Practical
Electricity

Practical Electricity

Second Edition

Nigel P. Cook

Prentice Hall

Upper Saddle River, New Jersey
Columbus, Ohio

Library of Congress Cataloging-in-Publication Data

Cook, Nigel P.,
    Practical electricity / Nigel P. Cook.—2nd ed.
      p. cm.
    ISBN 0-13-042047-6
      1. Electric engineering. 2. Electronics. I. Title.

TK146 .C625 2002
621.3—dc21
                                       2001036521

Editor in Chief: Stephen Helba
Product Manager: Scott J. Sambucci
Production Editor: Rex Davidson
Design Coordinator: Karrie Converse-Jones
Cover Designer: Tom Mack
Cover Photo: PhotoDisc
Illustrations: Rolin Graphics
Production Manager: Pat Tonneman
Project Management: Holly Henjum, Clarinda Publication Services

This book was set in Times Roman by The Clarinda Company and was printed and bound by Von Hoffmann Press, Inc. The cover was printed by The Lehigh Press, Inc.

Electronics Workbench® and MultiSim® are registered trademarks of Electronics Workbench.

10 9 8 7 6 5 4 3 2 1
ISBN: 0-13-042047-6

To Dawn, Candy, and Jon

Books by Nigel P. Cook

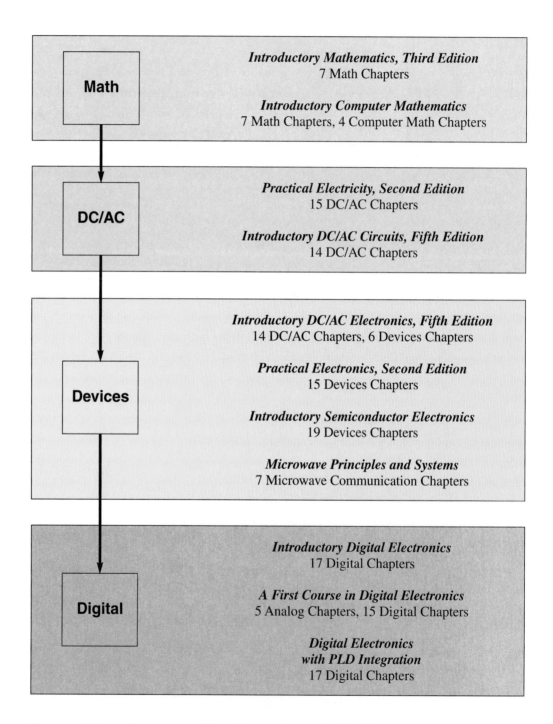

Math

Introductory Mathematics, Third Edition
7 Math Chapters

Introductory Computer Mathematics
7 Math Chapters, 4 Computer Math Chapters

DC/AC

Practical Electricity, Second Edition
15 DC/AC Chapters

Introductory DC/AC Circuits, Fifth Edition
14 DC/AC Chapters

Devices

Introductory DC/AC Electronics, Fifth Edition
14 DC/AC Chapters, 6 Devices Chapters

Practical Electronics, Second Edition
15 Devices Chapters

Introductory Semiconductor Electronics
19 Devices Chapters

Microwave Principles and Systems
7 Microwave Communication Chapters

Digital

Introductory Digital Electronics
17 Digital Chapters

A First Course in Digital Electronics
5 Analog Chapters, 15 Digital Chapters

***Digital Electronics
with PLD Integration***
17 Digital Chapters

For more information on any of the other textbooks by Nigel Cook, see his web page at www.prenhall.com/cook or ask your local Prentice Hall representative.

Preface

TO THE STUDENT

The early pioneers in electronics were intrigued by the mystery and wonder of a newly discovered science, whereas people today are attracted by its ability to lend its hand to any application and accomplish almost anything imaginable. If you analyze exactly how you feel at this stage, you will probably discover that you have mixed emotions about the journey ahead. On one hand, imagination, curiosity, and excitement are driving you on, while apprehension and reservations may be slowing you down. Your enthusiasm will overcome any indecision you have once you become actively involved in electronics and realize that it is as exciting as you ever expected it to be.

ORGANIZATION OF THE TEXTBOOK

This textbook has been divided into four basic parts. Chapters 1 through 3 introduce you to the world of electronics and the fundamentals of electricity. Chapters 4 through 7 cover direct current, or dc, circuits and Chapters 8 through 14 cover alternating current, or ac, circuits.

The material covered in this book has been logically divided and sequenced to provide a gradual progression from the known to the unknown, and from the simple to the complex.

ANCILLARIES ACCOMPANYING THIS TEXT

The following ancillaries accompanying this text provide extensive opportunity for further study and support:

- *Laboratory Manual.* Co-authored by Nigel Cook and Gary Lancaster, the lab manual offers numerous experiments designed to translate all of the textbook's theory into practical experimentation.

- *Companion Website,* located at http://www.prenhall.com/cook. Numerous interactive study questions are provided on this site to reinforce the concepts covered in the book.

- Many of the text's circuits have been rendered in Multisim from Electronics Workbench and are available on the Companion Website.

To complete the ancillary package, the following supplements are essential elements for any instructor using this text for a course:

- *Instructor's Solutions Manual*

- *Solutions Manual to Accompany Laboratory Manual*

- *PowerPoint*™ *Transparencies.* This CD-ROM includes a full set of **lecture presentations** as well as transparencies for all schematics appearing in the text.

- *Test Item File*

- *PH Test Manager*

DEVELOPMENT, CLASS TESTING, AND REVIEWING

The first phase of development for this manuscript was conducted in the classroom with students and instructors as critics. Each topic was class-tested by videotaping each lesson, and the results were then evaluated and implemented. This invaluable feedback enabled me to fine-tune my presentation of topics and instill understanding and confidence in the students.

The second phase of development was to forward a copy of the revised manuscript to several instructors at schools throughout the country. Their technical and topical critiques helped to mold the text into a more accurate form.

The third and final phase was to class-test the final revised manuscript and then commission the last technical review in the final stages of production.

ACKNOWLEDGMENTS

My appreciation and thanks are extended to the following instructors who have reviewed and contributed greatly to the development of this textbook: Venkata Anadu, Southwest Texas State University; Don Barrett, Jr., DeVry Institute of Technology; Charles Dale, Eastfield College; Robert Diffenderfer, DeVry Institute of Technology; Lynnette Garetz, Heald

College; Joe Gryniuk, Lake Washington Technical College; Jerry M. Manno, DeVry Institute of Technology; John Njemini, DeVry Institute of Technology; Gerald Schickman, Miami Dade Community College; George Sweiss, ITT Technical Institute; and Bradley J. Thompson, State University of New York College of Technology at Alfred.

Nigel P. Cook

Timeline Photo Credits

Contents

13

Transformers   344

Woolen Mill Makes Minis   344
Introduction   344

14

Resistive, Inductive, and Capacitive (*RLC*) Circuits   370

The Fairchildren   370
Introduction   370

Appendixes   399

Index   411

Introduction to Electronics

From World War II onward, no branch of science has contributed more to the development of the modern world than electronics. It has stimulated dramatic advances in the fields of communication, computing, consumer products, industrial automation, test and measurement, and health care. It has now become the largest single industry in the world, exceeding the automobile and oil industries, with annual sales of electronic systems exceeding $2 trillion.

Your Course in Electronics

Your future in the electronics industry begins with this text. To give you an idea of where you are going and what we will be covering, Figure I-1 acts as a sort of road map, breaking up your study of electronics into four basic steps.

Step 1: Basics of Electricity
Step 2: Electronic Components
Step 3: Electronic Circuits
Step 4: Electronic Systems

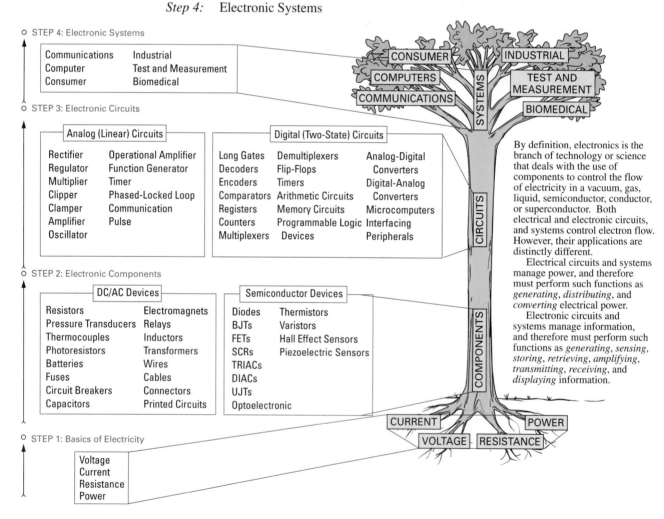

By definition, electronics is the branch of technology or science that deals with the use of components to control the flow of electricity in a vacuum, gas, liquid, semiconductor, conductor, or superconductor. Both electrical and electronic circuits, and systems control electron flow. However, their applications are distinctly different.

Electrical circuits and systems manage power, and therefore must perform such functions as *generating*, *distributing*, and *converting* electrical power.

Electronic circuits and systems manage information, and therefore must perform such functions as *generating*, *sensing*, *storing*, *retrieving*, *amplifying*, *transmitting*, *receiving*, and *displaying* information.

FIGURE I-1   The Steps Involved in Studying Electronics.

The main purpose of this introduction is not only to introduce you to the terms of the industry, but also to show you why the first two chapters in this text begin at the very beginning with "voltage and current," and then "resistance and power." **Components,** which are the basic electronic building blocks, were developed to control these four roots or properties, and when these devices are combined they form **circuits.** Moving up the tree to the six different branches of electronics, you will notice that just as components are the building blocks for circuits, circuits are in turn the building blocks for **systems.** Take a glance at the following pages, which will list many of the different types of components, circuits, and systems in the electronics industry.

Resistors

Fixed (chip, composition, deposited carbon, metal film, wirewound)
Resistive Networks (thick film, thin film)
Thermistors
Variable (rheostats, potentiometers, trimmers)

Capacitors

Fixed (ceramic, chips, electrolytic, film and paper, glass, mica)
Variable

Magnetic

Transformer (AF, RF, and power)
Inductors (coils and chokes)
Yokes
Flybacks
Relays (crystal can, general, reed,
    solid state, telephone type, time delay)

Transducers

Flow
Fluid Level
Motion (linear and angular)
Pressure (air, liquid, mechanical)
Temperature (thermocouples,
    thermistors)
Vibration

Switches and Keyboard

Dual-in-line
Keyboards (keypads and matrixes)
Lighted
Push Button
Rotary
Slide
Snap-action
Thumbwheel
Toggle

Electron Tubes

Cathode Ray
Power and Special Purpose (gas and vapor, klystrons,
    light and image sensing, magnetrons, TWTs)
TV Picture (color, monochrome)
Computer Monitors (text, graphics)

ELECTRONIC
COMPONENTS

CONSUMER   INDUSTRIAL
COMPUTERS   TEST AND
              MEASUREMENT
COMMUNICATIONS BIOMEDICAL
SYSTEMS
CIRCUITS
COMPONENTS
CURRENT   POWER
VOLTAGE RESISTANCE

Semiconductors

Discrete Semiconductors
Diodes (arrays, rectifiers
    signal, varactor,
    zeners, microwave)
Protection Devices
Thyristors
Transistors (bipolar, field
    effect, gallium arsenide)

Integrated Circuits
Custom and Semicustom
    (custom, gate arrays, PALs)
Linear (amplifiers, switches,
    communications, consumer,
    data conversion, interface,
    voltage reference)
Memories (RAM-dynamic and
    static, ROM-bipolar, CMOS
    EEROMs, EPROMs)
Microprocessors (bipolar, NMOS,
    CMOS)
Logic Families (bipolar, MOS)

Optoelectronic Devices
Imaging Arrays (CCD, MOS)
Laser Diodes
Light-emitting Diodes
    (infrared, visible)
Optically Coupled Isolators
Photoconductive Cells
Photodiodes
Phototransistors
Photovoltaic Cells

ELECTRONIC
CIRCUITS

Signal Generators and Timers

Sine-wave Generators (oscillators)
Square- and Pulse-wave Generators
Ramp- and Triangular-wave Generators

Power Supplies

Switching
Linear
Uninterruptible

Miscellaneous

Detectors and Mixers
Filters
Phase-locked Loops
Converters
Data Acquisition
Synthesizers

Digital Circuits

Oscillators and Generators
Gates and Flip-flops
Display Drivers
Counters and Dividers
Encoders and Decoders
Memories
Input/Output
Microprocessors

Latches
Registers
Multiplexers
Demultiplexers
Gate Arrays

Amplifiers

Bipolar Transistor
Field-effect Transistor
Operational Amplifier

COMMUNICATION
SYSTEMS

Electronic communications allow the
transmission and reception of information
between two points. Radio and television
are obvious communication devices,
broadcasting data or entertainment
between two points.

Telecommunications

Switching Systems
Data and Voice Switching
Voice-only Switching
Cellular Systems
Telephones
Corded
Cordless
Telephone/Video Equipment
Facsimile Terminals
Fiber Optic Communication
Systems

Television

Broadcast Equipment
CATV Equipment
CCTV Equipment

Satellite TV Equipment
HDTV (High Definition
TV) Equipment

Data Communications

Concentrators
Front-end Communications Processors
Message-switching Systems
Modems
Multiplexers
Network Controllers
Mixed Service (combining voice, data,
video, imaging)

Radio

Amateur (mobile and base stations)
Aviation Mobile and Ground Support Stations
Broadcast Equipment
Land Mobile (mobile and base stations)
Marine Mobile (ship and shore stations)
Microwave Systems
Satellite Systems
Radar and Sonar Systems

INTRODUCTION TO ELECTRONICS

CONSUMER

INDUSTRIAL

COMPUTERS

TEST AND MEASUREMENT

COMMUNICATIONS

BIOMEDICAL

SYSTEMS

CIRCUITS

COMPONENTS

COMPUTER SYSTEMS

The computer is proving to be one of the most useful of all systems. Its ability to process, store, and manipulate large groups of information at an extremely fast rate makes it ideal for almost any and every application. Systems vary in complexity and capability, ranging from the Cray super-computer to the home personal computer. The applications of word processing, record keeping, inventory, analysis, and accounting are but a few examples of data processing systems.

CURRENT

POWER

VOLTAGE

RESISTANCE

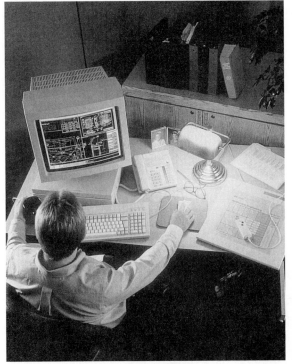

Data Terminals

CRT Terminals
   ASCII Terminals
     Graphics Terminals (color, monochrome)
Remote Batch Job Entry Terminals

Computer Systems

Microcomputers and Supermicrocomputers
Minicomputers (personal computers) and
   Superminicomputers (technical
   workstations, multiuser)
Mainframe Computers
Supercomputers

I/O Peripherals

Computer Microfilm
Digitizers
Graphics Tablets
Light Pens
Trackball and Mice
Optical Scanning Devices
Plotters
Printers
   Impact
   Nonimpact (laser, thermal, electrostatic, inkjet)

Data Storage Devices

Fixed Disk (14, 8, $5\frac{1}{4}$, and $3\frac{1}{2}$ in.)
Flexible Disk (8, $5\frac{1}{4}$, and $3\frac{1}{2}$ in.)
Optical Disk Drives (read-only, write once, erasable)
Cassette
Cartridge Magnetic Tape ($\frac{1}{4}$ in.)
Cartridge Tape Drives ($\frac{1}{2}$ in.)
Reel-type Magnetic Tape Drives

Audio Equipment

Car
Stereo Equipment
   Compact Systems (miniature components)
   Components (speakers, amps, turntables,
     tuners, tape decks)
Phonographs and Radio Phonographs
Radios (table, clock, portable)
Tape Players/Recorders
Compact Disk Players
Digital Tape Players

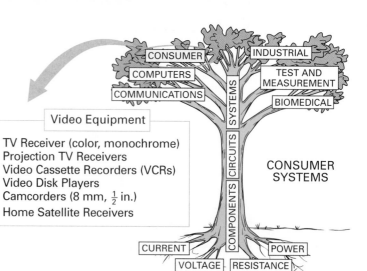

Video Equipment

TV Receiver (color, monochrome)
Projection TV Receivers
Video Cassette Recorders (VCRs)
Video Disk Players
Camcorders (8 mm, $\frac{1}{2}$ in.)
Home Satellite Receivers

CONSUMER
SYSTEMS

Personal

Calculators, Cameras, Watches
Telephone Answering Equipment
Personal Computers
Microwave Ovens
Musical Equipment and Instruments
Pacemakers and Hearing Aids
Alarms and Smoke Detectors

Automobile Electronics

Dashboard
Engine Monitoring and Analysis
Computer Navigation Systems
Alarms
Telephones

From the smart computer-controlled
automobiles, which provide navigational
information and monitor engine functions
and braking, to the compact disc players,
video camcorders, satellite TV receivers,
and wide-screen stereo TVs, this branch
of electronics provides us with
entertainment, information, safety, and,
in the case of the pacemaker, life.

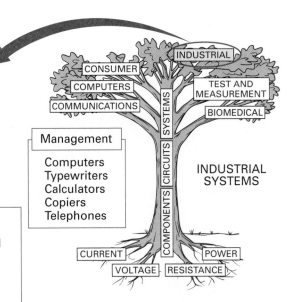

Management

Computers
Typewriters
Calculators
Copiers
Telephones

INDUSTRIAL SYSTEMS

Almost any industrial company can be divided into three basic sections, all of which utilize electronic equipment to perform their functions. The manufacturing section will typically use power, motor, and process control equipment, along with automatic insertion, inspection, and vision systems, for the fabrication of a product. The engineering section uses computers and test equipment for the design and testing of a product, while the management section uses electronic equipment such as computers, copiers, telephones, and so on.

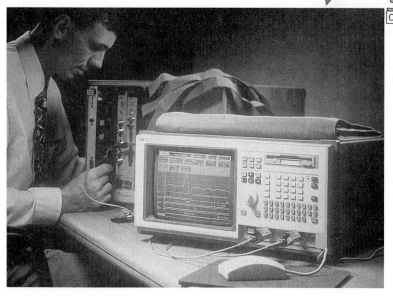

The rising complexity of electronic components, circuits, and equipment is causing a demand for sophisticated automatic test equipment for both the manufacturer and the customer to test their products.

Automated Test Systems
Active and Discrete Component Test Systems
Automated Field Service Testers
IC Testers (benchtop, general purpose, specialized)
Interconnect and Bare Printed Circuit-Board Testers
Loaded Printed Circuit-Board Testers (in-circuit, functional, combined)

General Test and Measurement Equipment

Amplifiers (lab)
Arbitrary Waveform Generators
Analog Voltmeters, Ammeters, and Multimeters
Audio Oscillators
Audio Waveform Analyzers and Distortion Meters
Calibrators and Standards
Dedicated IEEE–488 Bus Controllers
Digital Multimeters
Electronic Counters (RF, Microwave, Universal)
Frequency Synthesizers
Function Generators
Pulse/Timing Generators
Signal Generators (RF, Microwave)

Logic Analyzers
Microprocessor Development Systems
Modulation Analyzers
Noise-measuring Equipment
Oscilloscopes (Analog, Digital)
Panel Meters
Personal Computer (PC) Based Instruments
Recorders and Plotters
RF/Microwave Network Analyzers
RF/Microwave Power-measuring Equipment
Spectrum Analyzers
Stand-alone In-circuit Emulators
Temperature-measuring Instruments

Patient Care

Surgical Equipment
   Endoscope
   Laser Equipment
   Anesthesia Equipment
   Isotripters
   Electrocautery
   Microscopic Equipment

Monitoring Equipment (computer-controlled care units)
   Intensive Care Unit (ICU): Arrhythmia
      Computer Unit (heart wave, blood pressure,
      respiration rate, heart rate, blood gas)
   Cardiac Care Unit (CCU): Heart/Lung Machines and ICU Equipment
   Neonatal Intensive Care Unit (NICU): Temperature monitors (probes,
      incubators, open radiant warmers)

BIOMEDICAL SYSTEMS

Electronic equipment is used more and more within the biological and medical fields, which can be categorized simply as being either patient care or diagnostic equipment. In the operating room, the endoscope, which is an instrument used to examine the interior of a canal or hollow organ, and the laser, which is used to coagulate, cut, or vaporize tissue with extremely intense light, both reduce the use of invasive surgery. A large amount of monitoring equipment is used both in and out of operating rooms, and the equipment consists of generally large computer-controlled systems that can have a variety of modules inserted (based on the application) to monitor, on a continuous basis, body temperature, blood pressure, pulse rate, and so on. In the diagnostic group of equipment, the clinical laboratory test results are used as diagnostic tools. With the advances in automation and computerized information systems, multiple tests can be carried out at increased speeds. Diagnostic imaging, in which a computer constructs an image of a cross-sectional plane of the body, is probably one of the most interesting equipment areas.

Diagnostic

Diagnostic Equipment
   X-Ray (computed tomography)
   Magnetic Resonance Imaging (MRI)
   Diagnostic Sounder
   Electrocardiograph (EKG)
   Electromyograph
   Electroencephalograph (EEG)
   Coagulograph
   Ultrasound (computed sonography)
   Nuclear Medicine (isotopes, spectroscopy)

Clinical Laboratory
   Automated Clinical Analyzers
   Centrifuge Incubators
   Cell Counters

Development of an Electronic Product

The flow chart below shows the order, from top to bottom, in which an electronic product is developed from conception to shipping.

Product Conception

Product Definition

Design Breakdown

Hiring of Personnel

ENGINEERS

The engineer is responsible for designing the electronic system and supervising its development. During this process, the engineer relies heavily on the engineering technician to assist in the prototyping of the design and the diagnosis and repair of the problems.

Hardware Design | Software Design | Mechanical Design

Parts Purchasing

Documentation

Breadboard Design

Prototype Layout, Construction, & Test

Build Pre-production Units

TECHNICIANS

The technician is the expert in troubleshooting circuit and system malfunctions. Along with a thorough knowledge of all test equipment and how to use it to diagnose problems, the technician is also familiar with how to repair or replace faulty components. Referring to the color coding in this flow chart, you can see that technicians are involved in almost every step of product development.

Quality Engineering Test | Stock Components | Design Test Systems | Technician Training

ASSEMBLERS

The assembler is responsible for putting the electronic product together and is trained in the correct component handling and assembly techniques.

Chassis Fabrication

Component Insertion

Flow Soldering

Wire/Cable Hook-up

Final Assembly

Product Assembly

Product Test & Calibration

Quality Control Test

Product Release

Sales

Shipping

Customer

Customer Service

ENGINEERING TECHNICIAN

Here you can see an engineering technician breadboarding the design. From sketches supplied by the engineers, a breadboard model of the design is constructed. The breadboard model is an experimental arrangement of a circuit in which the components are temporarily attached to a flat board. In this arrangement, the components can be tested to prove the feasability of the circuit. A breadboard facilitates making easy changes when they are necessary.

Working under close supervision, the engineering technician performs all work assignments as given by all levels of engineers.

RESPONSIBILITIES:

• Breadboard electronic circuits from schematics.

• Test, evaluate, and document circuits and system performance under the engineer's direction.

• Check out, evaluate, and take data for the engineering proto-types including mechanical assembly of prototype circuits.

• Help to generate and maintain preliminary engineering docu-mentations. Assist engineers and senior Engineering Techni-cians in ERN documentation.

• Maintain the working station equipment and tools in orderly fashion.

• Support the engineers in all aspects of the development of new products.

REQUIREMENTS:

AS/AAS Degree in Electronics or equivalent plus 1–2 years technician experience. Ability to read color code, to solder properly, and to bond wires where the skill is required to com-plete breadboards and prototypes. Ability to use common machinery required to build prototype circuits. Working knowl-edge of common electronic components: TTL logic circuits, op-amps, capacitors, resistors, inductors, semiconductor devices.

Once the system is fully operational, it is calibrated by a calibration technician. The more complex problems are handled by the production test technicians seen in this photograph.

TECHNICIAN I

Working under close supervision, the production test technician performs all work assignments, and exercises limited decision-making.

RESPONSIBILITIES:

• Perform routine, simple operational tests and fault isolation on simple components, circuits, and systems for verification of product performance to well-defined specifications.

• May perform standard assembly operations and simple alignment of electronic components and assemblies.

• May set up simple test equipment to test performance of products to specifications.

REQUIREMENTS:

AS/AAS Degree in Electronic Technology or equivalent work experience.

TECHNICIAN II

Working under moderate supervision, the production test technician exercises general decision-making involving simple cause and effect relationships to identify trends and common problems.

RESPONSIBILITIES:

• Perform moderately complex operational tests and fault isolation on components, circuits, and systems for verification of product performance to well-defined specifications.

• Perform simple mathematical calculations to verify test measurements and product performance to well-defined specifications.

• Set up general test stations, utilizing varied test equipment, including some sophisticated equipment.

• Perform standard assembly operations.

REQUIREMENTS:

AS/AAS Degree in Electronic Technology or equivalent plus 2–4 years directly related experience. Demonstrated experience working mathematical formulas and equations. Working knowledge of counters, scopes, spectrum analyzers, and related industry standard test equipment.

CALIBRATION TECHNICIAN

The calibration technician shown here is undertaking a complete evaluation of the newly constructed breadboard's mechanical and electrical form, design, and performance.

Working under general supervision, the calibration technician interfaces with test equipment in a system environment requiring limited decision-making.

RESPONSIBILITIES:

• Work in an interactive mode with test station. Test, align, and calibrate products to defined specifications.

• May set up own test stations and those of other operators.

• Perform multiple alignments to get products to meet specifications.

• May perform other manufacturing-related tasks as required.

REQUIREMENTS:

1–2 years experience with test and measurement equipment, experience with multiple alignment and calibration of assemblies including test station setups. Able to follow written instructions and write clearly.

QUALITY ASSURANCE TECHNICIAN

The quality assurance (QA) technician takes one of the pre-production units through an extensive series of tests to determine whether it meets the standards listed. This technician is evaluating the new product as it is put through an extensive series of tests.

Working under direct supervision, the quality assurance technician performs functional tests on completed instruments.

RESPONSIBILITIES:

• Using established Acceptance Test Procedures, perform operational tests of all completed systems to ensure that all functional and electrical parameters are within specified limits.

• Perform visual inspection of all completed systems for cleanliness and absence of cosmetic defects.

• Reject all systems that do not meet specifications and/or established parameters of function and appearance.

• Make appropriate notations on the system history sheet.

• Maintain the QA Acceptance Log in accordance with current instructions.

• Refer questionable characteristics to supervisor.

REQUIREMENTS:

AS/AAS Degree in Electronics or equivalent including the use of test equipment. Must know color code and be able to distinguish between colors. Must have a working knowledge of related test equipment. Must know how to read and interpret drawings.

This photograph shows some in-house service technicians troubleshooting problems on returned units. Once the customer has received the electronic equipment, customer service provides assistance in maintenance and repair of the unit through direct in-house service or at service centers throughout the world.

Working under moderate supervision, the in-house service technician performs all work assignments given by lead tech or direct supervisor. Work necessary overtime as assigned by supervisors.

RESPONSIBILITIES:

- Utilizing all appropriate tools, troubleshoot and repair customer systems in a timely, quality manner, to the general component level.
- Working with basic test equipment, perform timely quality calibration of customer systems to specifications.
- Timely repair of QA rejects.
- Solder and desolder components where appropriate, meeting company standards.
- Aid marketing in solving customer problems via the telephone.
- When appropriate, instruct customers in the proper methods of calibration and repair of products.

REQUIREMENTS:

AS/AAS Degree in Electronics or equivalent plus 2–3 years experience troubleshooting analog and/or digital systems, at least 6–12 months of which should be in a service environment. Must be able to read and understand flow charts, block diagrams, schematics, and truth tables. Must be able to operate and utilize test equipment such as oscilloscopes, counters, voltmeters, and analyzers.

Ability to effectively communicate and work with customers.

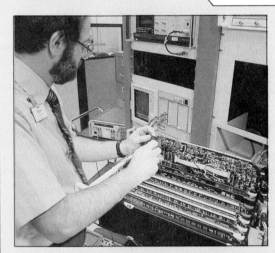

The field service technician seen here has been requested by the customer to make a service call on a malfunctioning unit that currently is under test.

Working under moderate supervision, the field service technician performs all work assignments given by the lead tech or direct supervisor. Work necessary overtime as assigned by supervisors.

RESPONSIBILITIES:

- Utilizing all appropriate tools, troubleshoot and repair customer systems in a timely, quality manner, to the general component level.
- Working with basic test equipment, perform timely quality calibration of customer systems to specifications.
- When appropriate, instruct others in proper soldering techniques meeting company standards.
- Timely repair of QA rejects.
- Aid marketing in solving customer problems via the telephone or at the customers' facility at marketings' discretion.
- When appropriate, aid marketing with sales applications.
- Help QA, Production, and Engineering in solving field problems.
- Evaluate manuals and other customer documents for errors or omissions.

REQUIREMENTS:

AS/AAS Degree in Electronics or equivalent plus 4–5 years experience troubleshooting analog and/or digital systems, at least 6–12 months of which should be in a service environment. Must be able to read and understand flow charts, block diagrams, schematics, and truth tables. A demonstrated ability to effectively communicate and work with customers, and suggest alternative applications for product utilization is also required.

Voltage and Current

1

Problem-Solver

Charles Proteus Steinmetz (1865–1923) was an outstanding electrical genius who specialized in mathematics, electrical engineering, and chemistry. His three greatest electrical contributions were his investigation and discovery of the law of hysteresis, his investigations in lightning, which resulted in his theory on traveling waves, and his discovery that complex numbers could be used to solve ac circuit problems. Solving problems was in fact his specialty, and on one occasion he was commissioned to troubleshoot a failure on a large company system that no one else had been able to repair. After studying the symptoms and schematics for a short time, he chalked an X on one of the metal cabinets, saying that this was where they would find the problem, and left. He was right, and the problem was remedied to the relief of the company executives; however, they were not pleased when they received a bill for $1000. When they demanded that Steinmetz itemize the charges, he replied—$1 for making the mark and $999 for knowing where to make the mark.

The strong message this vignette conveys is that you will get $1 for physical labor and $999 for mental labor, and this is a good example as to why you should continue in your pursuit of education.

Introduction

Before we begin, let me try to put you in the right frame of mind. As you proceed through this chapter and the succeeding chapters, it is imperative that you study every section, example, self-test evaluation point, and end-of-chapter question. If you cannot understand a particular section or example, go back and review the material that led up to the problem and make sure that you fully understand all the basics before you continue. Since each chapter builds on previous chapters, you may find that you need to return to an earlier chapter to refresh your understanding before moving on with the current chapter. This process of moving forward and then backtracking to refresh your understanding is very necessary and helps to engrave the material in your mind. Try never to skip a section or chapter because you feel that you already have a good understanding of the subject matter. If it is a basic topic that you have no problem with, read it anyway to refresh your understanding about the steps involved and the terminology because these may be used in a more complex operation in a later chapter.

In this chapter we will examine the smallest and most significant part of electricity and electronics—the **electron.** By having a good understanding of the electron and the atom, we will be able to obtain a clearer understanding of our four basic electrical quantities: **voltage, current, resistance,** and **power.** In this chapter we will be discussing the basic building blocks of matter, the electrical quantities of voltage and current, and the difference between a conductor and insulator.

1-1 THE STRUCTURE OF MATTER

All of the matter on the earth and in the air surrounding the earth can be classified as being either a solid, liquid, or gas. A total of approximately 107 different natural elements exist in, on, and around the earth. An **element,** by definition, is a substance consisting of only one type of atom; in other words, every element has its own distinctive atom, which makes it different from all the other elements. This **atom** is the smallest particle into which an element can be divided without losing its identity, and a group of identical atoms is called an element, shown in Figure 1-1.

1-1-1 *The Atom*

The word "atom" is a Greek word meaning a particle that is too small to be subdivided. At present, we cannot clearly see the atom; however, physicists and researchers do have the ability to record a picture as small as 12 billionths of an inch (about the diameter of one atom), and this image displays the atom as a white fuzzy ball.

In 1913, a Danish physicist, Neils Bohr, put forward a theory about the atom, and his basic model outlining the **subatomic** particles that make up the atom is still in use today and is illustrated in Figure 1-2. Bohr actually combined the ideas of Lord Rutherford's (1871–1937) nuclear atom with Max Planck's (1858–1947) and Albert Einstein's (1879–1955) quantum theory of radiation.

The three important particles of the atom are the *proton,* which has a positive charge, the *neutron,* which is neutral or has no charge, and the *electron,* which has a negative charge. Referring to Figure 1-2, you can see that the atom consists of a positively charged central mass called the *nucleus,* which is made up of protons and neutrons surrounded by a quantity of negatively charged orbiting electrons.

Table 1-1 lists the periodic table of the elements, in order of their atomic number. The **atomic number** of an atom describes the number of protons that exist within the nucleus.

The proton and the neutron are almost 2000 times heavier than the very small electron, so if we ignore the weight of the electron, we can use the fourth column in Table 1-1 (weight of an atom) to give us a clearer picture of the protons and neutrons within the atom's nucleus. For example, a hydrogen atom, shown in Figure 1-3(a), is the smallest of all atoms and has an atomic number of 1, which means that hydrogen has a one-proton nucleus. Helium, however [Figure 1-3(b)], is second on the table and has an atomic number of 2,

(a)

(b)

FIGURE 1-1   (a) Element: Many Similar Atoms. (b) Atom: Smallest Unit.

Electron
Smallest subatomic particle of negative charge that orbits the nucleus of the atom.

Voltage (*V* or *E*)
Term used to designate electrical pressure or the force that causes current to flow.

Current (*I*)
Measured in amperes or amps, it is the flow of electrons through a conductor.

Resistance
Symbolized *R* and measured in ohms (Ω), it is the opposition to current flow with the dissipation of energy in the form of heat.

Power
Amount of energy converted by a component or circuit in a unit of time, normally seconds. It is measured in units of watts (joules/second).

Element
There are 107 different natural chemical substances or elements that exist on the earth and can be categorized as either being a gas, solid, or liquid.

Atom
Smallest particle of an element.

Subatomic
Particles such as electrons, protons, and neutrons that are smaller than atoms.

Atomic Number
Number of positive charges or protons in the nucleus of an atom.

FIGURE 1-2   The Atom.

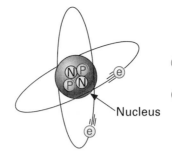

TIME LINE
The electron was first discovered by Jean Baptiste Perrin (1870-1942), a French physicist who was awarded the Nobel prize for physics.

P = Protons
N = Neutrons
e = Electrons

Nucleus

TABLE 1-1   Periodic Table of the Elements

ATOMIC NUMBER	ELEMENT NAME	SYMBOL	ATOMIC WEIGHT	ELECTRONS/SHELL K L M N O P Q							DISCOVERED	COMMENT
				K	L	M	N	O	P	Q		
1	Hydrogen	H	1.007	1							1766	Active gas
2	Helium	He	4.002	2							1895	Inert gas
3	Lithium	Li	6.941	2	1						1817	Solid
4	Beryllium	Be	9.01218	2	2						1798	Solid
5	Boron	B	10.81	2	3						1808	Solid
6	Carbon	C	12.011	2	4						Ancient	Semiconductor
7	Nitrogen	N	14.0067	2	5						1772	Gas
8	Oxygen	O	15.9994	2	6						1774	Gas
9	Fluorine	F	18.998403	2	7						1771	Active gas
10	Neon	Ne	20.179	2	8						1898	Inert gas
11	Sodium	Na	22.98977	2	8	1					1807	Solid
12	Magnesium	Mg	24.305	2	8	2					1755	Solid
13	Aluminum	Al	26.98154	2	8	3					1825	Metal conductor
14	Silicon	Si	28.0855	2	8	4					1823	Semiconductor
15	Phosphorus	P	30.97376	2	8	5					1669	Solid
16	Sulfur	S	32.06	2	8	6					Ancient	Solid
17	Chlorine	Cl	35.453	2	8	7					1774	Active gas
18	Argon	Ar	39.948	2	8	8					1894	Inert gas
19	Potassium	K	39.0983	2	8	8	1				1807	Solid
20	Calcium	Ca	40.08	2	8	8	2				1808	Solid
21	Scandium	Sc	44.9559	2	8	9	2				1879	Solid
22	Titanium	Ti	47.90	2	8	10	2				1791	Solid
23	Vanadium	V	50.9415	2	8	11	2				1831	Solid
24	Chromium	Cr	51.996	2	8	13	1				1798	Solid
25	Manganese	Mn	54.9380	2	8	13	2				1774	Solid
26	Iron	Fe	55.847	2	8	14	2				Ancient	Solid (magnetic)
27	Cobalt	Co	58.9332	2	8	15	2				1735	Solid
28	Nickel	Ni	58.70	2	8	16	2				1751	Solid
29	Copper	Cu	63.546	2	8	18	1				Ancient	Metal conductor
30	Zinc	Zn	65.38	2	8	18	3				1746	Solid
31	Gallium	Ga	69.72	2	8	18	4				1875	Liquid
32	Germanium	Ge	72.59	2	8	18	4				1886	Semiconductor
33	Arsenic	As	74.9216	2	8	18	5				1649	Solid
34	Selenium	Se	78.96	2	8	18	6				1818	Photosensitive
35	Bromine	Br	79.904	2	8	18	7				1898	Liquid
36	Krypton	Kr	83.80	2	8	18	8				1898	Inert gas
37	Rubidium	Rb	85.4678	2	8	18	8	1			1861	Solid
38	Strontium	Sr	87.62	2	8	18	8	2			1790	Solid

indicating that two protons are within the nucleus. The **atomic weight** of helium, however, is 4, meaning that two protons and two neutrons make up the atom's nucleus.

The number of neutrons within an atom's nucleus can therefore be calculated by subtracting the atomic number (protons) from the atomic weight (protons and neutrons). For example, Figure 1-4(a) illustrates a beryllium atom which has the following atomic number and weight.

Atomic Weight

The relative weight of a neutral atom of an element, based on a neutral oxygen atom having an atomic weight of 16.

Beryllium
Atomic number: 4 (protons)
Atomic weight: 9 (protons and neutrons)

TABLE 1-1 (*continued*)

ATOMIC NUMBER[a]	ELEMENT NAME	SYMBOL	ATOMIC WEIGHT	K	L	M	N	O	P	Q	DISCOVERED	COMMENT
39	Yttrium	Y	88.9059	2	8	18	9	2			1843	Solid
40	Zirconium	Zr	91.22	2	8	18	10	2			1789	Solid
41	Niobium	Nb	92.9064	2	8	18	12	1			1801	Solid
42	Molybdenum	Mo	95.94	2	8	18	13	1			1781	Solid
43	Technetium	Tc	98.0	2	8	18	14	1			1937	Solid
44	Ruthenium	Ru	101.07	2	8	18	15	1			1844	Solid
45	Rhodium	Rh	102.9055	2	8	18	16	1			1803	Solid
46	Palladium	Pd	106.4	2	8	18	18				1803	Solid
47	Silver	Ag	107.868	2	8	18	18	1			Ancient	Metal conductor
48	Cadmium	Cd	112.41	2	8	18	18	2			1803	Solid
49	Indium	In	114.82	2	8	18	18	3			1863	Solid
50	Tin	Sn	118.69	2	8	18	18	4			Ancient	Solid
51	Antimony	Sb	121.75	2	8	18	18	5			Ancient	Solid
52	Tellurium	Te	127.60	2	8	18	18	6			1783	Solid
53	Iodine	I	126.9045	2	8	18	18	7			1811	Solid
54	Xenon	Xe	131.30	2	8	18	18	8			1898	Inert gas
55	Cesium	Cs	132.9054	2	8	18	18	8	1		1803	Liquid
56	Barium	Ba	137.33	2	8	18	18	8	2		1808	Solid
57	Lanthanum	La	138.9055	2	8	18	18	9	2		1839	Solid
72	Hafnium	Hf	178.49	2	8	18	32	10	2		1923	Solid
73	Tantalum	Ta	180.9479	2	8	18	32	11	2		1802	Solid
74	Tungsten	W	183.85	2	8	18	32	12	2		1783	Solid
75	Rhenium	Re	186.207	2	8	18	32	13	2		1925	Solid
76	Osmium	Os	190.2	2	8	18	32	14	2		1804	Solid
77	Iridium	Ir	192.22	2	8	18	32	15	2		1804	Solid
78	Platinum	Pt	195.09	2	8	18	32	16	2		1735	Solid
79	Gold	Au	196.9665	2	8	18	32	18	1		Ancient	Solid
80	Mercury	Hg	200.59	2	8	18	32	18	2		Ancient	Liquid
81	Thallium	Tl	204.37	2	8	18	32	18	3		1861	Solid
82	Lead	Pb	207.2	2	8	18	32	18	4		Ancient	Solid
83	Bismuth	Bi	208.9804	2	8	18	32	18	5		1753	Solid
84	Polonium	Po	209.0	2	8	18	32	18	6		1898	Solid
85	Astatine	At	210.0	2	8	18	32	18	7		1945	Solid
86	Radon	Rn	222.0	2	8	18	32	18	8		1900	Inert gas
87	Francium	Fr	223.0	2	8	18	32	18	8	1	1945	Liquid
88	Radium	Ra	226.0254	2	8	18	32	18	8	2	1898	Solid
89	Actinium	Ac	227.0278	2	8	18	32	18	9	2	1899	Solid

[a]Rare earth series 58–71 and 90–107 have been omitted

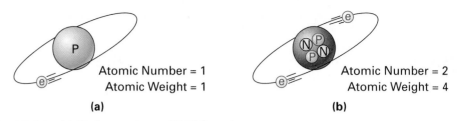

Atomic Number = 1
Atomic Weight = 1

(a)

Atomic Number = 2
Atomic Weight = 4

(b)

FIGURE 1-3   (a) Hydrogen Atom. (b) Helium Atom.

Neutral Atom

An atom in which the number of positive charges in the nucleus (protons) is equal to the number of negative charges (electrons) that surround the nucleus.

Shells or Bands

An orbital path containing a group of electrons that have a common energy level.

Subtracting the beryllium atom's weight from the beryllium atomic number, we can determine the number of neutrons in the beryllium atom's nucleus, as shown in Figure 1-4(a).

In most instances, atoms like the beryllium atom will not be drawn in the three-dimensional way shown in Figure 1-4(a). Figure 1-4(b) shows how a beryllium atom could be more easily drawn as a two-dimensional figure.

A **neutral atom** or *balanced atom* is one that has an equal number of protons and orbiting electrons, so the net positive proton charge is equal but opposite to the net negative electron charge, resulting in a balanced or neutral state. For example, Figure 1-5 illustrates a copper atom, which is the most commonly used metal in the field of electronics. It has an atomic number of 29, meaning that 29 protons and 29 electrons exist within the atom when it is in its neutral state.

Orbiting electrons travel around the nucleus at varying distances from the nucleus, and these orbital paths are known as **shells** or **bands.** The orbital shell nearest the nucleus is re-

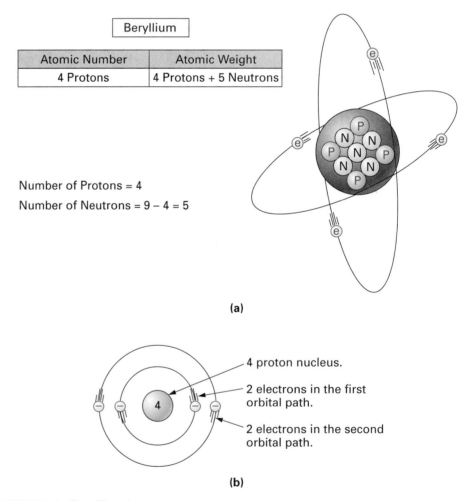

Beryllium	
Atomic Number	Atomic Weight
4 Protons	4 Protons + 5 Neutrons

Number of Protons = 4
Number of Neutrons = 9 − 4 = 5

(a)

4 proton nucleus.

2 electrons in the first orbital path.

2 electrons in the second orbital path.

(b)

FIGURE 1-4   Beryllium Atom.

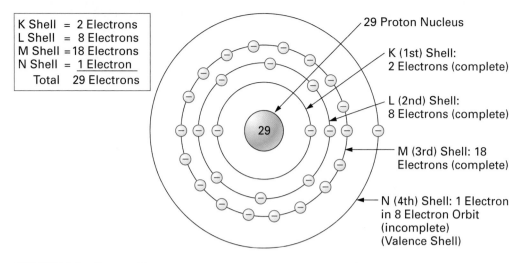

K Shell	= 2 Electrons
L Shell	= 8 Electrons
M Shell	= 18 Electrons
N Shell	= 1 Electron
Total	29 Electrons

29 Proton Nucleus

K (1st) Shell:
2 Electrons (complete)

L (2nd) Shell:
8 Electrons (complete)

M (3rd) Shell: 18
Electrons (complete)

N (4th) Shell: 1 Electron
in 8 Electron Orbit
(incomplete)
(Valence Shell)

FIGURE 1-5   Copper Atom.

ferred to as the first or K shell. The second is known as the L, the third is M, the fourth is N, the fifth is O, the sixth is P, and the seventh is referred to as the Q shell. There are seven shells available for electrons (K, L, M, N, O, P, and Q) around the nucleus, and each of these seven shells can only hold a certain number of electrons, as shown in Figure 1-6.

An atom's outermost electron-occupied shell is referred to as the **valence shell** or **ring,** and electrons in this shell are termed *valence electrons.* In the case of the copper atom, a single valence electron exists in the valence N shell.

All matter exists in one of three states: solid, liquid, or gas. The atoms of a solid are fixed in relation to one another but vibrate in a back-and-forth motion, unlike liquid atoms, which can flow over each other. The atoms of a gas move rapidly in all directions and collide with one another. The far-right column of Table 1-1 indicates whether the element is a gas, a solid, or a liquid.

Valence Shell or Ring
Outermost shell formed by electrons.

1-1-2   *Laws of Attraction and Repulsion*

For the sake of discussion and understanding, let us theoretically imagine that we are able to separate some positive and negative subatomic particles. Using these separated protons and

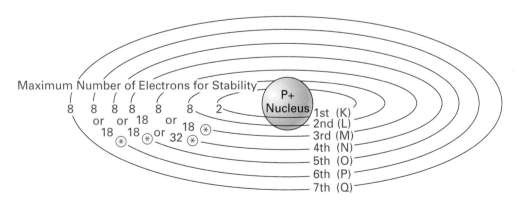

*The maximum number of electrons in these shells is dependent on the element's place in the periodic table.

FIGURE 1-6   Electrons and Shells.

P = Proton (positive)
e = Electron (negative)

FIGURE 1-7   Attraction and Repulsion. (a) Positive Repels Positive. (b) Negative Repels Negative. (c) Unlike Charges Attract.

electrons, let us carry out a few experiments, the results of which are illustrated in Figure 1-7. Studying Figure 1-7, you will notice that:

1. *Like charges* (positive and positive or negative and negative) repel one another.
2. *Unlike charges* (positive and negative or negative and positive) attract one another.

Orbiting negative electrons are therefore attracted toward the positive nucleus, which leads us to the question of why the electrons do not fly into the atom's nucleus. The answer is that the orbiting electrons remain in their stable orbit due to two equal but opposite forces. The centrifugal outward force exerted on the electrons due to the orbit counteracts the attractive inward force trying to pull the electrons toward the nucleus due to the unlike charges.

Due to their distance from the nucleus, valence electrons are described as being loosely bound to the atom. These electrons can easily be dislodged from their outer orbital shell by any external force, to become a **free electron.**

Free Electron

An electron that is not in any orbit around a nucleus.

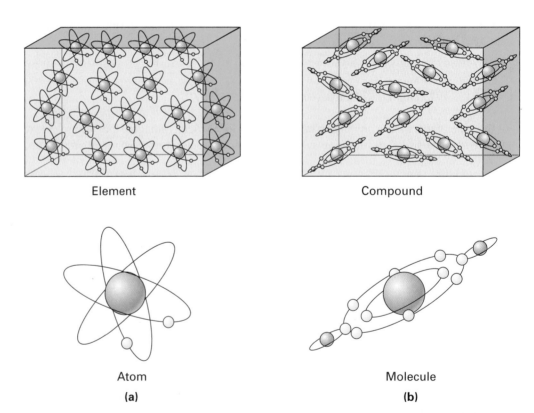

Element

Compound

Atom

(a)

Molecule

(b)

FIGURE 1-8   (a) An Element Is Made Up of Many Atoms. (b) A Compound Is Made Up of Many Molecules.

1-1-3 *The Molecule*

An atom is the smallest unit of a natural element, or an element is a substance consisting of a large number of the same atom. Combinations of elements are known as **compounds,** and the smallest unit of a compound is called a **molecule,** just as the smallest unit of an element is an atom. Figure 1-8 summarizes how elements are made up of atoms and compounds are made up of molecules.

    Water is an example of a liquid compound in which the molecule (H_2O) is a combination of an explosive gas (hydrogen) and a very vital gas (oxygen). Table salt is another example of a compound; here the molecule is made up of a highly poisonous gas atom (chlorine) and a potentially explosive solid atom (sodium). These examples of compounds each contain atoms that, when alone, are both poisonous and explosive, yet when combined the resulting substance is as ordinary and basic as water and salt.

> **Compound**
> A material composed of united separate elements.

> **Molecule**
> Smallest particle of a compound that still retains its chemical characteristics.

SELF-TEST EVALUATION POINT FOR SECTION 1-1

Use the following questions to test your understanding of Section 1-1:

1. Define the difference between an element and a compound.
2. Name the three subatomic particles that make up an atom.
3. What is the most commonly used metal in the field of electronics?
4. State the laws of attraction and repulsion.

1-2 CURRENT

The movement of electrons from one point to another is known as *electrical current.* Energy in the form of heat or light can cause an outer shell electron to be released from the valence shell of an atom. Once an electron is released, the atom is no longer electrically neutral and is called a **positive ion,** as it now has a net positive charge (more protons than electrons). The released electron tends to jump into a nearby atom, which will then have more electrons than protons and is referred to as a **negative ion.**

    Let us now take an example and see how electrons move from one point to another. Figure 1-9 shows a broken metal conductor between two charged objects. The metal conductor could be either gold, silver, or copper, but whichever it is, one common trait can be

> **Positive Ion**
> Atom that has lost one or more of its electrons and therefore has more protons than electrons, resulting in a net positive charge.

> **Negative Ion**
> Atom that has more than the normal neutral amount of electrons.

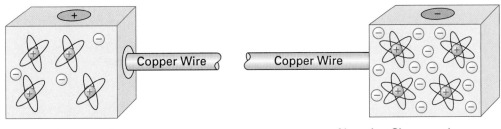

Positive Charge or Ions—
More Protons than Electrons
(absence of free electrons).

Negative Charge or Ions—
More Electrons than Protons
(abundance of free electrons).

FIGURE 1-9 **Positive and Negative Charges.**

noted: The valence electrons in the outermost shell are very loosely bound and can easily be pulled from their parent atom.

In Figure 1-10, the conductor between the two charges has been joined so that a path now exists for current flow. The negative ions on the right in Figure 1-10 have more electrons than protons, while the positive ions on the left in Figure 1-10 have fewer electrons than protons and so display a **positive charge.** The metal joining the two charges has its own atoms, which begin in the neutral condition.

Let us now concentrate on one of the negative ions. In Figure 1-10(a), the extra electrons in the outer shells of the negative ions on the right side will feel the attraction of the positive ions on the left side and the repulsion of the outer negative ions, or **negative charge.** This will cause an electron in a negative ion to jump away from its parent atom's orbit and land in an adjacent atom to the left within the metal wire conductor, as shown in Figure 1-10(b). This adjacent atom now has an extra electron and is called a negative ion, while the initial parent negative ion becomes a neutral atom, which will now receive an electron from one of the other negative ions, because their electrons are also feeling the attraction of the positive ions on the left side and the repulsion of the surrounding negative ions.

The electrons of the negative ion within the metal conductor feel the attraction of the positive ions, and eventually one of its electrons jumps to the left and into the adjacent atom,

Positive Charge

The charge that exists in a body which has fewer electrons than normal.

Negative Charge

An electric charge that has more electrons than protons.

TIME LINE

Ironically, Coulomb's law was not first discovered by Charles A. de Coulomb (1736-1806), but by Henry Cavendish, a wealthy scientist and philosopher. Cavendish did not publish his discovery, which he made several years before Coulomb discovered the law independently. James Clerk Maxwell published the scientific notebooks of Cavendish in 1879, describing his experiments and conclusions. However, about 100 years had passed, and Coulomb's name was firmly associated with the law. Many scientists demanded that the law be called Cavendish's law, while other scientists refused to change, stating that Coulomb was the discoverer because he made the law known promptly to the scientific community.

(a)

(b)

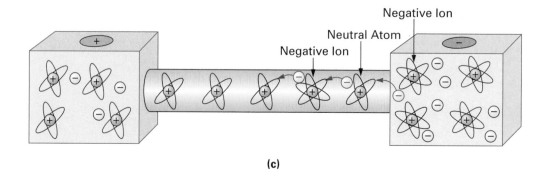

(c)

FIGURE 1-10   Electron Migration Due to Forces of Positive Attraction and Negative Repulsion on Electrons.

10

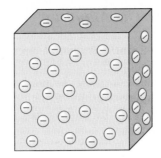

(a) 6.24×10^{18} Electrons
= 1 Coulomb of Charge

(b) 12.48×10^{18} Electrons
= 2 Coulombs of Charge

FIGURE 1-11   **(a) 1 C of Charge. (b) 2 C of Charge.**

as shown in Figure 1-10(c). This continual movement to the left will produce a stream of electrons flowing from right to left. Millions upon millions of atoms within the conductor pass a continuous movement of billions upon billions of electrons from right to left. This electron flow is known as **electric current.**

To summarize, we could say that as long as a force or pressure, produced by the positive charge and negative charge, exists it will cause electrons to flow from the negative to the positive terminal. The positive side has a deficiency of electrons and the negative side has an abundance, and so a continuous flow or migration of electrons takes place between the negative and positive terminal through our metal conducting wire. This electric current or electron flow is a measurable quantity, as will now be explained.

Electric Current (*I*)
Measured in amperes or amps, it is the flow of electrons through a conductor.

1-2-1   *Coulombs per Second*

There are 6.24×10^{18} electrons in 1 **coulomb of charge,** as illustrated in Figure 1-11. To calculate coulombs of charge (designated Q), we can use the formula

$$\text{charge, } Q = \frac{\text{total number of electrons } (n)}{6.24 \times 10^{18}}$$

where Q is the electric charge in coulombs.

Coulomb of Charge
Unit of electric charge. One coulomb equals 6.24×10^{18} electrons.

▢   **EXAMPLE:**

If a total of 3.75×10^{19} free electrons exist within a piece of metal conductor, how many coulombs (C) of charge would be within this conductor?

■   *Solution:*

By using the charge formula (Q) we can calculate the number of coulombs (C) in the conductor.

$$Q = \frac{n}{6.24 \times 10^{18}}$$
$$= \frac{3.75 \times 10^{19}}{6.24 \times 10^{18}}$$
$$= 6\ \text{C}$$

A total of 6 C of charge exists within the conductor.

In the calculator sequence you will see how the exponent key (E, EE, or EXP) on your calculator can be used.

CALCULATOR SEQUENCE

Step	Keypad Entry	Display Response
1.	③ . ⑦ ⑤ Ⓔ (Exponent) ① ⑨	3.75E19
2.	÷	
3.	⑥ . ② ④ Ⓔ ① ⑧	6.24E18
4.	=	6.0096

1-2-2   *The Ampere*

A coulomb is a **static** amount of electric charge. In electronics, we are more interested in electrons in motion. Coulombs and time are therefore combined to describe the number of electrons and the rate at which they flow. This relationship is called *current* (I) flow and has the unit of **amperes (A).** By definition, 1 ampere of current is said to be flowing if 6.24×10^{18} electrons (1 C) are drifting past a specific point on a conductor in 1 second of time. Stated as a formula:

$$\text{current } (I) = \frac{\text{coulombs } (Q)}{\text{time } (t)}$$

$$1 \text{ ampere} = 1 \text{ coulomb per } 1 \text{ second}$$

$$1 \text{ A} = \frac{1 \text{ C}}{1 \text{ s}}$$

In summary, 1 ampere equals a flow rate of 1 coulomb per second, and current is measured in amperes.

EXAMPLE:

If 5×10^{19} electrons pass a point in a conductor in 4 s, what is the amount of current flow in amperes?

Solution:

Current (I) is equal to Q/t. We must first convert electrons to coulombs.

$$Q = \frac{n}{6.24 \times 10^{18}}$$

$$= \frac{5 \times 10^{19}}{6.24 \times 10^{18}}$$

$$= 8 \text{ C}$$

Now, to calculate the amount of current, we use the formula

$$I = \frac{Q}{t}$$

$$= \frac{8 \text{ C}}{4 \text{ s}}$$

$$= 2 \text{ A}$$

This means that 2 A or 1.248×10^{19} electrons (2 C) are passing a specific point in the conductor every second.

CALCULATOR SEQUENCE

Step	Keypad Entry	Display Response
1.	5 E (exponent) 1 9	5E19
2.	÷	
3.	6 . 2 4 E 1 8	6.24E18
4.	=	8.012
5.	÷	
6.	4	2.003
7.	=	2.003

1-2-3   *Units of Current*

Current within electronic equipment is normally a value in milliamperes or microamperes and very rarely exceeds 1 ampere. Table 1-2 lists all the prefixes related to current. For example, 1 milliampere is one-thousandth of an ampere, which means that if 1 ampere were divided into 1000 parts, 1 part of the 1000 parts would be flowing through the circuit.

TABLE 1-2　Current Units

NAME	SYMBOL	VALUE
Picoampere	pA	$10^{-12} = \dfrac{1}{1,000,000,000,000}$
Nanoampere	nA	$10^{-9} = \dfrac{1}{1,000,000,000}$
Microampere	μA	$10^{-6} = \dfrac{1}{1,000,000}$
Milliampere	mA	$10^{-3} = \dfrac{1}{1000}$
Ampere	A	$10^{0} = 1$
Kiloampere	kA	$10^{3} = 1000$
Megaampere	MA	$10^{6} = 1,000,000$
Gigaampere	GA	$10^{9} = 1,000,000,000$
Teraampere	TA	$10^{12} = 1,000,000,000,000$

EXAMPLE:

Convert the following:

a. 0.003 A = _____ mA (milliamperes)

b. 0.07 mA = _____ μA (microamperes)

c. 7333 mA = _____ A (amperes)

d. 1275 μA = _____ mA (milliamperes)

Solution:

a. 0.003A = _____ mA. In this example, 0.003 A has to be converted so that it is represented in milliamperes (10^{-3} or ¹⁄₁₀₀₀ of an ampere). The basic algebraic rule to be remembered is that both expressions on either side of the equals must be equal.

LEFT		RIGHT	
Base	Multiplier	Base	Multiplier
0.003×10^{0}		= _____	$\times 10^{-3}$

The multiplier on the right in this example is going to be decreased 1000 times (10^{0} to 10^{-3}), so for the statement to balance the number on the right will have to be increased 1000 times; that is, the decimal point will have to be moved to the right three places (0.003 or 3). Therefore,

$$0.003 \times 10^{0} = 3 \times 10^{-3}$$

or

$$0.003 \text{ A} = 3 \times 10^{-3} \text{ A or 3 mA}$$

b. 0.07 mA = _____ μA. In this example the unit is going from milliamperes to microamperes (10^{-3} to 10^{-6}) or 1000 times smaller, so the number must be made 1000 times greater.

$$0.0\overset{\frown}{7}\overset{\frown}{0} \text{ or } 70.0$$

Therefore, 0.07 mA = 70 μA.

c. 7333 mA = _____ A. The unit is going from milliamperes to amperes, increasing 1000 times, so the number must decrease 1000 times.

$$7\overset{\frown}{3}\overset{\frown}{3}\overset{\frown}{3}. \text{ or } 7.333$$

Therefore, 7333 mA = 7.333 A.

d. 1275 μA = _____ mA. The unit is changing from microamperes to milliamperes, an increase of 1000 times, so the number must decrease by the same factor.

$$1\overset{\frown}{2}\overset{\frown}{7}\overset{\frown}{5}.0 \text{ or } 1.275$$

Therefore, 1275 μA = 1.275 mA.

1-2-4   *Conventional versus Electron Flow*

Electrons drift from a negative to a positive charge, as illustrated in Figure 1-12. As already stated, this current is known as **electron flow.**

In the eighteenth and nineteenth centuries, when very little was known about the atom, researchers believed that current was a flow of positive charges. Although this has now been proved incorrect, many texts still use **conventional current flow,** which is shown in Figure 1-13.

Whether conventional flow or electron flow is used, the same answers to problems, measurements, and designs are obtained. The key point to remember is that direction is not important, but the amount of current flow is.

Throughout this book we will be using electron flow so that we can relate back to the atom when necessary. If you wish to use conventional flow, just reverse the direction of the arrows. To avoid confusion, be consistent with your choice of flow.

FIGURE 1-12   Electron Current Flow.

TIME LINE

The galvanometer, which is used to measure electrical current, was named after Luigi Galvani (1737-1798), who conducted many experiments with electrical current, or, as it was known at the time, "galvanism."

FIGURE 1-13   Conventional Current Flow.

1-2-5   *How Is Current Measured?*

Ammeters (ampere meters) are used to measure the current flow within a circuit. Stepping through the sequence detailed in Figure 1-14, you will see how an ammeter is used to measure the value of current within a circuit.

In the simple circuit shown in Figure 1-14, an ON/OFF switch is being used to turn ON or OFF a small light bulb. One of the key points to remember with ammeters is that if you wish to measure the value of current flowing within a wire, the current path must be opened and the ammeter placed in the path so that it can sense and display the value.

Ammeter
Meter placed in the path of current flow to measure the amount.

SELF-TEST EVALUATION POINT FOR SECTION 1-2

Use the following questions to test your understanding of Section 1-2:

1. What is the unit of current?
2. Define current in relation to coulombs and time.
3. What is the difference between conventional and electron current flow?
4. What test instrument is used to measure current?

1-3   VOLTAGE

Voltage is the force or pressure exerted on electrons. Referring to Figure 1-15(a) and (b), you will notice two situations. Figure 1-15(a) shows highly concentrated positive and negative charges or potentials connected to one another by a copper wire. In this situation, a large potential difference or voltage is being applied across the copper atom's electrons. This force or voltage causes a large amount of copper atom electrons to move from right to left. On the other hand, Figure 1-15(b) illustrates a low concentration of positive and negative potentials, so a small voltage or pressure is being applied across the conductor, causing a small amount of force, and therefore current, to move from right to left.

In summary, we could say that a highly concentrated charge produces a high voltage, whereas a low concentrated charge produces a low voltage. Voltage is also appropriately

Light Bulb

Negative Charge

Positive Charge

ON/OFF Power Switch

8. View the reading shown on the ammeter's display, being sure to note the unit of measurement.

7. Turn ON the power to the circuit by closing the power switch.

6. Connect the meter probes to the circuit so that the ammeter is placed in the path of circuit current flow.

1. Turn OFF the power to the circuit by opening the power switch.

2. Create an opening in the circuit where the ammeter probes can be inserted.

3. Select AMPS (A) on the digital multimeter's function switch.

5. Plug the black test probe into the COM (common) input jack.

4. Plug the red test probe into the high-current (10 Amp) or low-current (300 milliamp) input jack depending on what value of current you expect. In electronic circuits current is generally a low value (300mA jack is used), whereas in electrical circuits current is generally a high value (10A jack is used).

NOTE: Connect the red test probe to the positive side of the circuit, and the black test probe to the negative side of the circuit. If you reverse the connections, the digital multimeter will display a negative sign.

FIGURE 1-14   Measuring Current with the Ammeter.

FIGURE 1-15 **(a) Large Potential Difference or Voltage. (b) Small Potential Difference or Voltage.**

known as the "electron moving force" or **electromotive force (emf),** and since two opposite potentials exist (one negative and one positive), the strength of the voltage can also be referred to as the amount of **potential difference (PD)** applied across the circuit. To compare, we can say that a large voltage, electromotive force, or potential difference exists across the copper conductor in Figure 1-15(a), while a small voltage, potential difference, or electromotive force is exerted across the conductor in Figure 1-15(b).

Voltage is the force, pressure, potential difference (PD), or electromotive force (emf) that causes electron flow or current and is symbolized by italic uppercase V. The unit for voltage is the volt, symbolized by roman uppercase V. This can become a bit confusing. For example, when the voltage applied to a circuit equals 5 volts, the circuit notation would appear as

$$V = 5 \text{ V}$$

Electromotive Force (emf)

Force that causes the motion of electrons due to a potential difference between two points.

Potential Difference (PD)

Voltage difference between two points, which will cause current to flow in a closed circuit.

You know the first V represents "voltage," not "volt," because 1 volt cannot equal 5 volts. To avoid confusion, some texts and circuits use E, symbolizing electromotive force, to represent voltage; for example,

$$E = 5\ \text{V}$$

In this text, we will maintain the original designation for voltage (V).

1-3-1   *Symbols*

A **battery,** like the one shown in Figure 1-16(a), converts chemical energy into electrical energy. At the positive terminal of the battery, positive charges or ions (atoms with more protons than electrons) are present, and at the negative terminal, negative charges or ions (atoms with more electrons than protons) are available to supply electrons for current flow within a circuit. A battery, therefore, chemically generates negative and positive ions at its respective terminals. The symbol for the battery is shown in Figure 1-16(b).

In Figure 1-17(a) you can see the schematic symbols for many of the devices discussed so far, along with their physical appearance.

In the circuit shown in Figure 1-17(b), a 9 V battery chemically generates positive and negative ions. The negative ions at the negative terminal force away the negative electrons, which are attracted by the positive charge or absence of electrons at the positive terminal. As the electrons proceed through the copper conductor wire, jumping from one atom to the next, they eventually reach the bulb. As they pass through the bulb, they cause it to glow. When emerging from the light bulb, the electrons travel through another connector cable and finally reach the positive terminal of the battery.

Studying Figure 1-17(b), you will notice two reasons why the circuit is drawn using symbols rather than illustrating the physical appearance:

1. A circuit with symbols can be drawn faster and more easily.
2. A circuit with symbols has less detail and clutter, and is therefore more easily comprehended since it has fewer distracting elements.

1-3-2   *Units of Voltage*

The unit for voltage is the volt (V). Voltage within electronic equipment is normally measured in volts, whereas heavy-duty industrial equipment normally requires high voltages that are generally measured in kilovolts (kV). Table 1-3 lists all the prefixes and values related to volts.

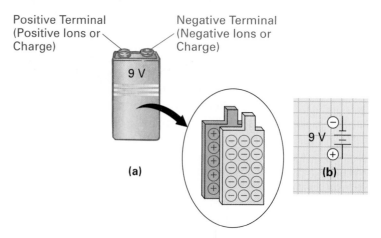

Positive Terminal
(Positive Ions or Charge)

Negative Terminal
(Negative Ions or Charge)

9 V

(a)

9 V

(b)

FIGURE 1-16   The Battery—A Source of Voltage. (a) Physical Appearance. (b) Schematic Symbol.

Component	Symbol	Name	Description
		Incandescent lamp	Incandescence: Release of visible radiation (light) by a heated object
		Connecting wire with end alligator clips	Used to connect different components
9 V		Battery	Source of voltage and current
	OPEN / CLOSED	Switch	A device used to open or close a current path
30.23	AM	Ammeter	Used to measure the current flow within a circuit

(a)

Schematic

Pictorial

(b)

FIGURE 1-17 (a) **Components.** (b) **Example Circuit.**

TABLE 1-3   Voltage Units

NAME	SYMBOL	VALUE
Picovolts	pV	$10^{-12} = \dfrac{1}{1,000,000,000,000}$
Nanovolts	nV	$10^{-9} = \dfrac{1}{1,000,000,000}$
Microvolts	μV	$10^{-6} = \dfrac{1}{1,000,000}$
Millivolts	mV	$10^{-3} = \dfrac{1}{1000}$
Volts	V	$10^{0} = 1$
Kilovolts	kV	$10^{3} = 1000$
Megavolts	MV	$10^{6} = 1,000,000$
Gigavolts	GV	$10^{9} = 1,000,000,000$
Teravolts	TV	$10^{12} = 1,000,000,000,000$

EXAMPLE:

Convert the following:

    a. 3000 V = _____ kV (kilovolts)

    b. 0.14 V = _____ mV (millivolts)

    c. 1500 kV = _____ MV (megavolts)

■ *Solution:*

    a. 3000 V = 3 kV or 3×10^{3} volts (multiplier ↑ 1000, number ↓ 1000)

    b. 0.14 V = 140 mV or 140×10^{-3} volt (multiplier ↓ 1000, number ↑ 1000)

    c. 1500 kV = 1.5 MV or 1.5×10^{6} volts (multiplier ↑ 1000, number ↓ 1000)

1-3-3   *How Is Voltage Measured?*

Voltmeter

Instrument designed to measure the voltage or potential difference. Its scale can be graduated in kilovolts, volts, or millivolts.

Voltmeters (voltage meters) are used to measure electrical pressure or voltage. Stepping through the sequence detailed in Figure 1-18, you will see how a voltmeter can be used to measure a battery's voltage.

In many instances, the voltmeter is used to measure the potential difference or voltage drop across a device, as shown in Figure 1-19(a) and (b). Figure 1-19(a) shows how to measure the voltage across light bulb 1 (L1), and Figure 1-19(b) shows how to measure the voltage across light bulb 2 (L2).

1-3-4   *Fluid Analogy of Current and Voltage*

An analogy is a comparison that describes the similarities between two otherwise different things. In this section, we will use a fluid analogy to reinforce your understanding of current and voltage.

5. View the reading shown on the voltmeter's display, being sure to note the unit of measurement.

12 V Battery

4. Touch the probe tips across a power source or across a device.

1. Select VOLTS (V) on the digital multimeter's function switch.

2. Plug the red test probe into the voltage (V) input jack.

3. Plug the black test probe into the common (COM) input jack.

NOTE: If test leads are reversed, a negative sign will show in the display.

(a)

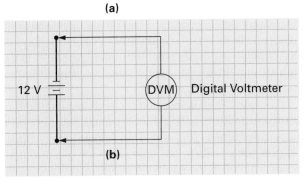

12 V ⎓ (DVM) Digital Voltmeter

(b)

FIGURE 1-18   **Using the Voltmeter to Measure Voltage.**

FIGURE 1-19 **Measuring the Voltage Drop across Components (a) Lamp 1 and (b) Lamp 2.**

In Figure 1-20(a), a system using a pump, pipes, and a waterwheel is being used to convert electrical energy into mechanical energy. When electrical energy is applied to the pump, it will operate and cause water to flow. The pump generates:

1. A high pressure at the outlet port, which pushes the water molecules out and into the system
2. A low pressure at the inlet port, which pulls the water molecules into the pump

The water current flow is in the direction indicated, and the high pressure or potential within the piping will be used to drive the waterwheel around, producing mechanical energy. The remaining water is attracted into the pump due to the suction or low pressure existing at the inlet port. In fact, the amount of water entering the inlet port is the same as the amount of water leaving the outlet port. It can therefore be said that the water flow rate is the same throughout the circuit. The only changing element is the pressure felt at different points throughout the system.

In Figure 1-20(b), an electric circuit containing a battery, conductors, and a bulb is being used to convert electrical energy into light energy. The battery generates a voltage just as the pump generates pressure. This voltage causes electrons to move through conductors, just as pressure causes water molecules to move through the piping. The amount of water flow is dependent on the pump's pressure, and the amount of current or electron flow is dependent on the battery's voltage. Water flow through the wheel can be compared to current flow through the bulb. The high pressure is lost in turning the wheel and producing mechanical energy, just as voltage is lost in producing light energy out of the bulb. We cannot say that pressure or voltage flows: Pressure and voltage are applied and cause water or current to flow, and it is this flow that is converted to mechanical energy in our fluid system and light energy in our electrical system. Voltage is the force of repulsion and attraction needed to

CHAPTER 1 / VOLTAGE AND CURRENT

(a) Fluid System

(b) Electrical System

(a) Fluid System		**(b)** Electrical System
Pump generates pressure, which is the water moving force.	Equivalent to (≅)	Battery generates voltage, which is the electron moving force.
Water current flow.		Electron current flow.
High pressure or potential.		High voltage or potential.
Low pressure or potential.		Low voltage or potential.

FIGURE 1-20 **Comparison between a Fluid System (a) and an Electrical System (b).**

cause current to flow through a circuit, and without this potential difference or pressure there cannot be current.

1-3-5   *Current Is Directly Proportional to Voltage*

Referring to the fluid system and then the electrical circuit in Figure 1-20, you can easily see that the flow is proportional to the pressure, which means: If the pump were to generate a greater pressure, a larger amount of water would flow through the system.

$$\text{pressure} \uparrow \qquad \text{water flow} \uparrow$$

Similarly, if a larger voltage were applied to the electrical circuit, this larger electron moving force (emf) would cause more electrons or current to flow through the circuit, as shown in Figure 1-21.

   Current (I) is therefore said to be *directly proportional (∝) to voltage (V)*, as a voltage decrease $(V \downarrow)$ results in a current decrease $(I \downarrow)$, and similarly, a voltage increase $(V \uparrow)$ causes a current increase $(I \uparrow)$.

Current is directly proportional to voltage $(I \propto V)$

$$V \downarrow \text{ causes a } I \downarrow$$

$$V \uparrow \text{ causes a } I \uparrow$$

As you can see from this example, directly **proportional** (∝) is a phrase which means that one term will change in proportion, or in size, relative to another term.

Proportional

A term used to describe the relationship between two quantities that have the same ratio.

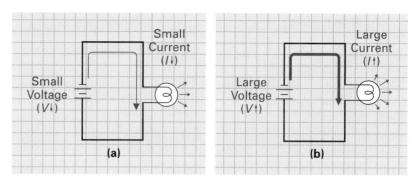

FIGURE 1-21 Current Is Directly Proportional to Voltage. (a) Small Voltage Produces a Small Current. (b) Large Voltage Produces a Large Current.

Use the following questions to test your understanding of Section 1-3:

1. What is the unit of voltage?

2. Convert 3 MV to kilovolts.

3. Which meter is used to measure voltage?

4. What is the relationship between current and voltage?

1-4   CONDUCTORS

A lightning bolt that sets fire to a tree and the operation of your calculator are both electrical results achieved by the flow of electrons. The only difference is that your calculator's circuits control the flow of electrons, while the lightning bolt is the uncontrolled flow of electrons. In electronics, a **conductor** is used to channel or control the path in which electrons flow.

Any material that passes current easily is called a conductor. These materials are said to have a "low resistance," which means that they will offer very little opposition to current flow. This characteristic can be explained by examining conductor atoms. As mentioned previously, the atom has a maximum of seven orbital paths known as shells, which are named K, L, M, N, O, P, and Q, stepping out toward the outermost or valence shell. Conductors are materials or natural elements whose valence electrons can easily be removed from their parent atoms. They are therefore said to be sources of free electrons, and these free electrons provide us with circuit current. The precious metals of silver and gold are the best conductors.

More specifically, a better conductor has:

Conductor

Length of wire whose properties are such that it will carry an electric current.

1. Electrons in shells farthest away from the nucleus; these electrons feel very little nucleus attraction and can be broken away from their atom quite easily.

2. More electrons per atom.

3. An incomplete valence shell. This means that the valence shell does not have in it the maximum possible number of electrons. If the atom had its valence ring complete (full), there would be no holes (absence of an electron) in that shell, so no encouragement for adjacent atom electrons to jump from their parent atom into the next atom would exist, preventing the chain reaction known as current.

Economy must be considered when choosing a conductor. Large quantities of conductors using precious metals are obviously going to send the cost of equipment beyond reach. The conductor must also satisfy some physical requirements, in that we must be able to shape it into wires of different sizes and easily bend it to allow us to connect one circuit to the next.

Copper is the most commonly used conductor, as it meets the following three requirements:

1. It is a good source of electrons.
2. It is inexpensive.
3. It is physically pliable.

Aluminum is also a very popular conductor, and although it does not possess as many free electrons as copper, it has the two advantages of being less expensive and lighter than copper.

1-4-1   *Conductance*

TIME LINE
Stephen Gray (1693-1736), an Englishman, discovered that certain substances would conduct electricity.

Conductance is the measure of how good a conductor is at carrying current. Conductance (symbolized G) is equal to the reciprocal of resistance (or opposition) and is measured in the unit siemens (S):

Conductance (G)
Measure of how well a circuit or path conducts or passes current. It is equal to the reciprocal of resistance.

$$\text{conductance } (G) = \frac{1}{\text{resistance } (R)}$$

Conductance (G) values are measured in siemens (S)
and resistance (R) in ohms (Ω).

This means that conductance is inversely proportional to resistance. For example, if the opposition to current flow (resistance) is low, the conductance is high and the material is said to have a good conductance.

$$\text{high conductance } G \uparrow = \frac{1}{R \downarrow \text{ (low resistance)}}$$

On the other hand, if the resistance of a conducting wire is high ($R \uparrow$), its conductance value is low ($G \downarrow$) and it is called a poor conductor. A good conductor therefore has a high conductance value and a very small resistance to current flow.

EXAMPLE:

A household electric blanket offers 25 ohms of resistance to current flow. Calculate the conductance of the electric blanket's heating element.

Solution:

$$\text{conductance} = \frac{1}{\text{resistance}}$$

$$G = \frac{1}{R} = \frac{1}{25 \text{ ohms}} = 0.04 \text{ siemens}$$

or 40 mS (millisiemens)

TIME LINE
Stephen Gray's lead was picked up in 1730 by Charles du Fay, a French experimenter, who believed that there were two types of electricity, which he called *vitreous* and *resinous* electricity.

To express how good or poor a conductor is, we must have some sort of reference point. The reference point we use is the best conductor, silver, which is assigned a relative conductivity value of 1.0. Table 1-4 lists other conductors and their relative conductivity values with respect to the best, silver. The formula for calculating relative conductivity is

$$\text{Relative Conductivity} = \frac{\text{Conductor's Relative Conductivity}}{\text{Reference Conductor's Relative Conductivity}}$$

TABLE 1-4 Relative Conductivity of Conductors

MATERIAL	CONDUCTIVITY (RELATIVE)
Silver	1.000
Copper	0.945
Gold	0.652
Aluminum	0.575
Tungsten	0.297
Nichrome	0.015

▢ EXAMPLE:

What is the relative conductivity of tungsten if copper is used as the reference conductor?

�◼ *Solution:*

CALCULATOR SEQUENCE

Step	Keypad Entry	Display Response
1.	0 . 2 9 7	0.297
2.	÷	
3.	0 . 9 4 5	0.945
4.	=	0.314

$$\text{tungsten} = 0.297$$
$$\text{copper} = 0.945$$
$$\text{relative conductivity} = \frac{\text{conductivity of conductor}}{\text{conductivity of reference}}$$
$$= \frac{0.297}{0.945}$$
$$= 0.314$$

▢ EXAMPLE:

What is the relative conductivity of silver if copper is used as the reference?

◼ *Solution:*

$$\text{relative conductivity} = \frac{\text{silver}}{\text{copper}}$$
$$= \frac{1.000}{0.945} = 1.058$$

SELF-TEST EVALUATION POINT FOR SECTION 1-4

Use the following questions to test your understanding of Section 1-4.

1. True or false: A conductor is a material used to block the flow of current.
2. What is the most commonly used conductor in the field of electronics?
3. Calculate the conductance of a 35 ohm heater element.

Insulator

A material that has few electrons per atom and those electrons are close to the nucleus and cannot be easily removed.

1-5 INSULATORS

Any material that offers a high resistance or opposition to current flow is called an **insulator.** Conductors permit the easy flow of current and so have good conductivity, while insulators allow small to almost no amount of free electrons to flow. Insulators can, with sufficient

TABLE 1-5   Breakdown Voltages of Certain Insulators

MATERIAL	BREAKDOWN STRENGTH (kV/cm)
Mica	2000
Glass	900
Teflon	600
Paper	500
Rubber	275
Bakelite	151
Oil	145
Porcelain	70
Air	30

pressure or voltage applied across them, "break down" and conduct current. This **breakdown voltage** must be great enough to dislodge the electrons from their close orbital shells (K, L shells) and release them as free electrons.

Insulators are also called "dielectrics," and the best insulator or dielectric should have the maximum possible resistance and conduct no current at all. The **dielectric strength** of an insulator indicates how good or bad an insulator is by indicating the voltage that will cause the insulating material to break down and conduct a large current. Table 1-5 lists some of the more popular insulators and the value of kilovolts that will cause a centimeter of insulator to break down. As an example, if 1 centimeter of paper is connected to a variable voltage source, a voltage of 500 kilovolts is needed to break down the paper and cause current to flow.

The following formula can be used to calculate the dielectric thickness needed to withstand a certain voltage.

Breakdown Voltage
The voltage at which breakdown of an insulator occurs.

Dielectric Strength
The maximum potential a material can withstand without rupture.

$$\text{dielectric thickness} = \frac{\text{voltage to insulate}}{\text{insulator's breakdown voltage}}$$

EXAMPLE:

What thickness of mica would be needed to withstand 16,000 V?

Solution:

$$\text{mica strength} = 2000 \text{ kV/cm}$$
$$\text{dielectric thickness} = \frac{16,000 \text{ V}}{2000 \text{ kV/cm}} = 0.008 \text{ cm}$$

CALCULATOR SEQUENCE

Step	Keypad Entry	Display Response
1.	[1] [6] [E](exponent) [3]	16E3
2.	[÷]	
3.	[2] [E] [6] (or 2000 E3)	2E6
4.	[=]	0.008

EXAMPLE:

What maximum voltage could 1 mm of air withstand?

Solution:

There are 10 mm in 1 cm. If air can withstand 30,000 V/cm, it can withstand 3000 V/mm.

Use the following questions to test your understanding of Section 1-5.

1. True or false: An insulator is a material used to block the flow of current.

2. What is considered to be the best insulator material?

3. Define *breakdown voltage.*

4. Would the conductance figure of a good insulator be large or small?

1-6   THE OPEN, CLOSED, AND SHORT CIRCUIT

In this section we will examine some of the terms used in association with circuits.

1-6-1   *Open Circuit (Open Switch)*

Figure 1-22 shows how an opened switch can produce an "open" in a circuit. The open prevents current flow as the extra electrons in the negative terminal of the battery cannot feel the attraction of the positive terminal due to the break in the path. The opened switch therefore has produced an **open circuit.**

Open Circuit

Break in the path of
current flow.

1-6-2   *Open Circuit (Open Component)*

In Figure 1-23, the switch is now closed so as to make a complete path in the circuit. An open circuit, however, could still exist due to the failure of one of the components in the circuit. In the example in Figure 1-23, the light bulb filament has burned out creating an open in the circuit, which prevents current flow. An open component therefore can also produce an open circuit.

1-6-3   *Closed Circuit (Closed Switch)*

If all of the devices in a circuit are operating properly and connected correctly, a closed switch produces a **closed circuit,** as shown in Figure 1-24. A closed circuit provides a complete path from the negative terminal of the battery to the positive, and so current will flow through the circuit.

Closed Circuit

Circuit having a complete
path for current to flow.

FIGURE 1-22   An Open Switch Causing an Open Circuit. (a) Pictorial. (b) Schematic.

FIGURE 1-23 **An Open Lamp Filament Causing an Open Circuit. (a) Pictorial. (b) Schematic.**

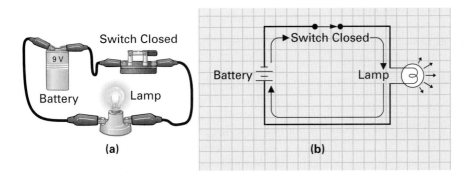

FIGURE 1-24 **Closed Switch Causing a Closed Circuit. (a) Pictorial. (b) Schematic.**

1-6-4 *Short Circuit (Shorted Component)*

A **short circuit** normally occurs when one point in a circuit is accidentally connected to another. Figure 1-25(a) illustrates how a short circuit could occur. In this example, a set of pliers was accidentally laid across the two contacts of the light bulb. Figure 1-25(b) shows the circuit's schematic diagram, with the effect of the short across the light bulb drawn in. In this instance, nearly all the current will flow through the metal of the pliers, which offers no resistance or opposition to the current flow. Since the light bulb does have some resistance or opposition, very little current will flow through the bulb, and therefore no light will be produced.

Short Circuit

Also called a short; it is a low-resistance connection between two points in a circuit, typically causing a large amount of current flow.

FIGURE 1-25 **Short Circuit Due to Pliers across a Lamp. (a) Pictorial. (b) Schematic.**

Use the following questions to test your understanding of Section 1-6.

1. Does current flow through an open circuit?

2. Does current flow through a short circuit?

3. Describe the difference between a closed circuit and a short circuit.

4. Describe the difference between an open switch and a closed switch.

REVIEW QUESTIONS

Multiple-Choice Questions

1. The most commonly used metal in the field of electronics is:

   a. Silver **b.** Copper **c.** Mica **d.** Gold

2. The smallest unit of an element is:

   a. A compound **c.** A molecule
   b. An atom **d.** A proton

3. The smallest unit of a compound is:

   a. An element **c.** An electron
   b. A neutron **d.** A molecule

4. A negative ion has:

   a. More protons than electrons
   b. More electrons than protons
   c. More neutrons than protons
   d. More neutrons than electrons

5. A positive ion has:

   a. Lost some of its electrons **c.** Lost neutrons
   b. Gained extra protons **d.** Gained more electrons

6. One coulomb of charge is equal to:

   a. 6.24×10^{18} electrons **c.** 6.24×10^{8} electrons
   b. 1018×10^{12} electrons **d.** 6.24×10^{81} electrons

7. If 14 C of charge passes by a point in 7 seconds, the current flow is said to be:

   a. 2 A **b.** 98 A **c.** 21 A **d.** 7 A

8. How many electrons are there within 16 C of charge?

   a. 9.98×10^{19} **b.** 14 **c.** 16 **d.** 10.73×10^{19}

9. Current is measured in:

   a. Volts **b.** Coulombs/second **c.** Ohms **d.** Siemens

10. Voltage is measured in units of:

    a. Amperes **b.** Ohms **c.** Siemens **d.** Volts

11. Another word used to describe voltage is:

    a. Potential difference **c.** Electromotive force (emf)
    b. Pressure **d.** All of the above

12. Conductors offer a _____ resistance to current flow.

    a. High **b.** Low **c.** Medium **d.** Maximum

13. Conductance is the measure of how good a conductor is at passing current, and is measured in:

    a. Siemens **b.** Volts **c.** Current **d.** Ohms

14. Insulators have:

    a. Electrons close to the nucleus
    b. Relatively few electrons per atom
    c. An almost complete valence shell
    d. All of the above
    e. None of the above

15. An open circuit will cause:

    a. No current flow **c.** A break in the circuit
    b. Maximum current flow **d.** Both (a) and (c)

Practice Problems

16. What is the value of conductance in siemens for a 100 ohm resistor?

17. Calculate the total number of electrons in 6.5 C of charge.

18. Calculate the amount of current in amperes passing through a conductor if 3 C of charge passes by a point in 4 s.

19. Convert the following:

    a. 0.014 A = _____ mA
    b. 1374 A = _____ kA
    c. 0.776 µA = _____ nA
    d. 0.91 mA = _____ µA

20. Convert the following:
    a. 1473 mV = _____ V
    b. 7143 V = _____ kV
    c. 0.139 kV = _____ V
    d. 0.390 MV = _____ kV

21. What is the relative conductivity of copper if silver is used as the reference conductor?

22. What minimum thickness of porcelain will withstand 24,000 V?

23. To insulate a circuit from 10 V, what insulator thickness would be needed if the insulator is rated at 750 kV/cm?

24. Convert the following:

    a. 2000 kV/cm to _____ meters
    b. 250 kV/cm to _____ mm

25. What maximum voltage may be placed across 35 mm of mica without it breaking down?

Resistance and Power

2

Genius of Chippewa Falls

In 1960, Seymour R. Cray, a young vice-president of engineering for Control Data Corporation, informed president William Norris that in order to build the world's most powerful computer he would need a small research lab built near his home. Norris would have shown any other employee the door, but Cray was his greatest asset, so in 1962 Cray moved into his lab, staffed by 34 people and nestled in the woods near his home overlooking the Chippewa River in Wisconsin. Eighteen months later the press was invited to view the 14- by 6-foot 6600 supercomputer that could execute 3 million instructions per second and contained 80 miles of circuitry and 350,000 transistors, which were so densely packed that a refrigeration cooling unit was needed due to the lack of airflow.

Cray left Control Data in 1972 and founded his own company, Cray Research. Four years later the $8.8 million Cray-1 scientific supercomputer outstripped the competition. It included some revolutionary design features, one of which is that since electronic signals cannot travel faster than the speed of light (1 foot per billionth of a second), the wire length should be kept as short as possible, because the longer the wire the longer it takes for a message to travel from one end to the other. With this in mind, Cray made sure that none of the supercomputer's conducting wires exceeded 4 feet in length.

In the summer of 1985, the Cray-2, Seymour Cray's latest design, was installed at Lawrence Livermore Laboratory. The Cray-2 was 12 times faster than the Cray-1, and its densely packed circuits are encased in clear Plexiglas and submerged in a bath of liquid coolant. The 60-year-old genius has moved on from his latest triumph, nicknamed "Bubbles," and is working on another revolution in the supercomputer field, because for Seymour Cray a triumph is merely a point of departure.

Introduction

Voltage, current, resistance, and power are the four basic properties of prime importance in our study of electronics. In Chapter 1, voltage and current were introduced, and in this chapter we will examine resistance and power.

2-1   WHAT IS RESISTANCE?

By definition, resistance is the opposition to current flow accompanied by the dissipation of heat. To help explain the concept of resistance, Figure 2-1 compares fluid opposition to electrical opposition.

In the fluid circuit in Figure 2-1(a), a valve has been opened almost completely, so a very small opposition to the water flow exists within the pipe. This small or low resistance within the pipe will not offer much opposition to water flow, so a large amount of water will flow through the pipe and gush from the outlet. In the electrical circuit in Figure 2-1(b), a small value **resistor** (which is symbolized by the zigzag line) has been placed in the circuit. The resistor is a device that is included within electrical and electronic circuits to oppose current flow by introducing a certain value of circuit **resistance.** In this circuit, a small value resistor has been included to provide very little resistance to the passage of current flow. This low resistance or small opposition will allow a large amount of current to flow through the conductor, as illustrated by the thick current line.

In the fluid circuit in Figure 2-1(c), the valve is almost completely closed, resulting in a high resistance or opposition to water flow, so only a trickle of water passes through the pipe and out from the outlet. In the electrical circuit in Figure 2-1(d), a large value resistor has been placed in the circuit, causing a large resistance to the passage of current. This high resistance allows only a small amount of current to flow through the conductor, as illustrated by the thin current line.

In the previous examples of low resistance and high resistance shown in Figure 2-1, you may have noticed that resistance and current are inversely proportional ($1/\infty$) to one another. This means that if resistance is high the current is low, and if resistance is low, the current is high.

Resistor

Device constructed of a material that opposes the flow of current by having some value of resistance.

Resistance

Symbolized R and measured in ohms (Ω), it is the opposition to current flow with the dissipation of energy in the form of heat.

TIME LINE

Ohm's law, the best known law in electrical circuits, was formulated by Georg S. Ohm (1787-1854), a German physicist. His law was so coldly received that his feelings were hurt and he resigned his teaching post. When his law was finally recognized, he was reinstated. In honor of his accomplishments, the unit of resistance is called the ohm.

(a)

(b)

FIGURE 2-1   The Opposite Relationship between Resistance and Current.

FIGURE 2-1 (continued) The Opposite Relationship between Resistance and Current.

> **Resistance (R) is Inversely Proportional $\left(\dfrac{1}{\propto}\right)$ to Current (I)**
>
> $$R \quad \frac{1}{\propto} \quad I$$

This relationship between resistance and current means that if resistance is increased by some value, current will be decreased by the same value (assuming a constant electrical pressure or voltage). For example, if resistance is doubled current is halved and similarly, if resistance is halved current is doubled.

$$R\uparrow\frac{1}{\propto}I\downarrow \qquad R\downarrow\frac{1}{\propto}I\uparrow$$

In Figure 2-1, the fluid analogy was used alongside the electrical circuit to help you understand the idea of low resistance and high resistance. In the following section, we will examine resistance in more detail, and define clearly exactly how much resistance exists within a circuit.

SELF-TEST EVALUATION POINT FOR SECTION 2-1

Use the following questions to test your understanding of Section 2-1:

1. What is resistance?
2. What is the difference between low and high resistance?
3. What is the relationship between current and resistance?
4. Would a very high resistance have a small or a large conductance figure?

2-2 THE OHM

Ohm

Unit of resistance, symbolized by the Greek capital letter omega (Ω).

As discussed in the previous chapter, current is measured in amperes and voltage is measured in volts. Resistance is measured in **ohms,** in honor of Georg Simon Ohm, who was the first to formulate the relationship between current, voltage, and resistance.

The larger the resistance, the larger the value of ohms and the more the resistor will oppose current flow. The ohm is given the symbol Ω, which is the Greek capital letter omega. By definition, 1 ohm is the value of resistance that will allow 1 ampere of current to flow through a circuit when a voltage of 1 volt is applied, as shown in Figure 2-2(a), where the resistor is drawn as a zigzag. In some schematics or circuit diagrams the resistor is drawn as a rectangular block, as shown in Figure 2-2(b). The pictorial of this circuit is shown in Figure 2-2(c).

The circuit in Figure 2-3 reinforces our understanding of the ohm. In this circuit, a 1 V battery is connected across a resistor, whose resistance can be either increased or decreased. As the resistance in the circuit is increased the current will decrease and, conversely, as the resistance of the resistor is decreased the circuit current will increase. So how much is 1 ohm of resistance? The answer to this can be explained by adjusting the resistance of the variable resistor. If the resistor is adjusted until exactly 1 amp of current is flowing around the circuit, the value of resistance offered by the resistor is referred to as 1 ohm (Ω).

FIGURE 2-2   One Ohm. (a) New and (b) Old Resistor Symbols. (c) Pictorial.

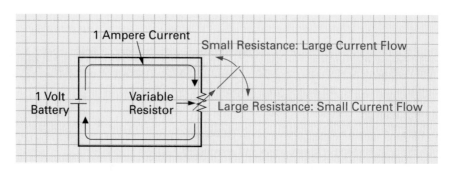

FIGURE 2-3   1 Ohm Allows 1 Amp to Flow When 1 Volt Is Applied.

2-2-1 *Ohm's Law*

Current flows in a circuit due to the electrical force or voltage applied. The amount of current flow in a circuit, however, is limited by the amount of resistance in the circuit. It can be said, therefore, that the amount of current flow around a circuit is dependent on both voltage and resistance. This relationship between the three electrical properties of current, voltage, and resistance was discovered by Georg Simon Ohm, a German physicist, in 1827. Published originally in 1826, **Ohm's law** states that the current flow in a circuit is directly proportional ($\propto$) to the source voltage applied and inversely proportional ($1/\propto$) to the resistance of the circuit.

Stated in mathematical form, Ohm arrived at this formula:

$$\text{current } (I) = \frac{\text{voltage } (V)}{\text{resistance } (R)}$$

$$\text{current } (I) \propto \text{voltage } (V)$$

$$\text{current } (I) \frac{1}{\propto} \text{resistance } (R)$$

Ohm's Law
Relationship between the three electrical properties of voltage, current, and resistance. Ohm's law states that the current flow within a circuit is directly proportional to the voltage applied across the circuit and inversely proportional to its resistance.

2-2-2 *The Ohm's Law Triangle*

Figure 2-4 shows how the three Ohm's law formulas can be placed within a triangle for easy memory recall.

2-2-3 *Current Is Proportional to Voltage (I $\propto$ V)*

Ohm's law states that the current flow in a circuit is proportional ($\propto$) to the circuit's applied voltage, and inversely proportional ($1/\propto$) to the circuit's resistance. In this section we will again use the fluid analogy to help you clearly see the first relationship described in Ohm's law.

Referring to the fluid system in Figure 2-5(a), you can see that an increase in the pump pressure will result in an increase in water flow. Similarly, in the electrical circuit in Figure 2-5(a), an increase in voltage (electron moving force) will exert a greater pressure on the circuit electrons and cause an increase in current flow. This point is reinforced with the Ohm's law formula:

$$I\uparrow = \frac{V\uparrow}{R} \qquad \begin{array}{l}\text{Both properties are above the line}\\ \text{and are therefore proportional.}\end{array}$$

In the two following examples, you will see how an increase in the voltage applied to a circuit results in a proportional increase in the circuit current.

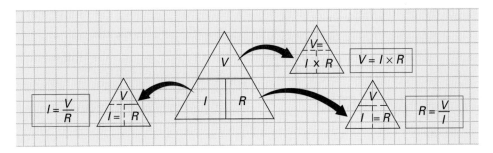

FIGURE 2-4   Ohm's Law Triangle.

(a)

(b)

FIGURE 2–5 **Current Flow Is Proportional to Voltage. (a) Pressure or Voltage Increase Causes a Water or Current Flow Increase (V↑, I ↑). (b) Pressure or Voltage Decrease Results in a Water or Current Flow Decrease (V↓, I ↓).**

EXAMPLE:

If the resistance in the circuit in Figure 2-6(a) remains constant at 1 Ω and the applied voltage equals 2 V, what is the value of current flowing through the circuit?

FIGURE 2-6

CHAPTER 2 / RESISTANCE AND POWER

Solution:

$$\text{current } (I) = \frac{\text{voltage } (V)}{\text{resistance } (R)}$$

$$= \frac{2 \text{ V}}{1 \text{ }\Omega}$$

$$= 2 \text{ A}$$

Step	Keypad Entry	Display Response
1.	[2]	2
2.	[÷]	
3.	[1]	1
4.	[=]	2

■ **EXAMPLE:**

If the voltage from the previous example is now doubled to 4 V as shown in Figure 2-6(b), what would be the change in current?

■ **Solution:**

$$\text{current } (I) = \frac{\text{voltage } (V)}{\text{resistance } (R)}$$

$$= \frac{4 \text{ V}}{1 \text{ }\Omega}$$

$$= 4 \text{ A}$$

On the other hand, the fluid system shown in Figure 2-5(b) shows how a decrease in pump pressure will result in a decrease in water flow. Similarly, the electrical circuit in Figure 2-5(b) shows how a decrease in the circuit's applied voltage will cause a decrease in circuit current. This is again reinforced by Ohm's law.

$$I\!\downarrow \; = \frac{V\!\downarrow}{R}$$

In the following two examples, you will see how a decrease in the applied voltage will result in a proportional decrease in circuit current.

■ **EXAMPLE:**

If a circuit has a resistance of 2 Ω and an applied voltage of 8 V, what would be the circuit current?

■ **Solution:**

$$\text{current } (I) = \frac{\text{voltage } (V)}{\text{resistance } (R)}$$

$$= \frac{8 \text{ V}}{2 \text{ }\Omega}$$

$$= 4 \text{ A}$$

■ **EXAMPLE:**

If the circuit described in the previous example were to have its applied voltage halved to 4 V, what would be the change in circuit current?

$$\text{current } (I) = \frac{\text{voltage } (V)}{\text{resistance } (R)}$$

$$= \frac{4 \text{ V}}{2 \text{ }\Omega}$$

$$= 2 \text{ A}$$

To summarize the relationship between circuit voltage and current, we can say that if the voltage were to double, the current within the circuit would also double (assuming the circuit's resistance were to remain at the same value). Similarly, if the voltage were halved, the current would also halve, making the two properties directly proportional to one another.

2-2-4 *The Current Is Inversely Proportional to Resistance* $\left(I \frac{1}{\propto} R \right)$

The second relationship in Ohm's law states that the circuit current is inversely proportional to the circuit's resistance. The fluid analogy is again used in Figure 2-7 to help you clearly see this second Ohm's law relationship.

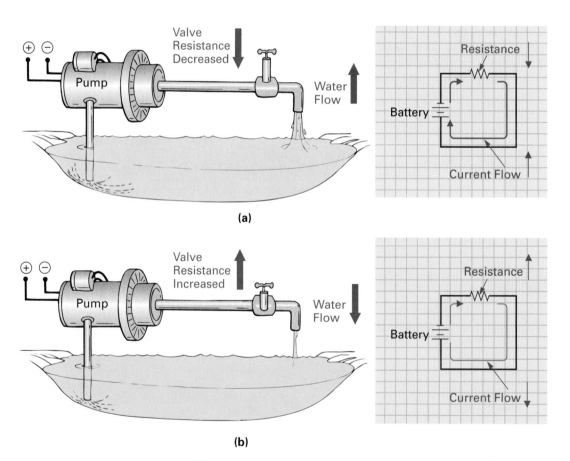

(a)

(b)

FIGURE 2-7 Current Is Inversely Proportional to Resistance. (a) Resistance Decrease Causes a Water or Current Flow Increase ($R\downarrow, I\uparrow$). (b) Resistance Increase Causes a Water or Current Flow Decrease ($R\uparrow, I\downarrow$).

In the fluid system in Figure 2-7(a), an open valve offers very little opposition to the water flow, and so a large water flow exists in a system with a low resistance. Similarly, in the electrical circuit in Figure 2-7(a), a small resistance ($R\downarrow$) allows a large current ($I\uparrow$) flow around the circuit.

This point is reinforced with the Ohm's law formula:

$$R\downarrow = \frac{V}{I\uparrow}$$ One property is above the line, the other is below the line, and therefore the two are inversely proportional.

On the other hand, Figure 2-7(b) shows how a large resistance ($R\uparrow$) will always result in a small current ($I\downarrow$).

$$R\uparrow = \frac{V}{I\downarrow}$$

To help reinforce this relationship, let us look at a couple of examples.

☐ EXAMPLE:

Figure 2-8 shows a circuit that has a constant 8 volts applied. If the resistance in this circuit were doubled from 2 ohms to 4 ohms, what would happen to the circuit current?

FIGURE 2-8

▪ *Solution:*

If the resistance in Figure 2-8 were doubled from 2 Ω to 4 Ω, the circuit current would halve from 4 A to:

$$\text{current } (I) = \frac{\text{voltage } (V)}{\text{resistance } (R)}$$
$$= \frac{8 \text{ V}}{4 \text{ } \Omega}$$
$$= 2 \text{ A}$$

☐ EXAMPLE:

If the resistance in Figure 2-8 were returned to its original value of 2 Ω, what would happen to the circuit current?

▪ *Solution:*

If the resistance in Figure 2-8 were halved from 4 Ω to 2 Ω, the circuit current would double from 2 A to:

$$\text{current } (I) = \frac{\text{voltage } (V)}{\text{resistance } (R)}$$
$$= \frac{8 \text{ V}}{2 \text{ } \Omega}$$
$$= 4 \text{ A}$$

To summarize the relationship between circuit current and resistance, we can say that if the resistance were to double, the current within the circuit would be halved (assuming the circuit's voltage were to remain at the same value). Similarly, if the circuit resistance were halved, the circuit current would double, confirming that current is inversely proportional to resistance.

SELF-TEST EVALUATION POINT FOR SECTION 2-2

Use the following questions to test your understanding of Section 2-2:

1. Define 1 ohm in relation to current and voltage.
2. Calculate I if $V = 24$ V and $R = 6$ Ω.
3. What is the Ohm's law triangle?
4. What is the relationship between: (a) current and voltage (b) current and resistance?
5. Calculate V if $I = 25$ mA and $R = 1$ kΩ.
6. Calculate R if $V = 12$ V and $I = 100$ μA.

2-3   CONDUCTORS AND THEIR RESISTANCE

If the current within a circuit needs to be reduced, a resistor is placed in the current path. Resistors are constructed using materials that are known to oppose current flow. Conductors, on the other hand, are not meant to offer any resistance or opposition to current flow. This, however, is not always the case since some are good conductors having a low resistance, while others are poor conductors having a high resistance. Conductance is the measure of how good a conductor is, and even the best conductors have some value of resistance. Up until this time, we have determined a conductor's resistance based on the circuit's electrical characteristics,

$$R = \frac{V}{I}$$

With this formula we could determine the conductor's resistance based on the voltage applied and circuit current. Another way to determine the resistance of a conductor is to examine the physical factors of the conductor.

2-3-1   *Physical Resistance Formula*

The total resistance of a typical conductor, like the one shown in Figure 2-9, is determined by four main physical factors:

1. The type of conducting material used
2. The conductor's cross-sectional area
3. The total length of the conductor
4. The temperature of the conductor

By combining all of the physical factors of a conductor, we can arrive at a formula for its resistance.

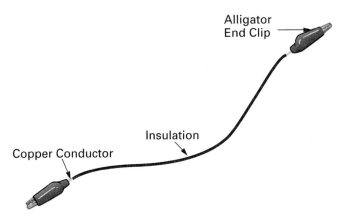

FIGURE 2-9 **Copper Conductor.**

TABLE 2-1 **Material Resistivity**

MATERIAL	RESISTIVITY (cmil/ft) IN OHMS
Silver	9.9
Copper	10.7
Gold	16.7
Aluminum	17.0
Tungsten	33.2
Zinc	37.4
Brass	42.0
Nickel	47.0
Platinum	60.2
Iron	70.0

$$R = \frac{\rho \times l}{a}$$

where R = resistance of conductor, in ohms
ρ = resistivity of conducting material
l = length of conductor, in feet
a = area of conductor, in circular mils

Resistivity, by definition, is the resistance (in ohms) that a certain length of conductive material (in feet) will offer to the flow of current. Table 2-1 lists the resistivity of the more commonly used conductors.

The resistance of a conductor, therefore, can be calculated based either on its physical factors or on electrical performance.

$$\frac{\rho \times l}{a} = R = \frac{V}{I}$$

Physical $\Longleftrightarrow$ Electrical

Resistivity
Measure of a material's resistance to current flow.

☐ **EXAMPLE:**

Calculate the resistance of 333 ft of copper conductor with a conductor area of 3257 cmil.

CALCULATOR SEQUENCE

Step	Keypad Entry	Display Response
1.	[1] [0] [.] [7]	10.7
2.	[×]	
3.	[3] [3] [3]	333
4.	[÷]	3563.1
5.	[3] [2] [5] [7]	3257
6.	[=]	1.09

$$R = \frac{\rho \times l}{a}$$

$$= \frac{10.7 \times 333}{3257}$$

$$= 1.09 \ \Omega$$

◾ EXAMPLE:

Calculate the resistance of 1274 ft of aluminum conductor with a diameter of 86.3 mils.

◾ *Solution:*

$$R = \frac{\rho \times l}{a}$$

If the diameter is equal to 86.3 mils, the circular mil area equals d^2:

$$86.3^2 = 7447.7 \ \text{cmil}$$

Referring to Table 2-1 you can see that the resistivity of aluminum is 17.0, and therefore

$$R = \frac{17 \times 1274}{7447.7}$$

$$= 2.9 \ \Omega$$

2-3-2   *Temperature Effects on Conductors*

When heat is applied to a conductor, the atoms within the conductor convert this thermal energy into mechanical energy or movement. These random moving atoms cause collisions between the directed electrons (current flow) and the adjacent atoms, resulting in an opposition to the current flow (resistance).

Metallic conductors are said to have a **positive temperature coefficient of resistance** (+ Temp. Coe. of R). This means that the greater the heat applied to the conductor, the greater the atom movement, causing more collisions of atoms to occur, and consequently the greater the conductor's resistance.

Positive Temperature Coefficient of Resistance

The rate of increase in resistance relative to an increase in temperature.

heat ↑   resistance ↑

2-3-3   *Maximum Conductor Current*

American Wire Gauge

American wire gauge (AWG) is a system of numerical designations of wire sizes, with the first being 0000 (the largest size) and then going to 000, 00, 0, 1, 2, 3, and so on up to the smallest sizes of 40 and above.

Any time that current flows through any conductor, a certain resistance or opposition is inherent in that conductor. This resistance will convert current to heat, and the heat further increases the conductor's resistance (+ Temp. Coe. of R), causing more heat to be generated due to the opposition. As a result, a conductor must be chosen carefully for each application so that it can carry the current without developing excessive heat. This is achieved by selecting a conductor with a greater cross-sectional area to decrease its resistance. The National Fire Protection Association has developed a set of standards known as the **American Wire Gauge** for all copper conductors, which lists their diameter, resistance, and maximum safe current in amperes. This table is given in Table 2-2. A rough guide for measuring wire size is shown in Figure 2-10.

TABLE 2-2 The American Wire Gauge (AWG) for Copper Conductor

AWG NUMBER[a]	DIAMETER (mils)	MAXIMUM CURRENT (A)	Ω/1000 ft
0000	460.0	230	0.0490
000	409.6	200	0.0618
00	364.8	175	0.0780
0	324.9	150	0.0983
1	289.3	130	0.1240
2	257.6	115	0.1563
3	229.4	100	0.1970
4	204.3	85	0.2485
5	181.9	75	0.3133
6	162.0	65	0.3951
7	144.3	55	0.4982
8	128.5	45	0.6282
9	114.4	40	0.7921
10	101.9	30	0.9981
11	90.74	25	1.260
12	80.81	20	1.588
13	71.96	17	2.003
14	64.08	15	2.525
15	57.07		3.184
16	50.82	6	4.016
17	45.26	Wires	5.064
18	40.30	of this 3	6.385
19	35.89	size	8.051
20	31.96	have	10.15
22	25.35	current	16.14
26	15.94	measured	40.81
30	10.03	in mA	103.21
40	3.145		1049.0

[a]The larger the AWG number, the smaller the size of the conductor.

FIGURE 2-10   Wire Gauge Size.

Conducting wires are normally covered with a plastic or rubber type of material, known as insulation, as shown in Figure 2-11. This insulation is used to protect the users and technicians from electrical shock and also to keep the conductor from physically contacting other conductors within the equipment. If the current through the conductor is too high, this insulation will burn due to the heat and may cause a fire hazard. This is why a conductor's maximum safe current value should not be exceeded.

FIGURE 2-11   Conductor with Insulator.

FIGURE 2-12   Wires and Cables. (a) Solid Wire. (b) Stranded Wire. (c) Braided Wire. (d) Printed Wire. (e) Coaxial Cable. (f) Twin Lead.

FIGURE 2-13   Connectors. (a) Temporary Connectors. (b) Plugs. (c) Lug and Binding Post. (d) Sockets.

CHAPTER 2 / RESISTANCE AND POWER

2-3-4 Conductor and Connector Types

A cable is made up of two or more wires. Figure 2-12 illustrates some of the different types of **wires** and **cables.** In Figure 2-12(a), (b), and (c), only one conductor exists within the insulation, so they are classified as wires, whereas the cables seen in Figure 2-12(e) and (f) have two conductors. The coaxial and twin-lead cables are most commonly used to connect TV signals into television sets. Figure 2-12(d) illustrates how conducting copper, silver, or gold strips are printed on a plastic insulating board and are used to connect components such as resistors that are mounted on the other side of the board.

Wires and cables have to connect from one point to another. Some are soldered directly, whereas others are attached to plugs that plug into sockets. A sample of connectors is shown in Figure 2-13.

Wire

Single solid or stranded group of conductors having a low resistance to current flow.

Cable

Group of two or more insulated wires.

SELF-TEST EVALUATION POINT FOR SECTION 2-3

Use the following questions to test your understanding of Section 2-3:

1. List the four factors that determine the total resistance of a conductor.
2. What is the relationship between the resistance of a conductor and its cross-sectional area, length, and resistivity?
3. True or false: Conductors are said to have a positive temperature coefficient.
4. True or false: The smaller the AWG number, the larger the size of the conductors.

2-4 ENERGY, WORK, AND POWER

The sun provides us with a consistent supply of energy in the form of light. Coal and oil are fossilized vegetation that grew, among other things, due to the sun, and are examples of **energy** that the earth has stored for millions of years. It can be said, then, that all energy begins from the sun. On the earth, energy is not created or destroyed; it is merely transformed from one form to another. The transforming of energy from one form to another is called **work.** The greater the energy transformed, the more work that is done.

The six basic forms of energy are light, heat, magnetic, chemical, electrical, and mechanical energy. The unit for energy is the **joule** (J).

Energy

Capacity to do work.

Work

Work is done anytime energy is transformed from one type to another, and the amount of work done is dependent on the amount of energy transformed.

Joule

The unit of work and energy.

EXAMPLE:

One person walks around a track and takes 5 minutes, while another person runs around the track and takes 50 seconds. Both were full of energy before they walked or ran around the track, and during their travels around the track they converted the chemical energy within their bodies into the mechanical energy of movement.

a. Who exerted the most energy?
b. Who did the most work?

Solution:

Both exerted the same amount of energy. The runner exerted all his energy (for example, 100 J) in the short time of 50 seconds, while the walker spaced his energy (100 J) over 5 minutes. Since they both did the same amount of work, the only difference between the runner and the walker is time, or the rate at which their energy was transformed.

TIME LINE

James P. Joule (1818-1889), an English physicist and self-taught scientist, conducted extensive research into the relationships between electrical, chemical, and mechanical effects, which led him to the discovery that one energy form can be converted into another. For this achievement, his name was given to the unit of energy, the joule.

Power

Amount of energy converted by a component or circuit in a unit of time, normally seconds. It is measured in units of watts (joules/second).

TIME LINE

The unit of power, the watt, was named after Scottish engineer and inventor James Watt (1736-1819) in honor of his advances in the field of science.

Watt (W)

Unit of electric power required to do work at a rate of 1 joule/second. One watt of power is expended when 1 ampere of direct current flows through a resistance of 1 ohm.

2-4-1   *Power*

Power (*P*) is the rate at which work is performed and is given the unit of **watt** (W), which is joules per second (J/s).

Returning to the example involving the two persons walking and running around the track, we could say that the number of joules of energy exerted in 1 second by the runner was far greater than the number of joules of energy exerted in 1 second by the walker, although the total energy exerted by both persons around the entire track was equal and therefore the same amount of work was done. Even though the same amount of energy was used, and therefore the same amount of work was done by the runner and the walker, the output power of each was different. The runner exerted a large value of joules/second or watts (high power output) in a short space of time, while the walker exerted only a small value of joules/second or watts (low power output) over a longer period of time.

Whether discussing a runner, walker, electric motor, heater, refrigerator, light bulb, or compact disk player—power is power. The output power, or power ratings, of electrical, electronic, or mechanical devices can be expressed in watts and describes the number of joules of energy converted every second. The output power of rotating machines is given in the unit *horsepower* (hp), the output power of heaters is given in the unit British thermal unit per hour (Btu/h), and the output power of cooling units is given in the unit *ton of refrigeration*. Despite the different names, they can all be expressed quite simply in the unit of watts. The conversions are as follows:

$$1 \text{ horsepower (hp)} = 746 \text{ W}$$
$$1 \text{ British thermal unit per hour (Btu/h)} = 0.293 \text{ W}$$
$$1 \text{ ton of refrigeration} = 3.52 \text{ kW (3520 W)}$$

Now we have an understanding of power, work, and energy. Let's reinforce our knowledge by introducing the energy formula and try some examples relating to electronics.

2-4-2   *Calculating Energy*

The amount of energy stored (*W*) is dependent on the coulombs of charge stored (*Q*) and the voltage (*V*).

$$W = Q \times V$$

where W = energy stored, in joules (J)
Q = coulombs of charge (1 coulomb = 6.24×10^{18} electrons)
V = voltage, in volts (V)

If you consider a battery as an example, you can probably better understand this formula. The battery's energy stored is dependent on how many coulombs of electrons it holds (current) and how much electrical pressure it is able to apply to these electrons (voltage).

☐   **EXAMPLE:**

How many coulombs of electrons would a 9 V battery have to store to have 63 J of energy?

Solution:

If $W = Q \times V$, then by transposition:

$$Q = \frac{W}{V}$$

coulombs of electrons $(Q) = \dfrac{\text{energy in joules } (W)}{\text{battery voltage } (V)}$

$$= \frac{63 \text{ J}}{9 \text{ V}}$$

$$= 7 \text{ C of electrons}$$

or

$$7 \times 6.24 \times 10^{18} = 4.36 \times 10^{19} \text{ electrons}$$

Step	Keypad Entry	Display Response
1.	6 3	63
2.	÷	
3.	9	9
4.	=	7
5.	×	7
6.	6 . 2 4 E 1 8	6.24E18
7.	=	4.36E19

2-4-3   *Calculating Power*

Power, in connection with electricity and electronics, is the rate (t) at which electric energy (W) is converted into some other form.

$$P = I \times V$$

where P = power, in watts (W)
I = current, in amperes (A)
V = voltage, in volts (V)

This formula states that the amount of power delivered to a device is dependent on the electrical pressure or voltage applied across the device and the current flowing through the device.

■ EXAMPLE:

3 Amperes

9 Volts

FIGURE 2-14   Calculating Power.

In regards to electrical and electronic circuits, power is the rate at which electric energy is converted into some other form. In the example in Figure 2–14, electric energy will be transformed into light and heat energy by the light bulb. Power has the unit of watts, which is the number of joules of energy transformed per second (J/s). If 27 J of electric energy is being transformed into light and heat per second, how many watts of power does the light bulb convert?

■ Solution:

$$\text{power} = \frac{\text{joules}}{\text{second}}$$

$$= \frac{27 \text{ J}}{1 \text{ s}}$$

$$= 27 \text{ W}$$

The power output in the previous example could have easily been calculated by merely multiplying current by voltage to arrive at the same result.

$$\text{power} = I \times V = 3\,\text{A} \times 9\,\text{V}$$
$$= 27\,\text{W}$$

We could say, therefore, that the light bulb dissipates 27 watts of power, or 27 joules of energy per second.

Studying the power formula $P = I \times V$, we can also say that 1 watt of power is expended when 1 ampere of current flows through a circuit that has 1 volt applied.

Like Ohm's law, we can transpose the power formula as follows:

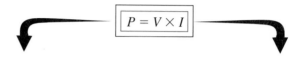

$$V = \frac{P}{I}$$

This formula can be used to calculate V when P and I are known.

$$I = \frac{P}{V}$$

This formula can be used to calculate I when P and V are known.

When trying to calculate power we may not always have the values of V and I available. For example, we may know only I and R, or V and R. The following shows how we can substitute terms in the $P = V \times I$ formula to arrive at alternative power formulas for wattage calculations.

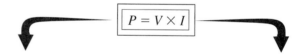

$$P = \frac{V^1}{R}$$

This formula can be used to calculate P when I and R are known.

$$P = I^2 \times R$$

This formula can be used to calculate P when I and R are known.

EXAMPLE:

In Figure 2-15 a 12 V battery is connected across a 36 Ω resistor. How much power does the resistor dissipate?

FIGURE 2-15

Solution:

Since V and R are known, the V2/R power formula should be used.

$$\text{power} = \frac{\text{voltage}^2}{\text{resistance}}$$

$$= \frac{12\ V^2}{36\ \Omega} = \frac{144}{36}$$

$$= 4W$$

Four joules of heat energy are being dissipated every second.

CALCULATOR SEQUENCE

Step	Keypad Entry	Display Response
1.	1 2	12
2.	X² (square key)	144
3.	÷	
4.	3 6	36
5.	=	4

2-4-4 Measuring Power

The multimeter can be used to measure power by following the procedure described in Figure 2-16. In this example, the power consumed by a car's music system is being determined. After the current measurement and voltage measurement have been taken, the product of the two values will have to be calculated, since power = current × voltage ($P = I \times V$).

EXAMPLE:

Calculate the power consumed by the car music system shown in Figure 2-16.

Solution:

Current drawn from battery by music system = 970 mA. Voltage applied to music system power input = 13.6 V.

$$P = I \times V = 970\ \text{mA} \times 13.6\ \text{V} = 13.2\ \text{watts}$$

By connecting a special current probe, as shown in Figure 2-17, some multimeters are able to make power measurements. In this configuration, the current probe senses current, the standard meter probes measure voltage, and the multimeter performs the multiplication process and displays the power reading.

2-4-5 The Kilowatt-Hour

You and I pay for our electric energy in a unit called the **kilowatt-hour** (kWh). The **kilowatt-hour meter,** shown in Figure 2-18, measures how many kilowatt-hours are consumed, and the electric company then charges accordingly.

So what is a kilowatt-hour? Well, since power is the rate at which energy is used, if we multiply power and time, we can calculate how much energy has been consumed.

energy consumed (W) = power (P) × time (t)

This formula uses the product of power (in watts) and time (in seconds or hours) and so we can use one of three units: the watt-second (Ws), watt-hour (Wh), or kilowatt-hour (kWh). The kilowatt-hour is most commonly used by electric companies, and by definition, a kilowatt-hour of energy is consumed when you use 1000 watts of power in 1 hour (1 kW in 1 h).

energy consumed (kWh) = power (kW) × time (h)

Kilowatt-Hour
1000 watts for 1 hour.

Kilowatt-Hour Meter
A meter used by electric companies to measure a customer's electric power use in kilowatt-hours.

1. First determine the current value being drawn by the device (follow "measuring current" procedure).

I = 970 mA

2. Then determine the voltage value being applied across device (follow "measuring voltage" procedure).

V = 13.6 V

Car AM/FM Radio and CD Player (14 Ω)

3. Multiply voltage value by current value to obtain power value ($P = V \times I$).

FIGURE 2-16   Measuring Power with a Multimeter.

Current Probe Clips onto Power Line and Senses Current (*I*).

Standard Meter Probes Measure Applied Voltage (*V*).

Meter's Internal Circuitry Performs Multiplication of *I* and *V* and Displays Wattage Result (*P*).

FIGURE 2-17   Configuring the Multimeter to Measure Power.

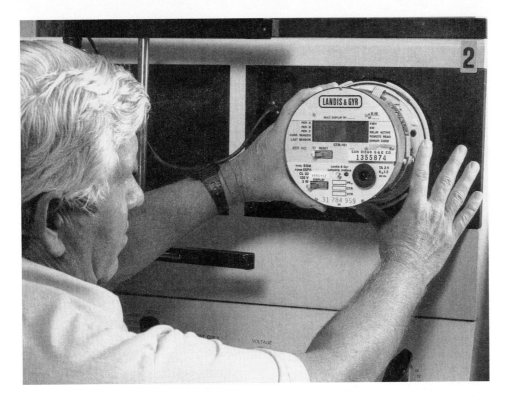

FIGURE 2-18   Digital Kilowatt-Hour Meter.

To see how this would apply, let us look at a couple of examples.

EXAMPLE:

If a 100 W light bulb is left on for 10 hours, how many kilowatt-hours will we be charged for?

Solution:

$$\begin{aligned}
\text{power consumed (kWh)} &= \text{power (kW)} \times \text{time (hours)} \\
&= 0.1 \text{ kW} \times 10 \text{ hours} \\
&\quad (100 \text{ W} = 0.1 \text{ kW}) \\
&= 1 \text{ kWh}
\end{aligned}$$

CALCULATOR SEQUENCE

Step	Keypad Entry	Display Response
1.	⓪ ⨀ ① Ⓔ ③	0.1E3
2.	ⓧ	
3.	① ⓪	10
4.	⊜	1E3

EXAMPLE:

Figure 2-19 illustrates a typical household electric heater and an equivalent electrical circuit. The heater has a resistance of 7 Ω and the electric company is charging 12 cents/kWh. Calculate:

a. The energy consumed by the heater

b. The cost of running the heater for 7 hours

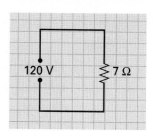

Electric Heater

120 V

120 V   7 Ω

FIGURE 2-19   An Electric Heater

■ *Solution:*

a. Power $(P) = \dfrac{V^2}{R} = \dfrac{120^2}{7}$

$$= \dfrac{14400 \text{ V}}{7} = \dfrac{14.4 \text{ kV}}{7}$$

$$= 2057 \text{ W (approximately 2 kilowatts or 2 kW)}$$

b. Energy consumed = power (kW) × time (hours)

$$= 2.057 \times 7 \text{ h}$$

$$= 14.4 \text{ kWh}$$

Cost = kWh × rate = 14.4 × 12 cents

$$= \$1.73$$

SELF-TEST EVALUATION POINT FOR SECTION 2-4

Use the following questions to test your understanding of Section 2-4:

1. List the six basic forms of energy.
2. What is the difference between energy, work, and power?
3. List the formulas for calculating energy and power.
4. What is 1 kilowatt-hour of energy?

REVIEW QUESTIONS

Multiple-Choice Questions

1. Resistance is measured in:
    a. Ohms   c. Amperes
    b. Volts   d. Siemens

2. Current is proportional to:
    a. Resistance   c. Both (a) and (b)
    b. Voltage   d. None of the above

3. Current is inversely proportional to:
    a. Resistance   c. Both (a) and (b)
    b. Voltage   d. None of the above

4. If the applied voltage is equal to 15 V and the circuit resistance equals 5 Ω, the total circuit current would be equal to:
    a. 4 A   b. 5 A   c. 3 A   d. 75 A

5. Calculate the applied voltage if 3 mA flows through a circuit resistance of 25 kΩ.
    a. 63 mV   b. 25 V   c. 77 μV   d. 75 V

6. Energy is measured in:
    a. Volts   b. Joules   c. Amperes   d. Watts

7. Work is measured in:
    a. Joules   c. Amperes
    b. Volts   d. Watts

8. Power is the rate at which energy is transformed and is measured in:
    a. Joules   c. Volts
    b. Watts   d. Amperes

9. A good conductor has a:

    a. Large conductance figure   **c.** Both (a) and (b)

    b. Small resistance figure   **d.** None of the above

10. AWG is an abbreviation for:

    a. Alternate Wire Gauge   **c.** American Wave Guide

    b. Alternating Wire Gauge **d.** American Wire Gauge

Practice Problems

11. An electric heater with a resistance of 6 Ω is connected across a 120 V wall outlet.

    a. Calculate the current flow.

    b. Draw the schematic diagram.

12. What source voltage would be needed to produce a current flow of 8 mA through a 16 kΩ resistor?

13. If an electric toaster draws 10 A when connected to a power outlet of 120 V, what is its resistance?

14. Calculate the power used in Problems 11, 12, and 13.

15. Calculate the current flowing through the following light bulbs when they are connected across 120 V:

    a. 300 W   **b.** 100 W   **c.** 60 W   **d.** 25 W

16. If an electric company charges 9 cents/kWh, calculate the cost for each light bulb in Problem 15 if on for 10 hours.

17. Indicate which of the following unit pairs is larger:

    a. Millivolts or volts

    b. Microamperes or milliamperes

    c. Kilowatts or watts

    d. Kilohms or megohms

18. Calculate the resistance of 200 ft of copper having a diameter of 80 mils.

19. What AWG size wire should be used to safely carry just over 15 A?

20. Calculate the voltage dropped across 1000 ft of No. 4 copper conductor when a current of 7.5 A is flowing through it.

21. Calculate the unknown resistance in a circuit when an ammeter indicates that a current of 12 mA is flowing and a voltmeter indicates 12 V.

22. What battery voltage would use 1000 J of energy to move 40 C of charge through a circuit?

23. Calculate the resistance of a light bulb that passes 500 mA of current when 120 V is applied. What is the bulb's wattage?

24. Which of the following circuits has the largest resistance and which has the smallest?

    a. $V = 120$ V, $I = 20$ mA   **c.** $V = 9$ V, $I = 100$ μA

    b. $V = 12$ V, $I = 2$ A   **d.** $V = 1.5$ V, $I = 4$ mA

25. Calculate the power dissipated in each circuit in Problem 54.

26. How many watts are dissipated if 5000 J of energy are consumed in 25 s?

27. Convert the following:

    a. 1000 W = _____ kW

    b. 0.345 W = _____ mW

    c. 1250×10^3 W = _____ MW

    d. 0.00125 W = _____ μW

28. What is the value of the resistor when a current of 4 A is causing 100 W to be dissipated?

29. How many kilowatt-hours of energy are consumed in each of the following:

    a. 7500 W in 1 hour   **c.** 127,000 W for half an hour

    b. 25 W for 6 hours

30. What is the maximum output power of a 12 V, 300 mA power supply?

Resistors

At the Core of Apple

**Steven Jobs (left), and
Stephen Wozniak**

Stephen Wozniak and Steven Jobs met at their Los Altos, California, high school, and became friends due to their common interest in electronics.

Wozniak was a conservative youth whose obsession with technology left little room for social relationships or studying; in fact, after one year at the University of Colorado, he dropped out with his academic record littered with F's. In contrast to this serious nature, Wozniak was renowned for his high-tech pranks, for which, on one occasion, he spent a night in juvenile hall for wiring up a fake bomb in a friend's locker. In another incident, he devised a way to place a free telephone call to the Pope at the Vatican, identifying himself as then Secretary of State Henry Kissinger.

Jobs, on the other hand, had interests outside electronics. He searched for intellectual, emotional, and spiritual stimulation. After one semester at Reed College, he dropped out to pursue an interest in Eastern religions, which led him to temples in India, searching for the meaning of life.

The first enterprise built by Wozniak and sold by Jobs was an illegal device called a blue box that could crash the telephone system. It generated a set of tones that fooled the computerized telephone switching systems and opened up free, long-distance circuits; these systems allowed "phone phreaks" to take long and illegal joy rides throughout the world's telephone networks.

By selling a Volkswagen van and a programmable calculator, Wozniak and Jobs raised the initial capital of $1300 to start a business in April 1976 called Apple Computer, since Jobs had a passion for the Beatles, who recorded under the Apple record label.

In just five years, Apple Computer grew faster than any other company in history, from a two-man assetless partnership building computers in a family home to a publicly traded corporation earning its youthful entrepreneurs, Jobs (27) and Wozniak (24), fortunes of nearly $200 million each.

Introduction

Resistance would seem to be an undesirable effect, as it reduces current flow and wastes energy as it dissipates heat. Resistors, however, are probably used more than any other component in electronics, and in this chapter we discuss all the different types.

3-1 RESISTOR TYPES

Conductors are used to connect one device to another, and although they offer a small amount of resistance, this resistance is not normally enough. In electronic circuits, additional resistance is normally needed to control the amount of current flow, and the component used to supply this additional resistance is called a **resistor.**

There are two basic types of resistors: fixed value and variable value. Fixed-value resistors have a resistance value that cannot be changed. On the other hand, variable-value resistors have a range of values that can be selected by adjusting a control.

3-1-1 *Fixed-Value Resistors*

In this section, we will examine some of the **fixed-value resistor** types.

Carbon Composition Resistors

This is the most common and least expensive type of fixed-value resistor, the appearance of which can be seen in Figure 3-1(a). It is constructed by placing a piece of resistive material, with embedded conductors at each end, within an insulating cylindrical molded case, as illustrated in Figure 3-1(b).

This resistive material is called carbon composition because powdered carbon and powdered insulator are bonded into a compound and used as the resistive material. By changing the ratio of powdered insulator to carbon, the value of resistance can be changed within the same area. For example, Figure 3-2 illustrates four fixed-value resistors, ranging from 2 Ω to 10 MΩ.

The 2 Ω resistor is exactly the same size as the 10 MΩ (10 million Ω) resistor. This is achieved by having more powdered carbon and less powdered insulator in the 2 W and less powdered carbon and more powdered insulator in the 10 MΩ. The color-coded rings or bands on the resistors in Figure 3-2 are a means of determining the value of resistance. This and other coding systems will be discussed later.

The physical size of the resistors lets the user know how much power in the form of heat can be **dissipated,** as shown in Figure 3-3. As you already know, resistance is the opposition to current flow, and this opposition causes heat to be generated whenever current is passing through a resistor. The amount of heat dissipated each second is measured in watts and each resistor has its own **wattage rating.** For example, a 2 watt size resistor can dissipate up to 2 joules of heat per second, whereas a ⅛ W size resistor can only dissipate up to ⅛

Resistor

Component made of a material that opposes the flow of current and therefore has some value of resistance.

Fixed-Value Resistor

A resistor whose value cannot be changed.

Carbon Composition Resistor

Fixed resistor consisting of carbon particles mixed with a binder, which is molded and then baked. Also called a composition resistor.

Dissipation

Release of electrical energy in the form of heat.

Wattage Rating

Maximum power a device can safely handle continuously.

(a)

(b)

FIGURE 3-1 Carbon Composition Resistors. (a) Appearance. (b) Construction. (Courtesy of Stackpole Electronics, Inc.)

Low Resistance

2 Ω

1 kΩ

Increasing Amount of Powdered Carbon

100 kΩ

Increasing Amount of Powdered Insulator

10 MΩ

High Resistance

FIGURE 3-2   Carbon Composition Resistor Ratios.

$\frac{1}{8}$ Watt $\frac{1}{4}$ Watt $\frac{1}{2}$ Watt   1 Watt   2 Watts

FIGURE 3-3   Resistor Wattage Rating Guide. All Resistor Silhouettes Are Drawn to Scale.

joule of heat per second. The key point to remember is that resistors in high current circuits should have a surface area that is large enough to dissipate away the heat faster than it is being generated. If the current passing through a resistor generates heat faster than the resistor can dissipate it, the resistor will burn up and no longer perform its function.

Another factor to consider when discussing resistors is their **tolerance.** Tolerance is the amount of deviation or error from the specified value. For example, a 1000 Ω (1 kΩ) resistor with a ±10% (plus and minus 10%) tolerance when manufactured could have a resistance anywhere between 900 and 1100 Ω.

Tolerance

Permissible deviation from a specified value, normally expressed as a percentage.

$$\pm 10\% \text{ of } 1000 = 100$$

$$10\% - \boxed{1000} + 10\%$$
$$\downarrow \qquad\qquad\qquad \downarrow$$
$$900 \qquad\qquad\quad 1100$$

This means that two identically marked resistors when measured could be from 900 to 1100 Ω, a difference of 200 Ω. In some applications, this may be acceptable. In other applications, where high precision is required, this deviation could be too large and so a more expensive, smaller tolerance resistor would have to be used.

EXAMPLE:

Calculate the amount of deviation of the following resistors:

    a. 2.2 kΩ ± 10%

    b. 5 MΩ ± 2%

    c. 3 Ω ± 1%

CALCULATOR SEQUENCE

Step	Keypad Entry	Display Response
1.	1 0	10
2.	%	10
3.	×	0.10
4.	2 . 2 E 3	2.2E3
5.	=	220

Solution:

a. 10% of 2.2 kΩ = 220 Ω. For + 10%, the value is

$$2200 + 220 \, \Omega = 2420 \, \Omega$$
$$= 2.42 \, k\Omega$$

For − 10%, the value is

$$2200 - 220 \, \Omega = 1980 \, \Omega$$
$$= 1.98 \, k\Omega$$

The resistor will measure anywhere from 1.98 kΩ to 2.42 kΩ.

b. 2% of 5 MΩ = 100 kΩ

$$5 \text{ M}\Omega + 100 \text{ k}\Omega = 5.1 \text{ M}\Omega$$

$$5 \text{ M}\Omega - 100 \text{ k}\Omega = 4.9 \text{ M}\Omega$$

Deviation = 4.9 MΩ to 5.1 MΩ

c. 1% of 3 Ω = 0.03 Ω or 30 milliohms (mΩ)

$$3 \Omega + 0.03 \Omega = 3.03 \Omega$$

$$3 \Omega - 0.03 \Omega = 2.97 \Omega$$

Deviation = 2.97 Ω to 3.03 Ω

Carbon Film Resistors

Figure 3-4(a) illustrates the physical appearance of **carbon film resistors.** This resistor type is constructed, as shown in Figure 3-4(b), by first depositing a thin layer or film of resistive material (a blend of carbon and insulator) on a ceramic (insulating) **substrate.** The film is then cut to form a helix or spiral. A greater ratio of carbon to insulator will achieve a low-resistance helix. On the other hand, a greater ratio of insulator to carbon will create a higher-resistance helix. Carbon film resistors have smaller tolerance figures (±5% to ±2%) and have a better temperature stability than carbon composition resistors.

Metal Film Resistors

Figure 3-5(a) illustrates the physical appearance of some typical **metal film resistors.** This resistor type is constructed by spraying a thin film of metal on a ceramic cylinder (or substrate) and then cutting the film, as shown in Figure 3-5(b). Metal film resistors have possibly the best tolerances commercially available of ±1% to ±0.1%. They also maintain a very stable resistance over a wide range of temperatures (good stability) and generate very little internal noise compared to any carbon resistor.

Carbon Film Resistor
Thin carbon film deposited on a ceramic form to create resistance.

Substrate
The mechanical insulating support on which a device is fabricated.

Metal Film Resistor
A resistor in which a film of a metal, metal oxide, or alloy is deposited on an insulating substrate.

(a)

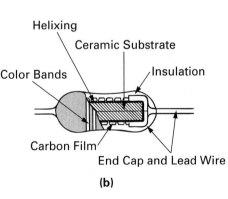

Helixing
Ceramic Substrate
Color Bands
Insulation
Carbon Film
End Cap and Lead Wire

(b)

FIGURE 3-4   Carbon Film Resistors. (a) Appearance. (b) Construction. (Courtesy of Stackpole Electronics, Inc.)

(a)

(b)

FIGURE 3-5 Metal Film Resistors. (a) Appearance. (b) Construction. (Courtesy of Stackpole Electronics, Inc.)

Wirewound Resistors

Figure 3-6(a) and (c) illustrate the appearance of different types of **wirewound resistors.** This resistor type is constructed, as can be seen in Figure 3-6(b), by wrapping a length of wire uniformly around a ceramic insulating core, with terminals making the connections at each end. The length and thickness of the wire are varied to change the resistance, which ranges from 1 Ω to 150 kΩ.

Since current flow is opposed by resistors and this opposition generates heat, the larger the physical size of the resistor, the greater the amount of heat that can be dissipated away

(a)

(b)   **(c)**

FIGURE 3-6 Wirewound Resistors. (a) Appearance. (b) Construction. (c) Other Types. (Courtesy of Stackpole Electronics, Inc.)

and so the greater the current that can be passed through the resistor. Wirewound resistors are generally used in applications requiring low resistance values, which means that the current and therefore power dissipated are high ($R\downarrow = V/I\uparrow$, $P\uparrow = I^2\uparrow R$). As a result, these resistors are designed to have large surface areas so that they can safely dissipate away the heat. The amount of heat that can be dissipated in normal conditions is measured in watts and is indicated on the resistor. If the resistor's heat dissipation is assisted, it can handle a higher current. For example, a 10 W resistor can be used to dissipate 20 W if air is blown across its surface by a cooling fan or if it is immersed in a coolant. Wirewound resistors typically have good tolerances of ±1%. Their large physical size and difficult manufacturing process, however, make them very expensive.

Metal Oxide Resistors

Figure 3-7(a) illustrates the physical appearance of some **metal oxide resistors.** This resistor type is constructed, as can be seen in Figure 3-7(b), by depositing an oxide of a metal such as tin onto an insulating substrate. The ratio of oxide (insulator) to tin (conductor) will determine the resistor's resistance.

Metal oxide resistors have excellent temperature stability, but are more expensive than carbon composition, carbon film, and metal film resistors.

Thick-Film Resistors

Figure 3-8 illustrates examples of **thick-film resistors.** Figure 3-8(a) and (b) show the two different types of resistor networks, called SIPs and DIPs. The **single in-line package** is so called because all its lead connections are in a single line, whereas the **dual in-line package** has two lines of connecting pins. The chip resistors shown in Figure 3-8(c) are small thick-film resistors that are approximately the size of a pencil lead.

The SIP and DIP resistor networks are constructed by first screening on the internal conducting strip (silver) that connects the external pins to the resistive material, and then screening on the thick film of resistive paste (bismuth/ruthenate), the blend of which will determine resistance value. The chip resistor uses the same resistive film paste material, which is deposited onto an insulating substrate with two conductive end terminations and protected by a glass coat.

The SIP and DIP resistor networks, once constructed, are trimmed by lasers to obtain close tolerances of typically ±2%. Resistance values ranging from 22 Ω to 2.2 MΩ are available, with a power rating of ½ W.

The chip resistor shown in Figure 3-8(c) is commercially available with resistance values from 10 Ω to 3.3 MΩ, a ±2% tolerance, and a ⅛ W heat dissipation capability. They are

(a)   (b)

FIGURE 3-7 **Metal Oxide Resistors. (a) Appearance. (b) Construction. (Courtesy of Stackpole Electronics, Inc.)**

FIGURE 3-8   Thick-Film Resistors. (a) SIPs: Single In-Line Packages. (b) DIPs: Dual In-Line Packages. (c) Chips. (Courtesy of Stackpole Electronics, Inc.)

ideally suited for applications requiring physically small sized resistors, as explained in the inset in Figure 3-8(c). The chip resistor is called a **surface mount technology** (SMT) device. The key advantage of SMT devices over "through-hole" devices is that a through-hole device needs both a hole in the printed circuit board (PCB) and a connecting pad around the hole. With the SMT device, no holes are needed since the package is soldered directly onto the surface of the PCB. Pads can be placed closer together. This results in a considerable space saving, as you can see in the inset in Figure 3-8(c) by comparing the through-hole resistor to the surface mount chip resistor.

3-1-2   *Variable-Value Resistors*

A **variable resistor** can have its resistance varied or changed while it is connected in a circuit. In certain applications, the ability to adjust the resistance of a resistor is needed. For example, the volume control on your television set makes use of a variable resistor to vary the amount of current passing to the speakers, and so change the volume of the sound.

Mechanically (User) Adjustable Variable Resistors

In this section we will discuss two variable resistor types, which will cause a change in resistance when a shaft is physically rotated.

Rheostat (two terminals: A and B).   Figure 3-9(a) shows the physical appearance of different rheostats, while Figure 3-9(b) shows the rheostat's schematic symbols. As can be seen in the construction of a circular rheostat in Figure 3-9(c), one terminal is connected to one side of a resistive track and the other terminal of this two-terminal device is connected to a movable wiper. As the wiper is moved away from the end of the track with the terminal, the resistance between the stationary end terminal and the mobile wiper terminal increases. This is summarized in Figure 3-9(d), where the wiper has been moved down by a clockwise rotation of the shaft. Current would have to flow through a large resistance as it travels from one terminal to the other. On the other hand, as the wiper is moved closer to the end of the track connected to the terminal, the resistance decreases. This is summarized in Figure 3-9(e), which shows that as the wiper is moved up, as a result of turning the shaft counterclockwise, current will see only a small resistance between the two terminals.

Rheostats come in many shapes and sizes, as can be seen in Figure 3-9(a). Some employ a straight-line motion to vary resistance, while others are classified as circular-motion rheostats. The resistive elements also vary; wirewound and carbon tracks are very popular. Cermet rheostats mix the ratio of ceramic (insulator) and metal (conductor) to produce different values of resistive tracks. A trimming rheostat is a miniature device used to change resistance by a small amount. Other circular-motion rheostats are available that require between two to ten turns to cover the full resistance range.

Potentiometer (three terminals: A, B, and C).   Figure 3-10(a) illustrates the physical appearance of a variety of potentiometers, also called pots (slang), while Figure 3-10(b) shows the potentiometer's schematic symbol. You will probably notice that the difference between a rheostat and potentiometer is the number of terminals; the rheostat has two terminals while the potentiometer has three. With the rheostat, there were only two terminals and the resistance between the wiper and terminal varied as the wiper was adjusted. Referring to the potentiometer shown in Figure 3-10(c), you can see that resistance can actually be measured across three separate combinations: between *A* and *B* (*X*), between *B* and *C* (*Y*), and between *C* and *A* (*Z*).

The only difference between the rheostat and the potentiometer in construction is the connection of a third terminal to the other end of the resistive track, as can be seen in Figure 3-10(d), which shows the single-turn potentiometer. Also illustrated in this section of the figure is the construction of a multiturn potentiometer in which a contact arm slides along a shaft and the resistive track is formed into a helix of 2 to 10 coils.

Surface Mount Technology

A method of installing tiny electronic components on the same side of a circuit board as the printed wiring pattern that interconnects them.

Variable Resistor

A resistor whose value can be changed.

Rheostat

Two-terminal variable resistor that, through mechanical turning of a shaft, can be used to vary its resistance and therefore its value of terminal to terminal current.

Potentiometer

Three-lead variable resistor that through mechanical turning of a shaft can be used to produce a variable voltage or potential.

(a) Physical Appearance

(b) Schematic Symbols

(c) Construction

(d) Increasing a Rheostat's Resistance

(e) Decreasing a Rheostat's Resistance

FIGURE 3-9 **Rheostat. (a) Physical Appearance. (b) Schematic Symbols. (c) Construction. (d) Increasing a Rheostat's Resistance. (e) Decreasing a Rheostat's Resistance.**

As shown in Figure 3-11, the resistance between A and B (X) and B and C (Y) will vary as the wiper's position is moved, as illustrated in Figure 3-11(a). If the user physically turns the shaft in a clockwise direction, the resistance between A and B increases, while the resistance between B and C decreases. Similarly, if the user mechanically turns the shaft counterclockwise, a decrease occurs between A and B and there is a resulting increase in resistance between B and C. This point is summarized in Figure 3-11(b).

Whether rheostat or potentiometer, the resistive track can be classified as having either a **linear** or a **tapered** (nonlinear) resistance. In Figure 3-12(a) we have taken a 1 kΩ rheostat and illustrated the resistance value changes between A and B for a linear and a tapered one-turn rheostat. The definition of linear is having an output that varies in direct proportion to the input. The input in this case is the user turning the shaft, and the output is the linearly increasing resistance between A and B.

Linear

Relationship between input and output in which the output varies in direct proportion to the input.

Tapered

Nonuniform distribution of resistance per unit length throughout the element.

CHAPTER 3 / RESISTORS

FIGURE 3-10 Potentiometer. (a) Physical Appearance. (b) Schematic Symbol. (c) Operation. (d) Construction.

With a tapered rheostat or potentiometer, the resistance varies nonuniformly along its resistor element, sometimes being greater or less for equal shaft movement at various points along the resistance element, as shown in the table in Figure 3-12(a). Figure 3-12(b) plots the position of the variable resistor's wiper against the resistance between the two output terminals, showing the difference between a linear increase and a nonlinear or tapered increase.

SELF-TEST EVALUATION POINT FOR SECTION 3-1

Use the following questions to test your understanding of Section 3-1:

1. List the six types of fixed-value resistors.
2. What is the difference between SIPs and DIPs?
3. Name the two types of mechanically adjustable variable resistors and state the difference between the two.
4. Describe a linear and a tapered potentiometer.

FIGURE 3-11 The Resistance Changes between the Terminals of a Potentiometer.

3-2 HOW IS RESISTANCE MEASURED?

Ohmmeter

Measurement device used to measure electric resistance.

Resistance is measured with an **ohmmeter.** Stepping through the procedure described in Figure 3-13, you will see how the ohmmeter is used to measure the resistance of a resistor.

Use the following questions to test your understanding of Section 3-2:

1. Name the instrument used to measure resistance.
2. Describe the five-step procedure that should be used to measure the resistance.

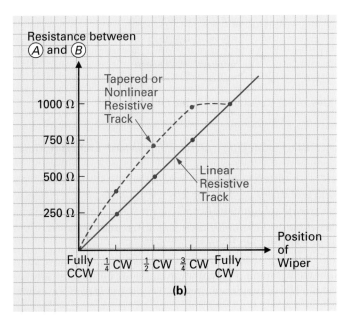

	Fully CCW	$\frac{1}{4}$ CW	$\frac{1}{2}$ CW	$\frac{3}{4}$ CW	Fully CW
Linear	0 Ω	250 Ω $\left(\begin{array}{c}\frac{1}{4}\text{ of}\\1000\text{ }\Omega\end{array}\right)$	500 Ω $\left(\begin{array}{c}\frac{1}{2}\text{ of}\\1000\text{ }\Omega\end{array}\right)$	750 Ω $\left(\begin{array}{c}\frac{3}{4}\text{ of}\\1000\text{ }\Omega\end{array}\right)$	1000 Ω
Tapered	0 Ω	350 Ω	625 Ω	900 Ω	1000 Ω

These values have been arbitrarily chosen to illustrate a nonlinear change.

(a)

(b)

FIGURE 3-12   **Linear versus Tapered Resistive Track.**

3-3   RESISTOR CODING

Manufacturers indicate the value and tolerance of resistors on the body of the component using either a color code (colored rings or bands) or printed alphanumerics (alphabet and numerals).

Resistors that are smaller in size tend to have the value and tolerance of the resistor encoded using colored rings, as shown in Figure 3-14(a). To determine the specifications of a color-coded resistor, therefore, you will have to decode the rings.

Referring to Figure 3-14(b), you can see that when the value, tolerance, and wattage of a resistor are printed on the body, no further explanation is needed.

1.
Turn off power to the circuit ⚠ and remove the component to be tested.

2.
Select resistance (Ω).

3.
Plug the black test probe into the COM input jack. Plug the red test probe into the Ω input jack.

4.
Connect the probe tips across the component or portion of the circuit for which you want to determine resistance.

5.
View the reading, being sure to note the unit of measurement — ohms (Ω), kilohms (kΩ), or megohms (MΩ).

FIGURE 3-13   **Measuring Resistance with the Ohmmeter.**

3-3-1   *General-Purpose and Precision Color Code*

There are basically two different types of fixed-value resistors: general purpose and precision. Resistors with tolerances of ±2% or less are classified as *precision resistors* and have five bands. Resistors with tolerances of ±5% or greater have four bands and are referred to as *general-purpose resistors.* The color code and differences between precision and general-purpose resistors are explained in Figure 3-15.

When you pick up a resistor, look for the bands that are nearer to one end. This end should be held in your left hand. If there are four bands on the resistor, follow the general-purpose resistor code. If five bands are present, follow the precision resistor code.

General-Purpose Resistor Code

1. The first band on either a general-purpose or precision resistor can never be black, and it is the first digit of the number.
2. The second band indicates the second digit of the number.
3. The third band specifies the multiplier to be applied to the number, which ranges from × ¹⁄₁₀₀ to 10,000,000.

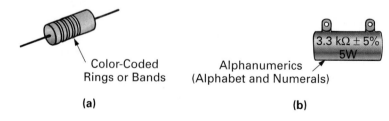

Color-Coded
Rings or Bands

Alphanumerics
(Alphabet and Numerals)

3.3 kΩ ± 5%
5W

(a)

(b)

FIGURE 3-14   **Resistor Value and Tolerance Markings.**

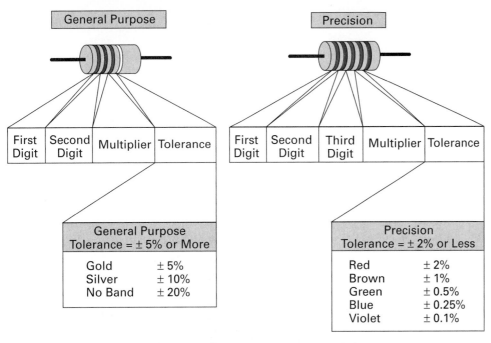

	Color	Digit Value	Multiplier	
Big	Black	0	1 One	1
Beautiful	Brown	1	10 One Zero	10
Roses	Red	2	100 Two Zeros	100
Occupy	Orange	3	1000 Three Zeros	1 k
Your	Yellow	4	10000 Four Zeros	10 k
Garden	Green	5	100000 Five Zeros	100 k
But	Blue	6	1000000 Six Zeros	1 M
Violets	Violet	7	10000000 Seven Zeros	10 M
Grow	Gray	8	–	
Wild	White	9	–	
So	Silver	–	10^{-2} or 0.01	1/100
Get some	Gold	–	10^{-1} or 0.1	1/10
Now	None	–		

FIGURE 3-15 General-Purpose and Precision Resistor Color Code.

4. The fourth band describes the tolerance or deviation from the specified resistance, which is ±5% or greater.

Precision Resistor Code

1. The first band, like the general-purpose resistor, is never black and is the first digit of the three-digit number.

2. The second band provides the second digit.

3. The third band indicates the third and final digit of the number.

4. The fourth band specifies the multiplier to be applied to the number.

5. The fifth and final band indicates the tolerance figure of the precision resistor, which is always less than ±2%, which is why precision resistors are more expensive than general-purpose resistors.

Band 1 = Green
Band 2 = Blue
Band 3 = Brown
Band 4 = Gold

FIGURE 3-16

■ **EXAMPLE:**

Figure 3-16 illustrates a ½ W resistor.

a. Is this a general-purpose or a precision resistor?

b. What is the resistor's value of resistance?

c. What tolerance does this resistor have, and what deviation plus and minus could occur to this value?

■ *Solution:*

a. General purpose (four bands)

b. green   blue × brown

     5   6 × 10   = 560 Ω

c. Tolerance band is gold, which is ±5%.

$$\text{deviation} = 5\% \text{ of } 560 = 28$$
$$560 + 28 = 588 \ \Omega$$
$$560 - 28 = 532 \ \Omega$$

The resistor could, when measured, be anywhere from 532 to 588 Ω.

■ **EXAMPLE:**

State the resistor's value and tolerance and whether it is general purpose or precision for the examples shown in Figure 3-17.

Orange/Green/Black/Silver   Green/Blue/Red/None   Red/Red/Green/Gold/Blue

(a)   (b)   (c)

FIGURE 3-17

■ *Solution:*

a. orange   green   black   silver   (four bands = general purpose)

     3   5   × 1   10%   = 35 Ω ± 10%

b. green   blue   red   none   (four bands = general purpose)

     5   6 × 100   20%   = 5.6 kΩ ± 20%

c. red   red   green   gold   blue   (five bands = precision)

    2   2   5   × 0.1   0.25% = 22.5 Ω ± 0.25%

3-3-2   *Other Resistor Identification Methods*

If a resistor color code is not found, some other form of marking will always be made on the resistor. The larger wirewound resistors and nearly all variable resistors normally have the resistance value, tolerance, and wattage printed on the resistor, as shown in Figure 3-18(a). SIP and DIP packages are normally coded with both letters and numerals, all of which have particular meaning, as shown in Figure 3-18(b). Chip resistors tend to be too small to have

Value = 5000 Ω
Tolerance = ± 10%
Power Rating = 5 watt

(a)

Number of Pins
Number of Resistors
Configuration of Resistors
6–5–2–330 — Value of Resistors
Dot Marks Pin 1

(b)

FIGURE 3-18   Typographical Value Indication. (a) Standard Variable and Fixed. (b) SIP and DIP.

any intelligible form of marking and so will have to be measured with an ohmmeter. They are packaged, as previously shown in Figure 3-8(c), in either polythene bags, paper tape, or plastic magazines, and in all these cases the value, tolerance, and wattage will be printed on the packaging.

SELF-TEST EVALUATION POINT FOR SECTION 3-3

Use the following questions to test your understanding of Section 3-3:

1. What tolerance differences occur between a general-purpose and a precision resistor?
2. A red/red/red/gold/blue resistor has what value and tolerance figure?
3. What color code would appear on a 4.7 kΩ, ±10% tolerance resistor?
4. Would a three-band resistor be considered a general-purpose type?

3-4   POWER DISSIPATION DUE TO RESISTANCE

The atoms of a resistor obstruct the flow of moving electrons (current), and this friction between the two causes heat to be generated. Current flow through any resistance therefore will always be accompanied by heat. In some instances, such as a heater or oven, the device has been designed specifically to generate heat. In other applications it may be necessary to include a resistor to reduce current flow, and the heat generated will be an unwanted side effect.

Any one of the three power formulas can be used to calculate the power dissipated by a resistance.

$$P = V \times I \qquad P = I^2 \times R \qquad P = \frac{V^2}{R}$$

However, since power is determined by the friction between current (I) and the circuit's resistance (R), the $P = I^2 \times R$ formula is more commonly used to calculate the heat generated.

EXAMPLE:

What wattage rating should a 33 Ω resistor have if the value of current is 100 mA?

Solution:

$$P = I^2 \times R$$
$$= (100 \text{ mA}^2) \times 33 \text{ }\Omega$$
$$= 0.33 \text{ W}$$

It would be safe therefore to use a ½ W (0.5 W) resistor.

EXAMPLE:

A coffee cup warming plate has a power rating of 120 V, 23 W. How much current is flowing through the heating element, and what is the heating element's resistance?

Solution:

$$P = V \times I \qquad I = \frac{P}{V}$$

$$I = \frac{23 \text{ W}}{120 \text{ V}} = 191.7 \text{ mA}$$

$$R = \frac{V}{I} = \frac{120 \text{ V}}{191.7 \text{ mA}} = 626 \text{ }\Omega$$

EXAMPLE:

To illustrate a point, let us work through a hypothetical example. Calculate the current drawn by a 100 W light bulb if it is first used in the home (and therefore connected to a 120 V source) and then used in the car (and therefore connected across a 12 V source).

Solution:

$$P = V \times I \qquad \text{therefore,} \qquad I = \frac{P}{V}$$

In the home,

$$I = \frac{P}{V} = \frac{100 \text{ W}}{120 \text{ V}} = 0.83 \text{ A}$$

In the car,

$$I = \frac{P}{V} = \frac{100 \text{ W}}{12 \text{ V}} = 8.33 \text{ A}$$

This example brings out a very important point in relation to voltage, current, and power. In the home, only a small current needs to be supplied when the applied voltage is large ($P = V\uparrow \times I\downarrow$). However, when the source voltage is small, as in the case of the car, a large current must be supplied in order to deliver the same amount of power ($P = V\downarrow \times I\uparrow$).

CHAPTER 3 / RESISTORS

Figure 3-19(a) illustrates the **filament resistor** within a glass bulb. This component is more commonly known as the household light bulb. The filament resistor is just a coil of wire that glows white hot when current is passed through it and, in so doing, dissipates both heat and light energy.

This **incandescent lamp** is an electric lamp in which electric current flowing through a filament of resistive material heats the filament until it glows and emits light. Figure 3-19(b) shows a test circuit for varying current through a small incandescent lamp. Potentiometer R_1 is used to vary the current flow through the lamp. The lamp is rated at 6 V, 60 mA. Using Ohm's law we can calculate the lamp's filament resistance at this rated voltage and current:

$$\text{filament resistance, } R = \frac{V}{I} = \frac{6 \text{ V}}{60 \text{ mA}} = 100 \text{ }\Omega$$

The filament material (tungsten, for example) is like all other conductors in that it has a positive temperature coefficient of resistance. Therefore, as the current through the filament increases, so does the temperature and so does the filament's resistance ($I\uparrow$, temperature $\uparrow$, $R\uparrow$). Consequently, when R_1's wiper is moved to the right so that it produces a high resistance ($R_1\uparrow$), the circuit current will be small ($I\downarrow$) and the lamp will glow dimly. Since the circuit current is small, the filament temperature will be small and so will the lamp's resistance ($I\downarrow$, temperature $\downarrow$, lamp resistance $\downarrow$). This small value of resistance is called the lamp's **cold resistance.** On the other hand, when R_1's wiper is moved to the left so that it produces a small resistance ($R\downarrow$), the circuit current will be large ($I\uparrow$), and the lamp will glow brightly. With the circuit current high, the filament temperature will be high and so will the lamp's resistance ($I\uparrow$, temperature $\uparrow$, lamp resistance $\uparrow$). This large value of resistance is called the lamp's **hot resistance.**

Figure 3-19(c) plots the filament voltage, which is being measured by the voltmeter, against the filament current, which is being measured by the ammeter. As you can see, an increase in current will cause a corresponding increase in filament resistance. Studying this graph, you may have also noticed that the lamp has been operated beyond its rated value of 6 V, 60 mA. Although the lamp can be operated beyond its rated value (for example, 10 V, 80 mA), its life expectancy will be decreased dramatically from several hundred hours to only a few hours.

A coil of wire similar to the one found in a light bulb, called a **ballast resistor,** is used to maintain a constant current despite variation in voltage. Since the coil of wire is a conductor and conductors have a positive temperature coefficient, an increase in voltage will cause a corresponding increase in current and therefore in the heat generated, which will result in an increase in the wire's resistance. This increase in resistance will decrease the initial current rise. Similarly, a decrease in voltage and therefore current ($I \propto V$) will result in a decrease in heat and wire resistance. This decrease in resistance will permit an increase in current to counteract the original decrease. Current is therefore regulated or maintained constant by the ballast resistor, despite variations in voltage.

Filament Resistor

The resistor in a light bulb or electron tube.

Incandescent Lamp

An electric lamp that generates light when an electric current is passed through its filament of resistance, causing it to heat to incandescence.

Cold Resistance

The resistance of a device when cold.

Hot Resistance

The resistance of a device when hot due to the generation of heat by electric current.

Ballast Resistor

A resistor that increases in resistance when current increases. It can therefore maintain a constant current despite variations in line voltage.

SELF-TEST EVALUATION POINT FOR SECTIONS 3-4 AND 3-5

Use the following questions to test your understanding of Sections 3-4 and 3-5:

1. Which formula is most commonly used to calculate the heat generated by a resistor?
2. An incandescent lamp has a ——————— (large/small) cold resistance and a ——————— (large/small) hot resistance.

Glass Envelope

Filament Resistor (Tungsten)

Gas

Connecting Wires

Metal screw base makes connection to one side of filament.

Small metal disk at base makes other connection to filament.

(a)

By changing the resistance of the potentiometer, the current flow through the lamp is varied.

R_1

Decrease Resistance | Increase Resistance

This ammeter measures the milliamperes of current flowing through the lamp.

AM

12 V

The voltmeter measures the voltage drop across the lamp.

VM

(b)

$R = V/I = 10\ \text{V}/80\ \text{mA} = 125\ \Omega$

$R = V/I = 6\ \text{V}/60\ \text{mA} = 100\ \Omega$ (Hot Resistance)

$R = V/I = 4\ \text{V}/50\ \text{mA} = 80\ \Omega$

$R = V/I = 1\ \text{V}/20\ \text{mA} = 50\ \Omega$ (Cold Resistance)

Filament Voltage (Volts)

Filament Current (mA)

(c)

FIGURE 3-19   Filament Resistor

3-6 TESTING RESISTORS

It is virtually impossible for fixed-value resistors to internally short-circuit (no resistance or zero ohms). Generally, the resistor's internal elements will begin to develop a higher resistance than its specified value (due to a partial internal open) or in some cases go completely open circuit (maximum resistance or infinite ohms).

Variable-value resistors have problems with the wiper making contact with the resistive track at all points. Faulty variable-value resistors in sound systems can normally be detected because they generate a scratchy noise whenever you adjust the volume control.

The ohmmeter is the ideal instrument for verifying whether or not a resistor is functioning correctly. When checking a resistor's resistance, here are some points that should be remembered.

1. The ohmmeter has its own internal power source (a battery), so always turn off the circuit power and remove from the circuit the resistor to be measured. If this is not done, you will not only obtain inaccurate readings, but you can damage the ohmmeter.

2. With an autoranging digital multimeter, the suitable ohms range is automatically selected by the ohmmeter. The range scales on a non-autoranging digital multimeter have to be selected, and often confuse people. For example, if a 100 kΩ range is chosen the highest reading that can be measured is 100 kΩ. If a 1 MΩ resistor is measured, the meter will indicate an infinite-ohms reading and you may be misled into believing that you have found the problem (an open resistor). To overcome this problem, always start on the highest range and then work down to a lower range for a more accurate reading.

3. Another point to keep in mind is tolerance. For example, if a suspected faulty 1 kΩ (1000 Ω) resistor is measured with an ohmmeter and reads 1.2 kΩ (1200 Ω), it could be within tolerance if no tolerance band is present on the resistor's body (±20%). A 1 kΩ resistor with ±20% tolerance could measure anywhere between 800 and 1200 Ω.

4. The ohmmeter's internal battery voltage is really too small to deliver an electrical shock. You should, however, avoid touching the bare metal parts of the probes or resistor leads, as your body resistance of approximately 50 kΩ will affect your meter reading.

SELF-TEST EVALUATION POINT FOR SECTION 3-6

Use the following questions to test your understanding of Section 3-6:

1. What problem normally occurs with resistors? (open/shorts)
2. What instrument is used to verify a resistor's resistance?

REVIEW QUESTIONS

Multiple-Choice Questions

1. The most common type of fixed-value resistor is the:
   a. Wirewound   c. Film
   b. Carbon composition d. Metal oxide

2. A SIP package has:
   a. A single line of connectors   c. No connectors
   b. A double line of connectors   d. None of the above

3. The most commonly used mechanically adjustable variable-value resistor is the:

   a. Rheostat   c. Potentiometer
   b. Thermistor   d. Photoresistor

4. A _____ -band resistor is called a general-purpose resistor, while a _____ -band resistor is known as a precision resistor.

   a. Four, five   c. One, four
   b. Three, four   d. Two, seven

5. What would be the power dissipated by a 2 kΩ carbon composition resistor when a current of 20 mA is flowing through it?

    a. 40 W   **c.** 1.25 W
    b. 0.8 W   **d.** None of the above

6. A rheostat is a _____ terminal device, while the potentiometer is a _____ terminal device.

    a. 2, 3   **c.** 2, 4
    b. 1, 2   **d.** 3, 2

7. Which of the following is *not* true?

    a. Precision resistors have a tolerance of > 5%.
    b. General-purpose resistors can be recognized because they have either three or four bands.
    c. The fifth band indicates the tolerance of a precision resistor.
    d. The third band of a general-purpose resistor specifies the multiplier.

8. The first band on either a general-purpose or a precision resistor can never be:

    a. Brown   **c.** Black
    b. Red or black stripe **d.** Red

9. The term *infinite ohms* describes:

    a. A small finite resistance
    b. A resistance so large that a value cannot be placed on it
    c. A resistance between maximum and minimum
    d. None of the above

10. A 10% tolerance, 2.7 MΩ carbon composition resistor measures 2.99 MΩ when checked with an ohmmeter. It is:

    a. Within tolerance   **d.** Both (a) and (c)
    b. Outside tolerance   **e.** Both (a) and (b)
    c. Faulty

Practice Problems

11. If a 5.6 kΩ resistor has a tolerance of ±10%, what would be the allowable deviation in resistance above and below 5.6 kΩ?

12. (a) If a current of 50 mA is flowing through a 10 kΩ, 25 W resistor, how much power is the resistor dissipating? (b) Can the current be increased, and if so by how much?

13. What minimum wattage size could be used if a 12 Ω wire-wound resistor were connected across a 12 V supply?

14. Calculate the resistance deviation for the tolerances of the resistors listed below

    a. 1.2 MΩ, ±10%
    b. 10 Ω, ±5%
    c. 27 kΩ, ±20%
    d. 273 kΩ, ±0.5%
    e. Orange, orange, black
    f. Red, red, green, red, red
    g. Brown, black, orange, gold
    h. White, brown, brown, silver

15. If a 1 kΩ rheostat has 50 V across it, calculate the different values of current if the rheostat is varied in 100 Ω steps. Referring to Figure 3-19, first calculate and insert the values into the table in Figure 3-20(a), and then plot these values in the graph in Figure 3-20(b).

Troubleshooting Questions

16. Briefly describe some of the problems that can occur with fixed and variable resistors.

17. Describe why the tolerance of a resistor can make you think that it has a problem when in fact it does not.

18. If a 20 kΩ, 1-W, ±10% tolerance resistor measures 20.39 kΩ, is it in or out of tolerance?

19. If 33 μA is flowing through a carbon film 4.7 kΩ, ¼ W, ±5% resistor, will it burn up?

20. What points should be remembered when using the ohmmeter to verify a resistor's value?

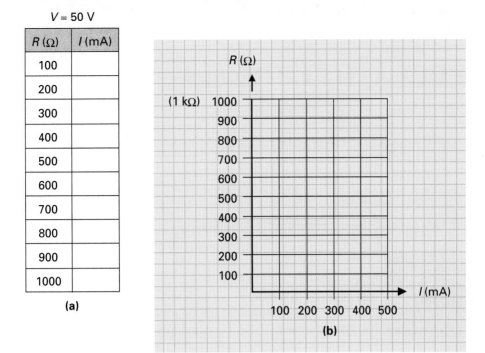

FIGURE 3-20   Plotting a Rheostat's Current against Resistance.

Direct Current (DC)

4

The First Computer Bug

Mathematician Grace Murray Hopper, an extremely independent U.S. naval officer, was assigned to the Bureau of Ordnance Computation Project at Harvard during World War II. As Hopper recalled, "We were not programmers in those days, the word had not yet come over from England. We were 'coders'," and with her colleagues she was assigned to compute ballistic firing tables on the Harvard Mark 1 computer. In carrying out this task, Hopper developed programming method fundamentals that are still in use.

Hopper is also credited, on a less important note, with creating a term frequently used today falling under the category of computer jargon. During the hot summer of 1945, the computer developed a mysterious problem. Upon investigation, Hopper discovered that a moth had somehow strayed into the computer and prevented the operation of one of the thousands of electromechanical relay switches. In her usual meticulous manner, Hopper removed the remains and taped and entered it into the logbook. In her own words, "From then on, when an officer came in to ask if we were accomplishing anything, we told him we were 'debugging' the computer," a term that is still used to describe the process of finding problems in a computer program.

Introduction

Direct current, abbreviated dc, is a flow of continuous current in only one direction and it is what we have been dealing with up to this point. A dc current is produced when a dc voltage source, such as a battery, is connected across a closed circuit. The requirement for a dc voltage source is that its output voltage polarity remain constant. It is the fixed polarity of a dc voltage source that will produce a unidirectional (one direction) current through a circuit. In this chapter, we will see how a dc voltage can be generated using several different methods. We will also examine fuses, circuit breakers, switches, and the protoboard.

4-1 SOURCE AND LOAD

Figure 4-1(a) illustrates a dc circuit with a battery connected across a light bulb. In this circuit, the potential energy of the battery (voltage) produces kinetic energy (current) that is used to produce light energy from the bulb. The battery is the **source** in this circuit and the bulb is the **load.** By definition, a load is a device that absorbs the energy being supplied and converts it into the desired form. The **load resistance** will determine how hard the voltage source has to work. For example, the bulb filament has a resistance of 300 Ω. The connecting wires have a combined resistance of 0.02 Ω (20 mΩ), and since this value is so small compared to the filament's resistance, it will be ignored, as it will have no noticeable effect. Consequently, a 300 Ω load resistance will permit 10 mA of **load current** to flow when a 3 V source is connected across the dc circuit ($I = V/R = 3 \text{ V}/300 \ \Omega = 10 \text{ mA}$). If you were to change the filament resistance by using a different bulb, the circuit current would be changed. To be specific, if the load resistance is increased, the load current decreases ($I \downarrow = V/R \uparrow$). Conversely, if the load resistance is decreased, the load current increases ($I \uparrow = V/R \downarrow$), assuming of course that the voltage source remains constant.

Figure 4-2 illustrates a heater, light bulb, motor, computer, robot, television, and microwave oven. All these and all other devices or pieces of equipment are connected to some source, such as a battery or dc power supply. This supply or source does not see all the circuitry and internal workings of the equipment; it simply sees the whole device or piece of equipment as having some value of resistance. The resistance of the equipment is referred to as the device's *load resistance,* and it is this value of load resistance and the value of *source voltage* that determines the value of *load current.*

Source
Device that supplies the signal power or electric energy to a load.

Load
A component, circuit, or piece of equipment connected to the source. The load resistance will determine the load current.

Load Resistance
The resistance of the load.

Load Current
The current that is present in the load.

◻ EXAMPLE:

A small portable radio offers a 390 Ω load resistance to a 9 V battery source. What current will flow in this circuit?

FIGURE 4-1 Load on a Voltage Source. (a) Pictorial. (b) Schematic.

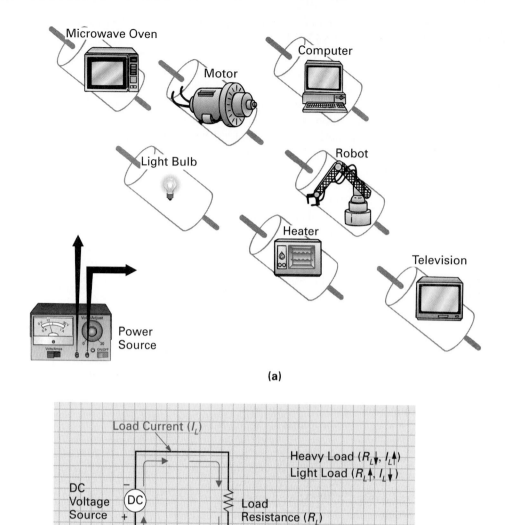

FIGURE 4-2 (a) Load Resistance of Equipment. (b) Schematic Symbols and Terms.

■ *Solution:*

$$\text{current } (I) = \frac{\text{voltage } (V)}{\text{resistance } (R)}$$

$$= \frac{9 \text{ V}}{390 \text{ }\Omega}$$

$$= 23 \text{ mA}$$

In summary, the phrase *load resistance* describes the device or equipment's circuit resistance, whereas the phrase *load current* describes the amount of current drawn by the device or equipment. A device that causes a large load current to flow (due to its small load resistance) is called a large or heavy load because it is heavily loading down or working the supply or source. On the other hand, a device that causes a small load current to flow (due to its large load resistance) is referred to as a small or light load because it is only lightly loading down or working the supply or source. It can be said, therefore, that load current and load resistance are inversely proportional to one another.

Use the following questions to test your understanding of Section 4-1:

1. Define *load resistance, load current, dc voltage source,* and *dc current.*
2. True or false: A large load current will flow when a large load resistance is connected across the supply.
3. What is the relationship between load resistance and load current?
4. Describe a heavy load resistance.

4-2 DIRECT-CURRENT SOURCES

The six basic forms of energy are mechanical, heat, light, magnetic, chemical, and electrical. To produce direct current (electrical energy), one of the other forms of energy must be used as a source and then converted or transformed into electrical energy in the form of a dc voltage that can be used to produce direct current.

4-2-1 *Mechanically Generated DC:*
   Mechanical ————————→ Electrical

Mechanical energy in the form of **pressure** can be used to generate electrical energy. Quartz and Rochelle salts are solid compounds that will produce electricity when a pressure is applied, as shown in Figure 4-3. Quartz is a natural or artificially grown crystal composed of silicon dioxide, from which thin slabs or plates are carefully cut and ground. Rochelle salt is a crystal of sodium potassium tartrate. They are both piezoelectric crystals. **Piezoelectric effect** is the generation of a voltage between the opposite faces of a crystal as a result of pressure being applied. This effect is made use of in most modern cigarette lighters. A voltage is produced between the opposite sides of a crystal (usually a crystalline lead compound) when it is either struck, pressured, or twisted. The resulting voltage is discharged across a gap to ignite the gas, which is released simultaneously by the operating button. The flame then burns for as long as the button is depressed. Piezoelectric effect also describes the reverse of what has just been discussed, as it was discovered that crystals would not only generate a voltage when mechanical force was applied, but would generate mechanical force when a voltage was applied.

   Figure 4-3(e) illustrates the physical appearance and operation of a piezoresistive pressure sensor. Piezoresistance is described as a change in resistance due to a change in the applied pressure. Unlike quartz, which generates a voltage, the piezoresistive sensor changes its resistance based on the applied pressure. Piezoresistive sensors need an excitation voltage applied, and it is this value that will determine the output voltage range. As an application, this sensor could be used to measure either the cooling system, hydraulic transmission, or fuel injection pressure in an automobile, which all vary in the 0 to 100 psi (pounds per square inch) range. The piezoresistive transducer would receive its excitation voltage from the car's battery (12 V) and produce a 0 to 100 mV output voltage, which would be sent to the car's computer, where the data would be analyzed and, if necessary, acted upon.

Pressure
The application of force to something by something else in direct contact with it.

Piezoelectric Effect
The generation of a voltage between the opposite sides of a piezoelectric crystal as a result of pressure or twisting. Also, the reverse effect in which the application of voltage to opposite sides causes deformation to occur at the frequency of the applied voltage.

4-2-2 *Thermally Generated DC:*
   Heat ————————→ Electrical

Heat can be used to generate an electric charge, as illustrated in Figure 4-4(a). When two dissimilar metals, such as copper and iron, are welded together and heated, an electrical charge is produced. This junction is called a **thermocouple,** and the heat causes the release

Thermocouple
Temperature transducer consisting of two dissimilar metals welded together at one end to form a junction that when heated will generate a voltage.

FIGURE 4-3   Direct Current from Pressure Using Crystals. (a) Quartz Crystal. (b) Physical Appearance. (c) Construction. (d) Operation. (e) Piezoresistive Sensor.

Two Metals	Temperature Range (°F)	Voltage Generated for Temperature Range (mV)
Iron vs. Copper–Nickel	32 to 1382	0 to 42.3
Nickel–Chromium vs. Nickel–Aluminum	–328 to 2282	–5.9 to 50.6
Platinum–6% Rhodium vs. Platinum–30% Rhodium	32 to 3092	0 to 12.4

(b)

FIGURE 4-4   **Direct Current from Heat Using the Thermocouple. (a) Operation.
(b) Typical Thermocouples.**

of electrons from their parent atoms, resulting in a meter deflection indicating the generation of a charge. The size of the charge is proportional to the temperature difference between the two metals. Figure 4-4(b) lists some typical thermocouples and their voltages generated when heat is applied, and shows the physical appearance of two types.

4-2-3   *Optically Generated DC: Light ⟶ Electrical*

A **photovoltaic cell,** shown in Figure 4-5(a), also makes use of two dissimilar materials in its transformation of light to electrical energy. Figure 4-5(b) illustrates the construction and operation of a photovoltaic cell, which is also called a photoelectric cell or a **solar cell.**

A light-sensitive metal or semiconductor is placed behind a transparent piece of dissimilar metal. When the light illuminates the light-sensitive material, a charge is generated, causing current to flow and the meter to indicate current flow. This phenomenon is known as **photovoltaic action.**

To obtain a reasonable amount of power, solar cells are interconnected so that they work collectively on solar panels. These solar panels are used to power calculators, emergency freeway phones, and satellites. The power obtained on average is about 100 milliwatts per square centimeter.

Photovoltaic Cell

A solid-state device that generates electricity when light is applied. Also called a solar cell.

Solar Cell

Photovoltaic cell that converts light into electric energy. They are especially useful as a power source for space vehicles.

Photovoltaic Action

The development of a voltage across the junction of two dissimilar materials when light is applied.

(a)

(b)

FIGURE 4-5 **Direct Current from Light Using the Solar Cell. (a) Physical Appearance.
(b) Construction and Operation.**

4-2-4 *Magnetically Generated DC:*
Magnetic ———————→ Electrical

A magnet like the one in Figure 4-6(a) can be used to generate electrical current by movement of the magnet past a conductor, as shown in Figure 4-6(b). When the magnet is stationary, no voltage is induced into the wire and therefore the light bulb is OFF. If the magnet
is moved so that the invisible magnetic lines of force cut through the wire conductor, a voltage is induced in the conductor, and this voltage produces a current in the circuit causing the

FIGURE 4-6 **Direct Current from Magnetism. (a) Stationary Magnet. (b) Moving Magnet.**

CHAPTER 4 / DIRECT CURRENT (DC)

light bulb to emit light. An energy conversion from magnetic to electric can be achieved by moving either the magnet past the wire or the wire past the magnet.

To produce a continuous supply of electric current, the magnet or wire must be constantly in motion. The device that achieves this is the dc electric generator.

4-2-5   *Chemically Generated DC:*
Chemical ————→ Electrical

The battery is a chemical voltage source that contains one or more voltaic cells. The **voltaic cell** was discovered by Alessandro Volta, an Italian physicist, in 1800. Figure 4-7 shows the three basic components within a battery's cell, which are a negative plate, a positive plate, and an electrolyte.

Cell Operation

Two dissimilar, separated metal plates such as copper and zinc are placed within a container that is filled with a substance known as the **electrolyte,** usually an acid. A chemical reaction causes electrons to pass through the electrolyte as they are repelled from one plate and attracted to the other. This action causes a large number of electrons to collect on one plate (negative plate), and an absence or deficiency of electrons to exist on the opposite plate (positive plate). The electrolyte therefore acts on the two plates and transforms chemical energy into electrical energy, which can be taken from the cell at its two output terminals as an electrical current flow.

If nothing is connected across the battery, as shown in Figure 4-8(a), a chemical reaction between the electrolyte and the negative electrode produces free electrons that travel from atom to atom but are held in the negative electrode. On the other hand, if a light bulb is connected between the negative and positive electrodes, as shown in Figure 4-8(b), the mutual repulsion of the free electrons at the negative electrode combined with the attraction of the positive electrode will cause a migration of free electrons (current flow) through the light bulb, causing its filament to produce light.

Primary Cells

The chemical reaction within the cell will eventually dissolve the negative plate. This discharging process also results in hydrogen gas bubbles forming around the positive plate, causing a resistance between the two plates, known as the battery's internal resistance, to

Voltaic Cell

A battery cell having two unlike metal electrodes immersed in a solution that chemically interacts with the plates to produce an emf.

Electrolyte

Electrically conducting liquid (wet) or paste (dry).

FIGURE 4-7   A Battery's Cell.

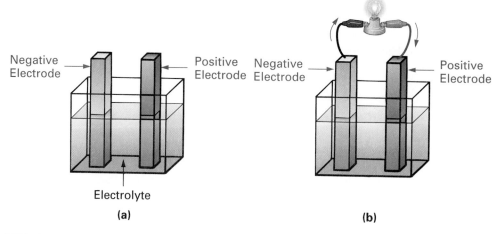

Negative Electrode   Positive Electrode   Negative Electrode   Positive Electrode

Electrolyte

(a)   **(b)**

FIGURE 4-8   **Cell Operation.**

increase (called *polarization*). To counteract this problem, all dry cells have a chemical within them known as a depolarizer agent, which reduces the buildup of these gas bubbles. With time, however, the depolarizer's effectiveness will be reduced, and the battery's internal resistance will increase as the battery reaches a completely discharged condition. These types of batteries are known as **primary cells** (first time, last time) because they discharge once and must then be discarded. Almost all primary cells have their electrolyte in paste form, which is why they are also referred to as **dry cells.** Wet cells have an electrolyte that is in liquid form.

Primary cell types.   The shelf life of a battery is the length of time a battery can remain on the shelf (in storage) and still retain its usability. Most primary cells will deteriorate in a three-year period to approximately 80% of their original capacity. Of all the types of primary cells, five dominate the market as far as applications and sales are concerned. These are the carbon–zinc, alkaline–manganese, mercury, silver oxide, and lithium types, all of which are described and illustrated in Table 4-1. The cell voltage in Table 4-1 describes the voltage produced by one *cell* (one set of plates). If two or more sets of plates are installed in one package, the component is called a *battery*.

Secondary Cells

Secondary cells operate on the same basic principle as that of primary cells. In this case, however, the plates are not eaten away or dissolved, they only undergo a chemical change during discharge. Once discharged, the secondary cell can have the chemical change that occurred during discharge reversed by recharging, resulting once more in a fully charged cell. Restoring a secondary cell to the charged condition is achieved by sending a current through the cell in the direction opposite to that of the discharge current, as illustrated in Figure 4-9.

In Figure 4-9(a), we have first connected a light bulb across the battery, and free electrons are being supplied by the negative electrode and moving through the light bulb or load and back to the positive electrode. After a certain amount of time, any secondary cell will run down or will have discharged to such an extent that no usable value of current can be produced. The surfaces of the plates are changed, and the electrolyte lacks the necessary chemicals.

In Figure 4-9(b), during the recharging process, we use a battery charger, which reverses the chemical process of discharge by forcing electrons back into the cell and restoring the battery to its charged condition. The battery charger voltage is normally set to about 115% of the battery voltage. To use an example, a battery charger connected across a 12 V battery would be set to 115% of 12 V, or 13.8 V. If this voltage is set too high, an excessive current can result, causing the battery to overheat.

TABLE 4-1 Primary Cell Types

Battery type	Features	Cell Voltage (V)	Applications	Construction
Carbon–zinc (also called Leclanché cell in honor of its inventor)	Most popular due to its low cost. Cylindrical D and C cells are most commonly used. Flat cells are stacked in series to obtain voltages greater than 1.5 V, as in the case of the 9 V battery	1.5	Portable radios, tape players, televisions, toys.	Has a zinc case (negative), a carbon center rod (positive), with ammonium chloride as the electrolyte and manganese dioxide as the depolarizing agent
Alkaline–maganese (also called alkaline cell)	Has three times the shelf life and capacity of carbon zinc. Cylindrical or miniature cell sizes are available	1.4	Portable radios, tape players, televisions, toys; large capacity is worth investment	The negative plate is granular zinc mixed with an electrolyte (alkaline); the positive plate is a polarizer in contact with the outer metal can
Mercury (also called mercuric–oxide)	Higher energy density than preceding two with good shelf life and small size; used in low-power applications. Flat or cylindrical cells are available	1.35 and 1.4	Watches, hearing aids, pacemakers, cameras, test equipment	A zinc anode and a mercuric oxide cathode in a potassium hydroxide electrolyte
Silver–oxide	High capacity; however, the material used (silver) makes it the most costly; can supply high currents for short periods of time; used in low-power applications	1.5	Watches, hearing aids, pacemakers, cameras	Contains a cathode of silver oxide, an alkaline electrolyte, and zinc anode
Lithium	High discharge rate and long shelf life; light weight and higher output voltage; the familiar cylindrical and flat cell are available	1.9	Liquid crystal watches, semi-conductor memories, hand-held calculators, sensor circuits	Contains a lithium anode, porous carbon cathode, and sulfur dioxide electrolyte

Construction diagrams:

Metal Cap (+ Terminal)

Carbon Rod (+ Electrode)
Electrolyte (Paste)
Zinc Can (– Electrode)

(a) Carbon–Zinc

9 V Flat cell has 6 stacked 1.5 V cells   9 V

Anode
Cathode

(b) Alkaline–Manganese

Alkaline Miniature Cell

Amalgamated Zinc Anode
Mercuric Oxide Depolarizing Cathode
Potassium Hydroxide Electrolyte

(c) Mercury

Anode
Cathode

(d) Silver-Oxide

Cathode
Solid Electrolyte
Anode

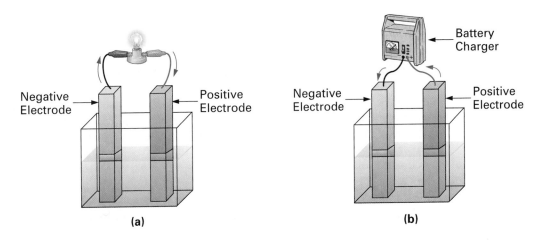

Negative Electrode   Positive Electrode

Battery Charger

Negative Electrode   Positive Electrode

(a)   (b)

FIGURE 4-9 Secondary Cell Operation. (a) Discharge. (b) Charge.

Secondary cells are often referred to as **wet cells** because the electrolyte is not normally in paste form (dry) as in the primary cell, but is in liquid (wet) form and is free to move and flow within the container.

Secondary cell types. Of all the secondary cells on the market today, the lead–acid and nickel–cadmium are the two most popular, and are illustrated and described in Table 4-2. Other secondary cell types include the nickel–metal hydride and lithium ion, which are used in cellular telephones, camcorders, and other portable electronic systems.

Ampere-Hours

If you multiply the amount of current, in amperes, that can be supplied for a given time frame in hours, a value of ampere-hours is obtained.

Secondary cell capacity. Capacity (C) is measured by the amount of **ampere-hours** (Ah) a battery can supply during discharge. For example, if a battery has a discharge capacity of 10 ampere-hours, the battery could supply 1 ampere for 10 hours, 10 amperes for 1 hour, 5 amperes for 2 hours, and so on, during discharge. Automobile batteries (12 V) will typically have an ampere-hour rating of between 100 and 300 Ah.

Ampere-hour units are actually specifying the coulombs of charge in the battery. For example, if a lead–acid car battery was rated at 150 Ah, this value would have to be converted to ampere-seconds to determine coulombs of charge, as 1 ampere-second (As) is equal to 1 coulomb. Ampere-hours are easily converted to ampere-seconds simply by multiplying ampere-hours by 3600, which is the number of seconds in 1 hour. In our example, a 150 Ah battery will have a charge of 54×10^4 coulombs.

EXAMPLE:

A Ni-Cd battery is rated at 300 Ah and is fully discharged.

    a. How many coulombs of charge must the battery charger put into the battery to restore it to full charge?

    b. If the charger is supplying a charging current of 3 A, how long will it take the battery to fully charge?

    c. Once fully charged, the battery is connected across a load that is pulling a current of 30 A. How long will it take until the battery is fully discharged?

Solution:

    a. The same amount that was taken out: 300 Ah $\times$ 3600 = 1080 $\times$ 10^3 C

    b. $\dfrac{300 \text{ Ah}}{3 \text{ A}}$ = 100 hours until charged

    c. $\dfrac{300 \text{ Ah}}{30 \text{ A}}$ = 10 hours until discharged

---

4-2-6   *Electrically Generated DC*

Up until now we have only discussed electrical energy in the form of direct current (dc). However, as we will discover in Chapter 8, electrical energy is more readily available in the form of alternating current (ac). In fact, that is what arrives at the wall receptacle at your home and workplace. We can use a dc power supply, like the one seen in Figure 4-10, to convert the ac electrical energy at the wall outlet to dc electrical energy for experiments.

Power in the laboratory can be obtained from a battery, but since ac power is so accessible, dc power supplies are used. The power supply has advantages over the battery as a source of dc voltage in that it can quickly provide an accurate voltage that can easily be varied by a control on the front panel, and it never runs down.

During your electronic studies, you will use a power supply in the lab to provide different voltages for various experiments. In Figure 4-10 you can see how a dc power supply is being used to supply 12 V to a 12 kΩ resistor.

Table 4-2 Secondary Cell Types.

Battery type	Features	Applications	Construction
Lead–acid (lead cell)	High cycle (discharge–charge) life and high current capacity; 2.1 V per cell; must be operated upright if a wet lead–acid. Lead–acid gel cells can be used in any position	*Gel Cell Batteries* televisions, recorders, robots, alarms, and tools *Wet Cell Batteries* Starting source for automobiles and power for robotics equipment	The electrodes are lead oxide immersed in an electrolyte of dilute sulfuric acid Wet-type lead–acid battery with 6 internal 2.1 V cells making 12.6 V car battery *Positive Plate* · *Insulating Separators (porous rubber)* · *Negative Plate* · *Car Battery*
Nickel–cadmium (Ni–Cd)	High capacity and high cost; three times discharge current of lead–acid for same amp-hour rating; can be sealed and operated in any position, so ideal for drills and other such portable equipment; 1.2 V per cell. Nickel cadmium batteries suffer from "memory effect." Short discharge and charge cycles cause the batteries' capacity to drop, because the cell appears to "remember" the lower capacity.	*Ni–Cd Gel Cell Cell* televisions, radios, toys, recorders, shavers, toothbrushes *Ni–Cd Wet Cell Batteries* Starter source for jet engines (resembles lead–acid car battery in appearance). Robotics equipment	The positive plate is made of cadmium, the negative plate of nickel hydroxide, and the electrolyte used is potassium hydroxide *Negative Sintered Plate* · *Separator* · *Nickel-plated Steel Jacket* · *Positive Sintered (applied in powered form) Plate*
Nickel–metal Hydride (Ni–MH) and Lithium Ion	Discharge characteristics are very similar to those of the nickel–cadmium cell. The cell's nominal voltage is 1.2 V, and it does not suffer from memory effect.	Biggest applications include cellular phones, portable computers, camcorders, and other portable electronic systems	*(+) Positive Terminal* · *Positive Electrode* · *Separator* · *Negative Electrode* · *(–) Negative Terminal*

FIGURE 4-10   **Direct Current from Alternating Current.**

SELF-TEST EVALUATION POINT FOR SECTION 4-2

Use the following questions to test your understanding of Section 4-2:

1. What is a piezoelectric device?
2. What is a thermocouple?
3. What device converts light into a direct current?
4. What device is used to convert magnetic energy into electrical energy?
5. List the three basic components in a battery.

4-3   EQUIPMENT PROTECTION

Current must be monitored and not be allowed to exceed a safe level so as to protect users from shock, to protect the equipment from damage, and to prevent fire hazards. There are two basic types of protective devices: fuses and circuit breakers. Let's now discuss each of these in more detail.

4-3-1   *Fuses*

Fuse

This circuit- or equipment-protecting device consists of a short, thin piece of wire that melts and breaks the current path if the current exceeds a rated damaging level.

A **fuse** is an equipment protection device that consists of a thin wire link within a casing. Figure 4-11(a) shows the physical appearance of a fuse and Figure 4-11(b) shows the fuse's schematic symbol.

To protect a system, a fuse needs to disconnect power from the unit the moment current exceeds a safe value. To explain how this happens, Figure 4-11(c) shows how a fuse is mounted in the rear of a dc power supply. All fuses have a current rating, and this rating indicates what value of current will generate enough heat to melt the thin wire link within the fuse. In the example in Figure 4-11(c), if the current is less than the fuse rating of 2 amps, the metal link of the fuse will remain intact. If, on the other hand, the current through the thin metal element exceeds the current rating of 2 A, the excessive current will create enough heat to melt the element and "open" or "blow" the fuse, thus disconnecting the dc power supply from the 120 volt source and protecting it from damage.

FIGURE 4-11   Fuses.

One important point to remember is that if the current increases to a damaging level, the fuse will open and protect the dc power supply. This implies that something went wrong with the 120 V source and it started to supply too much current. This is almost never the case. What actually happens is that, as with all equipment, eventually something internally breaks down. This can cause the overall load resistance of the piece of equipment to increase or decrease. An increase in load resistance $(R_L\uparrow)$ means that the source sees a higher resistance in the current path and therefore a small current would flow from the source ($I\downarrow$), and the user would be aware of the problem due to the nonoperation of the equipment. If an internal equipment breakdown causes the equipment's load resistance to decrease ($R_L\downarrow$), the source will see a smaller circuit resistance and will supply a heavier circuit current ($I\uparrow$), which could severely damage the equipment before the user had time to turn it off. Fuse protection is needed to disconnect the current automatically in this situation, to protect the equipment from damage.

Fuse elements come in various shapes and sizes so as to produce either quick heating and then melting (fast blow) or delayed heating and then melting (slow blow), as illustrated in Figure 4-11(d). The reason for the variety is based in the application differences. When turned on, some pieces of equipment will be of such low resistance that a short momentary current surge, sometimes in the region of four times the fuse's current rating, will result for the first couple of seconds. If a fast-blow fuse were placed in the circuit, it would blow at the instant the equipment was turned on, even though the equipment did not need to be protected. The system merely needs a large amount of current initially to start up. A slow-blow fuse would be ideal in this application, as it would permit the initial heavy current and would begin to heat up and yet not blow due to the delay. Once the surge had ended, the current would decrease, and the fuse would still be intact (some slow-blow fuses will allow a 400% overload current to flow for a few seconds).

On the other hand, some equipment cannot take any increase in current without being damaged. The slow blow would not be good in this application, as it would allow an increase of current to pass to the equipment for too long a period of time. The fast blow would be ideal in this application, since it would disconnect the current instantly if the current rating of the fuse were exceeded, and so prevent equipment damage. The automobile was the first application for a fuse in a glass holder, and so a size standard named AG (automobile glass) was established. This physical size standard is still used today, and is listed in Figure 4-11(d).

Fuses also have a voltage rating that indicates the maximum circuit voltage that can be applied across the fuse by the circuit in which the fuse resides. This rating, which is important after the fuse has blown, prevents arcing across the blown fuse contacts, as illustrated in Figure 4-11(e). Once the fuse has blown, the circuit's positive and negative voltages are now connected across the fuse contacts. If the voltage is too great, an arc can jump across the gap, causing a sudden surge of current, damaging the equipment connected.

Fuses are mounted within fuse holders and normally placed at the back of the equipment for easy access, as was shown in Figure 4-11(c). When replacing blown fuses, you must be sure that the equipment is turned OFF and disconnected from the source. If you do not remove power you could easily get a shock, since the entire source voltage appears across the fuse as was shown in Figure 4-11(e). Also, make sure that a fuse with the correct current and voltage rating is used.

A good fuse should have a resistance of 0 Ω when it is checked with an ohmmeter. In circuit and with power ON, the fuse should have no voltage drop across it because it is just a piece of wire. A blown or burned-out fuse will, when removed, read infinite ohms, and when in its holder and power ON, will have the full applied voltage across its two terminals.

4-3-2 *Circuit Breaker*

Circuit Breaker

Reusable fuse. This device will open a current-carrying path without damaging itself once the current value exceeds its maximum current rating.

In many appliances and in a home electrical system, **circuit breakers** are used in place of fuses to prevent damaging current. A circuit breaker can open a current-carrying circuit without damaging itself, then be manually reset and used repeatedly, unlike a fuse, which must be replaced when it blows. We can say, therefore, that a circuit breaker is a reusable fuse. Figure 4-12 illustrates a circuit breaker's schematic symbol, physical appearance, and a typical application in which it is used to protect a television.

In this section we will examine the three basic types of circuit breakers, which are thermal type, magnetic type, and thermomagnetic type.

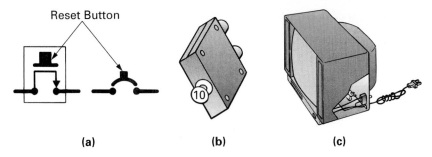

FIGURE 4-12   Circuit Breaker. (a) Symbol. (b) Appearance. (c) Typical Application.

Thermal Circuit Breakers

The operation of this type of circuit breaker depends on temperature expansion due to electrical heating. Figure 4-13 illustrates the construction of a thermal circuit breaker. A U-shaped bimetallic (two-metal) strip is attached to the housing of the circuit breaker. This strip is composed of a layer of brass on one side and a layer of steel on the other. The current arrives at terminal A, enters the right side of the bimetallic U-shaped strip, leaves the left side, and travels to the upper contact, which is engaging the bottom contact, where it leaves and exits the circuit breaker from terminal B.

If current were to begin to increase beyond the circuit breaker's current rating, heating of the bimetallic strip would occur. As with all metals, heat will cause the strip to expand. Some metals expand more than others, and this measurement is called the *thermal expansion coefficient*. In this situation, brass will expand more than steel, resulting in the lower end of the bimetallic strip bending to the right. This action allows a catch to release a pivoted arm. The arm's right side will be pulled down by the tension of a spring, lifting the left side of the arm and the attached top contact. The result is to separate the top contact surface from the bottom contact surface, thereby opening the current path and protecting the circuit from the excessive current. This action is also known as *tripping* the breaker. To reset the breaker, the reset button must be pressed to close the contacts. If the problem in the system still exists, however, the breaker will just trip once more, due to the excessive current.

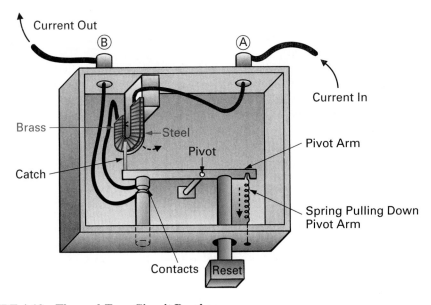

FIGURE 4-13   Thermal-Type Circuit Breaker.

Magnetic Circuit Breakers

Earlier in this chapter, we discussed how a generator converts a magnetic energy input into an electrical energy output. The magnetic type of circuit breaker makes use of a device that performs the opposite function. An electromagnet has a coil of insulated wire wrapped around a soft iron core, and it will generate a magnetic field whenever a current flows through its coil. The magnetic-type circuit breaker depends on the response of an electromagnet to break the circuit for protection.

The construction of a magnetic circuit breaker is shown in Figure 4-14, and it will operate as follows. A small level of current flowing through the coil of the electromagnet will provide only a small amount of magnetic pull to the left on the iron arm. This magnetic force cannot overcome the pull to the right being generated by spring A. This safe value of current will therefore be allowed to pass from the A terminal, through the coil to the top contact, out of the bottom contact, and then exit the breaker at terminal B.

If the current exceeds the current rating of the circuit breaker, an increase in current through the coil of the electromagnet generates a greater magnetic force on the vertical arm, which pulls the top half of the vertical arm to the left and the lower half below the pivot to the right. This action releases the catch holding the horizontal arm, allowing spring B to pull the right side of the lower arm down, opening the contact and disconnecting or tripping the breaker. The reset button must now be pressed to close the contacts. Once again, however, if the problem still exists, the breaker will continue to trip, as the excessive circuit current still exists.

Thermomagnetic Circuit Breakers

Figure 4-15(a) illustrates the typical residential-type thermomagnetic circuit breaker, and Figure 4-15(b) shows the physical appearance and schematic of a typical home breaker box. The thermal circuit breaker is similar to a slow-blow fuse in that it is ideal for passing momentary surges without tripping because of the delay caused by heating of the bimetallic strip. The magnetic circuit breaker, on the other hand, is most similar to a fast-blow fuse in that it is tripped immediately when an increase of current occurs.

The thermomagnetic circuit breaker combines the advantages of both previously mentioned circuit breakers by incorporating both a bimetallic strip and an electromagnet in its actuating mechanism. For currents at and below the current rating, the circuit breaker is a short circuit and connects the current source to the load. For momentary small overcurrent surges, the electromagnet is activated, but does not have enough force to trip the breaker. If

FIGURE 4-14   Magnetic-Type Circuit Breaker.

CHAPTER 4 / DIRECT CURRENT (DC)

FIGURE 4-15 **Thermomagnetic-Type Circuit Breaker.**

overcurrent continues, however, the bimetallic strip will have heated and the combined forces of the bimetallic strip and electromagnet will trip the breaker and protect the load by disconnecting the source. Momentary large overcurrent surges will supply enough current to the electromagnet to cause it to trip the breaker independently of the bimetallic strip and so disconnect the source from the load.

4-4 SWITCHES

As we have seen previously, a *switch* is a device that completes (closes) or breaks (opens) the path of current, as seen in Figure 4-16. All mechanical switches can be classified into one of the eight categories shown in Figure 4-17. Many different variations of these eight different classifications exist.

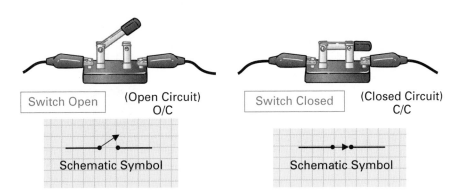

FIGURE 4-16 **Open and Closed Switches.**

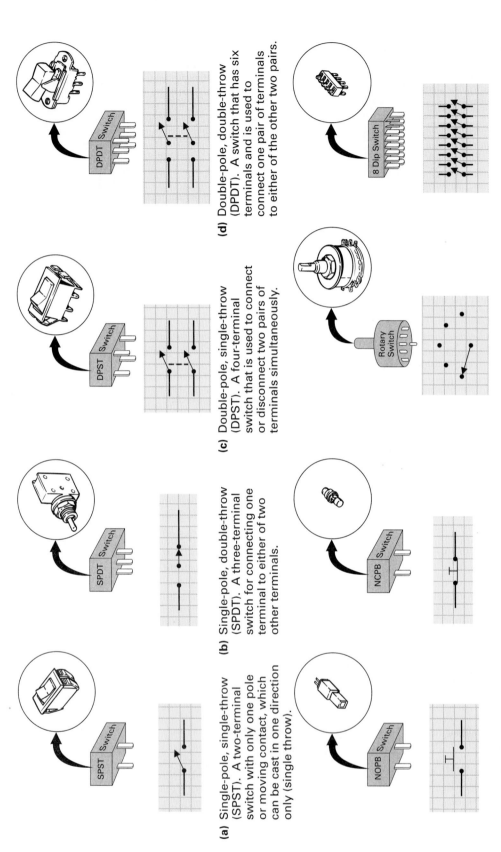

(a) Single-pole, single-throw (SPST). A two-terminal switch with only one pole or moving contact, which can be cast in one direction only (single throw).

(b) Single-pole, double-throw (SPDT). A three-terminal switch for connecting one terminal to either of two other terminals.

(c) Double-pole, single-throw (DPST). A four-terminal switch that is used to connect or disconnect two pairs of terminals simultaneously.

(d) Double-pole, double-throw (DPDT). A switch that has six terminals and is used to connect one pair of terminals to either of the other two pairs.

(e) Normally open push button (NOPB). A switch that will make contact and pass current when it is pressed.

(f) Normally closed push button (NCPB). A switch that normally makes contact and passes current, but disconnects or opens when pressed.

(g) Rotary. An electromechanical device that is capable of making (closing contacts) or breaking (opening contacts) in a circuit.

(h) Dual-in-line package (DIP) switch. A DIP switch is a group of separate miniature switches within two (dual) rows of external connecting pins or terminals.

Single pole: one moving contact.   Single throw: pole can be thrown or cast in only one direction.
Double pole: two moving contacts.   Double throw: pole can be thrown or cast in two directions.

FIGURE 4-17   Eight Basic Types of Switches.

Use the following questions to test your understanding of Sections 4-3 and 4-4:

1. What do the current and voltage ratings of a fuse indicate?
2. What is the difference between a slow-blow and a fast-blow fuse?
3. List the three types of circuit breakers.
4. List the eight basic types of switches.

4-5   PROTOBOARDS

The solderless prototyping board (**protoboard**) or breadboard is designed to accommodate the many experiments described in this textbook's associated lab manual. This protoboard will hold and interconnect resistors, capacitors, inductors, and many other components, as well as provide electrical power. Figure 4-18(a) shows an experimental circuit wired up on a protoboard.

Protoboard

An experimental arrangement of a circuit on a board. Also called breadboard.

(a)

Top View of Protoboard

Conductive Strips within Protoboard

This Bus Line Is Electrically Connected from this End of the Board to the Other by a Metal Strip within the Protoboard

Bus Strips

These Five Holes Are All Electrically Connected by a Metal Strip within the Protoboard

(b)

FIGURE 4-18   **Experimenting with the Protoboard.**

Figure 4-18(b) shows the top view of a basic protoboard. As you can see by the cross section on the right side, electrical connector strips are within the protoboard. These conductive strips make a connection between the five hole groups. The bus strips have an electrical connector strip running from end to end, as shown in the cross section. They are usually connected to a power supply as seen in the example circuit in Figure 4-19(a). In this circuit, you can see that the positive supply voltage is connected to the upper bus strip, while the negative supply (ground) is connected to the lower bus strip. These power supply "rails" can then be connected to a circuit formed on the protoboard with hookup wire, as shown in Figure 4-19(a). In this example, three resistors are connected end-to-end, as shown in the schematic diagram in Figure 4-19(b).

Figure 4-20 illustrates how the multimeter could be used to make voltage, current, and resistance measurements of a circuit constructed on the protoboard.

Other protoboards may vary slightly as far as layout, but you should be able to determine the pattern of conductive strips by making a few resistance checks with an ohmmeter.

FIGURE 4-19   Constructing Circuits on the Protoboard.

CHAPTER 4 / DIRECT CURRENT (DC)

FIGURE 4-20 **Making Measurements of a Circuit on the Protoboard.**

Use the following questions to test your understanding of Section 4-5:

1. What is the purpose of the protoboard?
2. What is generally connected to the bus strips?

REVIEW QUESTIONS

Multiple-Choice Questions

1. Direct current is:
   a. A reversing of current continually in a circuit
   b. A flow of current in only one direction
   c. Produced by an ac voltage
   d. None of the above

2. If load resistance were doubled, load current would
   _____.
   a. Halve b. Double c. Triple d. Remain the same

3. Quartz is a solid compound that will produce electricity when _____ is applied.
   a. Friction   c. A magnetic field
   b. An electrolyte   d. Pressure

4. Direct current is generated thermally by use of a:
   a. Crystal   c. Thermistor
   b. Thermocouple   d. None of the above

5. An application of a photovoltaic cell would be:
   a. A satellite power source
   b. To turn on and off security lights
   c. Both (a) and (b)
   d. None of the above

6. Magnetic flux lines _____ the north pole and _____ at the south pole.
   a. Arrive at, leave   c. Are never present at, always present
   b. Leave, arrive   d. None of the above

7. Which type of battery can be used for only one discharge (and cannot be rejuvenated)?
   a. A lead–acid cell   c. A nickel–cadmium cell
   b. A secondary cell   d. A primary cell

8. A battery with a capacity of 12 ampere-hours could supply:
   a. 12 A for 1 hour   c. 6 A for 2 hours
   b. 3 A for 4 hours   d. All of the above

9. The dc power supply converts:
   a. ac to dc   c. High dc to a low dc
   b. dc to ac   d. Low ac to a high ac

10. A fuse's current rating states the:
    a. Maximum amount of current allowed to pass
    b. Minimum amount of current allowed to pass
    c. Maximum permissible circuit voltage
    d. None of the above

11. A thermal-type circuit breaker is equivalent to a _____ fuse, whereas a magnetic-type circuit breaker is equivalent to a _____ fuse.
    a. Slow-blow, slow-blow   c. Slow-blow, fast-blow
    b. Fast-blow, slow-blow   d. Glass case, ceramic case

12. The type of circuit breaker found in the home is the:
    a. Thermal type   c. Magnetic type
    b. Thermomagnetic type   d. Carbon–zinc type

13. A single-pole, double-throw switch would have _____ terminals.
    a. Two b. Three c. Four d. Five

14. A switch will very simply _____ (open) or _____ (close) a path of current.
    a. Break, make   c. s/c, o/c
    b. make, break   d. o/c, break

15. With switches, the word pole describes a:
    a. Stationary contact   c. Path of current
    b. Moving contact   d. None of the above

Practice Problems

16. If a 50 W light bulb is connected across a 120 V source and the following fuses are available, which should be used to protect the circuit?

    a. 0.5 A/120 V (fast blow)   **c.** 1 A/12 V (slow blow)

    b. ¾ A/120 V (fast blow)   **d.** 0.5 A/120 V (slow blow)

17. Draw a diagram and indicate the polarities of a 12 V lead–acid battery being charged by a 15 V battery charger. If this battery is rated at 150 Ah:

a. How many coulombs of charge will be stored in the fully charged condition?

b. How long will it take the battery to charge if a charging current of 5 A is flowing?

18. If a 6 V nickel–cadmium battery discharges at a rate of 3.5 A in 4 hours:

a. Calculate its ampere-hour and ampere-second rating.

b. How long will it take the battery to discharge if a current of 2 A is drawn?

Series DC Circuits

5

Introduction

A series circuit, by definition, is the connecting of components end to end in a circuit to provide a single path for the current. This is true not only for resistors, but also for other components that can be connected in series. In all cases, however, the components are connected in succession or strung together one after another so that only one path for current exists between the negative ($-$) and positive ($+$) terminals of the supply.

5-1   COMPONENTS IN SERIES

Figure 5-1 illustrates five examples of **series** resistive **circuits.** In all five examples, you will notice that the resistors are connected "in-line" with one another so that the current through the first resistor must pass through the second resistor, and the current through the second resistor must pass through the third, and so on.

Series Circuit

Circuit in which the components are connected end to end so that current has only one path to follow throughout the circuit.

▢ EXAMPLE:

In Figure 5-2(a), seven resistors are laid out on a table top. Using a protoboard, connect all the resistors in series, starting at R_1, and proceeding in numerical order through the resistors until reaching R_7. After completing the circuit, connect the series circuit to a dc power supply.

▢ *Solution:*

In Figure 5-2(b), you can see that all the resistors are now connected in series (end-to-end), and the current has only one path to follow from negative to positive.

▢ EXAMPLE:

Figure 5-3(a) shows four 1.5 V cells and three lamps. Using wires, connect all of the cells in series to create a 6 V battery source. Then connect all of the three lamps in series with one another, and finally, connect the 6 V battery source across the three-series-connected-lamp load.

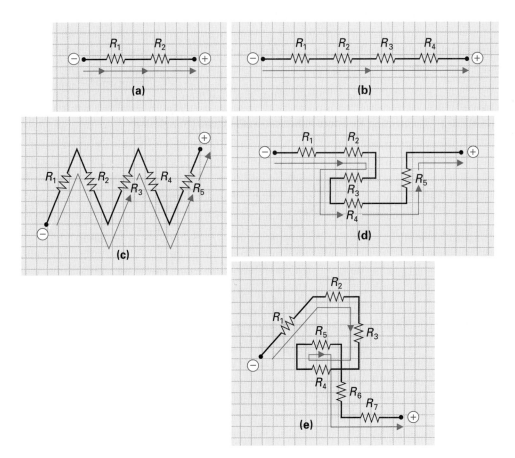

FIGURE 5-1   Five Series Resistive Circuits.

(a) Problem

(b) Solution

FIGURE 5-2   Connecting Resistors in Series. (a) Problem. (b) Solution.

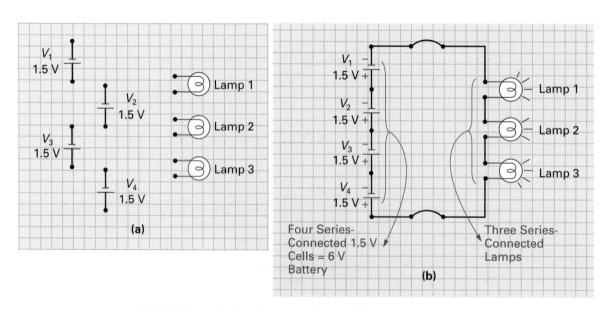

FIGURE 5-3   Series-Connected Cells and Lamps.

■ *Solution:*

In Figure 5-3(b) you can see the final circuit containing a source, made up of four series-connected 1.5 V cells, and a load, consisting of three series-connected lamps. As explained in Chapter 4, when cells are connected in series, the total voltage (V_T) will be equal to the sum of all the cell voltages:

$$V_T = V_1 + V_2 + V_3 + V_4 = 1.5\,\text{V} + 1.5\,\text{V} + 1.5\,\text{V} + 1.5\,\text{V} = 6\,\text{V}$$

5-2 CURRENT IN A SERIES CIRCUIT

The current in a series circuit has only one path to follow and cannot divert in any other direction. The current through a series circuit, therefore, is the same throughout that circuit.

Returning once again to the water analogy, you can see in Figure 5-4(a) that if 2 gallons of water per second are being supplied by the pump, 2 gallons per second must be pulled into the pump. If the rate at which water is leaving and arriving at the pump is the same, 2 gallons of water per second must be flowing throughout the circuit. It can be said, therefore, that the same value of water flow exists throughout a series-connected fluid system. This rule will always remain true, for if the valves were adjusted to double the opposition to flow, then half the flow, or 1 gallon of water per second, would be leaving the pump and flowing throughout the system.

Similarly, with the electronic series circuit shown in Figure 5-4(b), there is a total of 2 A leaving and 2 A arriving at the battery, and so the same value of current exists throughout the series-connected electronic circuit. If the circuit's resistance were changed, a new value of series circuit current would be present throughout the circuit. For example, if the resistance of the circuit was doubled, then half the current, or 1 A, will leave the battery, but that

FIGURE 5-4   Series Circuit Current. (a) Fluid System. (b) Electric System.

same value of 1 A will flow throughout the entire circuit. This series circuit current characteristic can be stated mathematically as

$$I_T = I_1 = I_2 = I_3 = \cdots$$

Total current = current through R_1 = current through R_2 = current through R_3, and so on.

EXAMPLE:

In Figure 5-5(a) and (b) on the previous page, a total current (I_T) of 1 A is flowing out of the
negative terminal of the dc power supply, through two end-to-end resistors R_1 and R_2, and
returning to the positive terminal of the dc power supply. Calculate:

a. The current through R_1 (I_1)

b. The current through R_2 (I_2)

**FIGURE 5-5   Total Current Example.
(a) Schematic. (b) Protoboard Circuit.
(c) Circuit Analysis Table.**

CIRCUIT ANALYSIS TABLE

Resistance $R = V/I$	Current $I = V/R$	Voltage $V = I \times R$	Power $P = V \times I$
$R_1 = 5\,\Omega$	$I_1 = 1$ A		
$R_2 = 15\,\Omega$	$I_2 = 1$ A		

$I_T = 1$ A

(c)

CHAPTER 5 / SERIES DC CIRCUITS

Solution:

Since R_1 and R_2 are connected in series, the current through both will be the same as the circuit current, which is equal to 1 A.

$$I_T = I_1 = I_2$$
$$1\ A = 1\ A = 1\ A$$

The details of this circuit are summarized in the analysis table shown in Figure 5-5(c).

SELF-TEST EVALUATION POINT FOR SECTIONS 5-1 AND 5-2

Use the following questions to test your understanding of Sections 5-1 and 5-2.

1. What is a series circuit?
2. What is the current flow through each of eight series-connected 8 Ω resistors if 8 A total current is flowing out of a battery?

5-3   RESISTANCE IN A SERIES CIRCUIT

Resistance is the opposition to current flow, and in a series circuit every resistor in series offers opposition to the current flow. In the water analogy of Figure 5-4, the total resistance or opposition to water flow is the sum of the two individual valve opposition values. Like the battery, the pump senses the total opposition in the circuit offered by all the valves or resistors, and the amount of current that flows is dependent on this resistance or opposition.

The total resistance in a series-connected electronic resistive circuit is thus equal to the sum of all the individual resistances, as shown in Figure 5-6(a) through (d). An equivalent circuit can be drawn for each of the circuits in Figure 5-6(b), (c), and (d) with one resistor of a value equal to the sum of all the series resistance values.

No matter how many resistors are connected in series, the total resistance or opposition to current flow is always equal to the sum of all the resistor values. This formula can be stated mathematically as

$$R_T = R_1 + R_2 + R_3 + \cdots$$

Total resistance = value of R_1 + value of R_2 + value of R_3, and so on.

Total resistance (R_T) is the only opposition a source can sense. It does not see the individual separate resistors, but one **equivalent resistance.** Based on the source's voltage and the circuit's total resistance, a value of current will be produced to flow through the circuit (Ohm's law, $I = V/R$).

Equivalent Resistance
(R_{eq})
Total resistance of all the individual resistances in a circuit.

◻ **EXAMPLE:**

Referring to Figure 5-7(a) and (b), calculate

  a. The circuit's total resistance
  b. The current flowing through R_2

FIGURE 5-6   Total or Equivalent Resistance.

■ *Solution:*

    a. $R_T = R_1 + R_2 + R_3 + R_4$
            $= 25\ \Omega + 20\ \Omega + 33\ \Omega + 10\ \Omega$
            $= 88\ \Omega$

    b. $I_T = I_1 = I_2 = I_3 = I_4$. Therefore, $I_2 = I_T = 3$ A.

The details of this circuit are summarized in the analysis table shown in Figure 5-7(c).

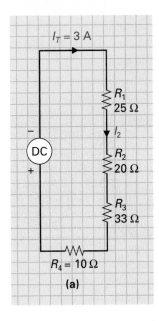

$I_T = 3$ A

R_1
25 Ω

I_2

DC

R_2
20 Ω

R_3
33 Ω

$R_4 = 10$ Ω

(a)

DC Power Supply

(b)

CIRCUIT ANALYSIS TABLE			
Resistance $R = V/I$	Current $I = V/R$	Voltage $V = I \times R$	Power $P = V \times I$
$R_1 = 25$ Ω	$I_1 = 3$ A		
$R_2 = 20$ Ω	$I_2 = 3$ A		
$R_3 = 33$ Ω	$I_3 = 3$ A		
$R_4 = 10$ Ω	$I_4 = 3$ A		

$R_T = 88$ Ω $I_T = 3$ A

(c)

FIGURE 5-7   Total Resistance Example. (a) Schematic. (b) Protoboard Circuit. (c) Circuit Analysis Table.

EXAMPLE:

Figure 5-8(a) and (b) shows how a single-pole three-position switch is being used to provide three different lamp brightness levels. In position ① R_1 is placed in series with the lamp, in position ② R_2 is placed in series with the lamp, and in position ③ R_3 is placed in series with the lamp. If the lamp has a resistance of 75 Ω, calculate the three values of current for each switch position.

Solution:

$$\text{Position 1: } R_T = R_1 + R_{\text{lamp}}$$
$$= 25 \text{ Ω} + 75 \text{ Ω} = 100 \text{ Ω}$$
$$I_T = \frac{V_T}{R_T} = \frac{12 \text{ V}}{100 \text{ Ω}} = 120 \text{ mA}$$

$$\text{Position 2: } R_T = R_2 + R_{\text{lamp}} = 50 \text{ Ω} + 75 \text{ Ω} = 125 \text{ Ω}$$
$$I_T = \frac{V_T}{R_T} = \frac{12 \text{ V}}{125 \text{ Ω}} = 96 \text{ mA}$$

$$\text{Position 3: } R_T = R_3 + R_{\text{lamp}} = 75 \text{ Ω} + 75 \text{ Ω} = 150 \text{ Ω}$$
$$I_T = \frac{V_T}{R_T} = \frac{12 \text{ V}}{150 \text{ Ω}} = 80 \text{ mA}$$

CIRCUIT ANALYSIS TABLE

	Resistance $R = V/I$	Current $I = V/R$	Voltage $V = I \times R$	Power $P = V \times I$
$S_1 = \text{①}$	$R_1 = 25\ \Omega$	$I_1 = 120\ \text{mA}$		
$S_1 = \text{②}$	$R_2 = 50\ \Omega$	$I_2 = 96\ \text{mA}$		
$S_1 = \text{③}$	$R_3 = 75\ \Omega$	$I_3 = 80\ \text{mA}$		

$R_{L1} = 75\ \Omega$

(c)

FIGURE 5-8 Three-Position Switch Controlling Lamp Brightness. (a) Schematic. (b) Protoboard Circuit. (c) Circuit Analysis Table.

The details of this circuit are summarized in the analysis table shown in Figure 5-8(c).

SELF-TEST EVALUATION POINT FOR SECTION 5-3

Use the following questions to test your understanding of Section 5-3.

1. State the total resistance formula for a series circuit.
2. Calculate R_T if $R_1 = 2\ \text{k}\Omega$, $R_2 = 3\ \text{k}\Omega$, and $R_3 = 4700\ \Omega$.

5-4 VOLTAGE IN A SERIES CIRCUIT

A potential difference or voltage drop will occur across each resistor in a series circuit when current is flowing. The amount of voltage drop is dependent on the value of the resistor and the amount of current flow. This idea of potential difference or voltage drop is best explained by returning to the water analogy. In Figure 5-9(a), you can see that the high pressure from the pump's outlet is present on the left side of the valve. On the right side of the valve, however, the high pressure is no longer present. The high potential that exists on the left of the

Pressure or Potential Difference

High Pressure   High Pressure Low Pressure
Outlet

Pump

Low Pressure
Inlet

(a)

High Potential
or Voltage

High Voltage

Potential
Difference

Low Voltage

Low Potential
or Voltage

(b)

**FIGURE 5-9   Series Circuit Voltage. (a) Fluid Analogy of Potential Difference.
(b) Electrical Potential Difference.**

valve is not present on the right, so a potential or pressure difference is said to exist across the valve.

Similarly, with the electronic circuit shown in Figure 5-9(b), the battery produces a high voltage or potential that is present at the top of the resistor. The high voltage that exists at the top of the resistor, however, is not present at the bottom. Therefore, a potential difference or voltage drop is said to occur across the resistor. This voltage drop that exists across resistors can be found by utilizing Ohm's law: $V = I \times R$.

EXAMPLE:

Referring to Figure 5-10, calculate:

a. Total resistance (R_T)

b. Amount of series current flowing throughout the circuit (I_T)

c. Voltage drop across R_1

d. Voltage drop across R_2

e. Voltage drop across R_3

Solution:

a. Total resistance $(R_T) = R_1 + R_2 + R_3$
$$= 20\ \Omega + 30\ \Omega + 50\ \Omega$$
$$= 100\ \Omega$$

b. Total current $(I_T) = \dfrac{V_{source}}{R_T}$
$$= \dfrac{100\ \text{V}}{100\ \Omega}$$
$$= 1\ \text{A}$$

FIGURE 5-10   Series Circuit Example. (a) Schematic. (b) Protoboard Circuit.

The same current will flow through the complete series circuit, so the current through R_1 will equal 1 A, the current through R_2 will equal 1 A, and the current through R_3 will equal 1 A.

c. Voltage across R_1 (V_{R1}) $= I_1 \times R_1$
$$= 1\,A \times 20\,\Omega$$
$$= 20\,V$$

d. Voltage across R_2 (V_{R2}) $= I_2 \times R_2$
$$= 1\,A \times 30\,\Omega$$
$$= 30\,V$$

e. Voltage across R_3 (V_{R3}) $= I_3 \times R_3$
$$= 1\,A \times 50\,\Omega$$
$$= 50\,V$$

Figure 5-11(a) shows the schematic and Figure 5-11(b) shows the analysis table for this example, with all of the calculated data inserted. As you can see, the 20 Ω resistor drops 20 V, the 30 Ω resistor has 30 V across it, and the 50 Ω resistor has dropped 50 V. From this example, you will notice that the larger the resistor value, the larger the voltage drop. Resistance and voltage drops are consequently proportional to one another.

The Voltage Drop across a Device is $\propto$ to Its Resistance

(a)

CIRCUIT ANALYSIS TABLE			
Resistance $R = V/I$	Current $I = V/R$	Voltage $V = I \times R$	Power $P = V \times I$
$R_1 = 20\ \Omega$	$I_1 = 1\ A$	$V_{R1} = 20\ V$	
$R_2 = 30\ \Omega$	$I_2 = 1\ A$	$V_{R2} = 30\ V$	
$R_3 = 50\ \Omega$	$I_3 = 1\ A$	$V_{R3} = 50\ V$	

$R_T = 100\ \Omega \qquad I_T = 1\ A \quad V_T = V_S = 100\ V$

(b)

FIGURE 5-11   **Voltage Drop and Resistance. (a) Schematic. (b) Circuit Analysis Table.**

$$V_{\text{drop}} \uparrow = I(\text{constant}) \times R \uparrow$$
$$V_{\text{drop}} \downarrow = I(\text{constant}) \times R \downarrow$$

Another interesting point you may have noticed from Figure 5-11 is that, if you were to add up all the voltage drops around a series circuit, they would equal the source (V_S) applied:

total voltage applied (V_S or V_T) $= V_{R1} + V_{R2} + V_{R3} + \cdots$

In the example in Figure 5-11, you can see that this is true, since

$$100\ V = 20\ V + 30\ V + 50\ V$$
$$100\ V = 100\ V$$

The series circuit has in fact divided up the applied voltage, and it appears proportionally across all the individual resistors. This characteristic was first observed by Gustav Kirchhoff in 1847. In honor of his discovery, this effect is known as **Kirchhoff's voltage law,** which states: The sum of the voltage drops in a series circuit is equal to the total voltage applied.

To summarize the effects of current, resistance, and voltage in a series circuit so far, we can say that:

1. The current in a series circuit has only one path to follow.

2. The value of current in a series circuit is the same throughout the entire circuit.

3. The total resistance in a series circuit is equal to the sum of all the resistances.

4. Resistance and voltage drops in a series circuit are proportional to one another, so a large resistance will have a large voltage drop and a small resistance will have a small voltage drop.

5. The sum of the voltage drops in a series circuit is equal to the total voltage applied.

TIME LINE

German physicist Gustav Robert Kirchhoff (1824-1879) extended Ohm's law and developed two important laws of his own, known as Kirchhoff's voltage law and Kirchhoff's current law.

Kirchhoff's Voltage Law

The algebraic sum of the voltage drops in a closed path circuit is equal to the algebraic sum of the source voltage applied.

EXAMPLE:

First, calculate the voltage drop across the resistor R_1 in the circuit in Figure 5-12(a) for a re-sistance of 4 Ω. Then, change the 4 Ω resistor to a 2 Ω resistor and recalculate the voltage drop across the new resistance value. Use a constant source of 4 V.

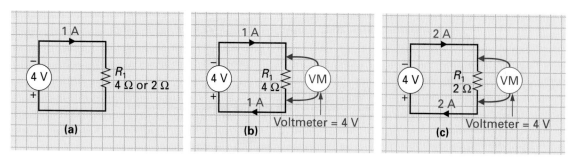

FIGURE 5-12   Single Resistor Circuit.

■   *Solution:*

Referring to Figure 5-12(b), you can see that when $R_1 = 4$ Ω, the voltage across R_1 can be calculated by using Ohm's law and is equal to

$$V_{R1} = I_1 \times R_1$$
$$V_{R1} = 1 \text{ A} \times 4 \text{ Ω}$$
$$= 4 \text{ V}$$

If the resistance is now changed to 2 Ω as shown in Figure 5-12(c), the current flow within the circuit will be equal to

$$I = \frac{V_S}{R} = \frac{4 \text{ V}}{2 \text{ Ω}} = 2 \text{ A}$$

The voltage dropped across the 2 Ω resistor will still be equal to

$$V_{R1} = I_1 \times R_1$$
$$= 2 \text{ A} \times 2 \text{ Ω}$$
$$= 4 \text{ V}$$

As you can see from this example, if only one resistor is connected in a series circuit, the entire applied voltage appears across this resistor. The value of this single resistor will determine the amount of current flow through the circuit and this value of circuit current will remain the same throughout.

▨   **EXAMPLE:**

Referring to Figure 5-13(a), calculate the following, and then draw the circuit schematic again with all of the new values inserted.

    a. Total circuit resistance

    b. Value of circuit current (I_T)

    c. Voltage drop across each resistor

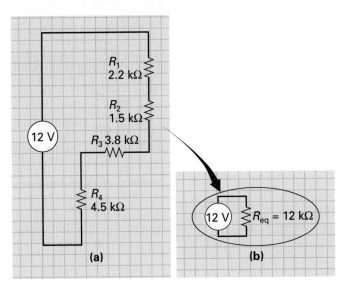

FIGURE 5-13   Series Circuit Example.

Solution:

a. $R_T = R_1 + R_2 + R_3 + R_4$
   $= 2.2 \text{ k}\Omega + 1.5 \text{ k}\Omega + 3.8 \text{ k}\Omega + 4.5 \text{ k}\Omega$
   $= (2.2 \times 10^3) + (1.5 \times 10^3) + (3.8 \times 10^3) + (4.5 \times 10^3)$
   $= 12 \text{ k}\Omega$ [Figure 5-13(b)]

b. $I_T = \dfrac{V_S}{R_T} = \dfrac{12 \text{ V}}{12 \text{ k}\Omega} = 1 \text{ mA}$

c. Voltage drop across each resistor:

$$V_{R1} = I_T \times R_1$$
$$= 1 \text{ mA} \times 2.2 \text{ k}\Omega$$
$$= 2.2 \text{ V}$$
$$V_{R2} = I_T \times R_2$$
$$= 1 \text{ mA} \times 1.5 \text{ k}\Omega$$
$$= 1.5 \text{ V}$$
$$V_{R3} = I_T \times R_3$$
$$= 1 \text{ mA} \times 3.8 \text{ k}\Omega$$
$$= 3.8 \text{ V}$$
$$V_{R4} = I_T \times R_4$$
$$= 1 \text{ mA} \times 4.5 \text{ k}\Omega$$
$$= 4.5 \text{ V}$$

Figure 5-14(a) shows the schematic diagram for this example with all of the values inserted, and Figure 5-14(b) shows this circuit's analysis table.

5-4-1   *Fixed Voltage Divider*

A series-connected circuit is often referred to as a *voltage-divider circuit,* because the total voltage applied (V_T) or source voltage (V_S) is divided and dropped proportionally across all the resistors in a series circuit. The amount of voltage dropped across a resistor is proportional to the value of resistance, and so a larger resistance will have a larger voltage drop, while a smaller resistance will have a smaller voltage drop.

FIGURE 5-14 Series Circuit Example with All Values Inserted.
(a) Schematic.
(b) Circuit Analysis Table.

(a)

CIRCUIT ANALYSIS TABLE

Resistance $R = V/I$	Current $I = V/R$	Voltage $V = I \times R$	Power $P = V \times I$
$R_1 = 2.2\ \text{k}\Omega$	$I_1 = 1\ \text{mA}$	$V_{R1} = 2.2\ \text{V}$	
$R_2 = 1.5\ \text{k}\Omega$	$I_2 = 1\ \text{mA}$	$V_{R2} = 1.5\ \text{V}$	
$R_3 = 3.8\ \text{k}\Omega$	$I_3 = 1\ \text{mA}$	$V_{R3} = 3.8\ \text{V}$	
$R_4 = 4.5\ \text{k}\Omega$	$I_4 = 1\ \text{mA}$	$V_{R4} = 4.5\ \text{V}$	

$R_T = 12\ \text{k}\Omega$ $I_T = 1\ \text{mA}$ $V_T = V_S = 12\ \text{V}$

(b)

The voltage dropped across a resistor is normally a factor that needs to be calculated. The voltage-divider formula allows you to calculate the voltage drop across any resistor without having to first calculate the value of circuit current. This formula is stated as

$$V_X = \frac{R_X}{R_T} \times V_S$$

where V_X = voltage dropped across selected resistor
R_X = selected resistor's value
R_T = total series circuit resistance
V_S = source or applied voltage

Figure 5-15 illustrates a circuit from a previous example. To calculate the voltage drop across each resistor, we would normally have to:

1. Calculate the total resistance by adding up all the resistance values.
2. Once we have the total resistance and source voltage (V_S), we could then calculate current.
3. Having calculated the current flowing through each resistor, we could then use the current value to calculate the voltage dropped across any one of the four resistors merely by multiplying current by the individual resistance value.

The voltage-divider formula allows us to bypass the last two steps in this procedure. If we know total resistance, supply voltage, and the individual resistance values, we can calculate the voltage drop across the resistor without having to calculate steps 2 and 3. For example, what would be the voltage dropped across R_2 and R_4 for the circuit shown in Figure 5-15? The voltage dropped across R_2 is

$$V_{R2} = \frac{R_2}{R_T} \times V_S$$

$$= \frac{1.5\ \text{k}\Omega}{12\ \text{k}\Omega} \times 12\ \text{V}$$

$$= 1.5\ \text{V}$$

FIGURE 5-15 Series Circuit Example.

The voltage dropped across R_4 is

$$V_{R4} = \frac{R_4}{R_T} \times V_S$$

$$= \frac{4.5 \text{ k}\Omega}{12 \text{ k}\Omega} \times 12 \text{ V}$$

$$= 4.5 \text{ V}$$

CALCULATOR SEQUENCE

Step	Keypad Entry	Display Response
1.	①　.　⑤　Ⓔ　③	1.5E3
2.	⊕	
3.	①　②　Ⓔ　③	12E3
4.	✕	0.125
5.	①　②	12
6.	⊜	1.5

The voltage-divider formula could also be used to find the voltage drop across two or more series-connected resistors. For example, referring again to the example circuit shown in Figure 5-15, what would be the voltage dropped across R_2 and R_3 combined? The voltage across $R_2 + R_3$ can be calculated using the voltage-divider formula as follows:

$$V_{R2} \text{ and } V_{R3} = \frac{R_2 + R_3}{R_T} \times V_S$$

$$= \frac{5.3 \text{ k}\Omega}{12 \text{ k}\Omega} \times 12 \text{ V}$$

$$= 5.3 \text{ V}$$

As can be seen in Figure 5-16, the voltage drop across R_2 and R_3 is 5.3 V.

EXAMPLE:

Referring to Figure 5-17(a), calculate the voltage drop across:

a. R_1, R_2, and R_3 separately

b. R_2 and R_3 combined

c. R_1, R_2, and R_3 combined

Solution:

a. The voltage drop across a resistor is proportional to the resistance value. The total resistance (R_T) in this circuit is 100 Ω or 100% of R_T. R_1 is 20% of the total resistance, so 20%

FIGURE 5-16   Series Circuit Example.

of the source voltage will appear across R_1. R_2 is 30% of the total resistance, so 30% of the source voltage will appear across R_2. R_3 is 50% of the total resistance, so 50% of the source voltage will appear across R_3. This was a very simple problem in which the figures worked out very neatly. The voltage-divider formula achieves the very same thing by calculating the ratio of the resistance value to the total resistance. This percentage is then multiplied by the source voltage in order to find the desired resistor's voltage drop:

$$V_{R1} = \frac{R_1}{R_T} \times V_S$$
$$= \frac{20 \ \Omega}{100 \ \Omega} \times V_S$$
$$= 0.2 \times 100 \text{ V} \qquad (20\% \text{ of } 100 \text{ V})$$
$$= 20 \text{ V}$$

CIRCUIT ANALYSIS TABLE

Resistance $R = V/I$	Current $I = V/R$	Voltage $V = I \times R$	Power $P = V \times I$
$R_1 = 20 \ \Omega$		$V_{R1} = 20$ V	
$R_2 = 30 \ \Omega$		$V_{R2} = 30$ V	
$R_3 = 50 \ \Omega$		$V_{R3} = 50$ V	
$R_T = 100 \ \Omega$		$V_T = V_S = 100$ V	

(b)

(a)

FIGURE 5-17   Series Circuit Example. (a) Schematic. (b) Circuit Analysis Table.

$$V_{R2} = \frac{R_2}{R_T} \times V_S$$

$$= \frac{30\ \Omega}{100\ \Omega} \times V_S$$

$$= 0.3 \times 100\ \text{V} \qquad (30\% \text{ of } 100\ \text{V})$$

$$= 30\ \text{V}$$

$$V_{R3} = \frac{R_3}{R_T} \times V_S$$

$$= \frac{50\ \Omega}{100\ \Omega} \times V_S$$

$$= 0.5 \times 100\ \text{V} \qquad (50\% \text{ of } 100\ \text{V})$$

$$= 50\ \text{V}$$

b. Voltage dropped across R_2 and $R_3 = 30 + 50 = 80$ V.

c. Voltage dropped across R_1, R_2, and $R_3 = 20 + 30 + 50 = 100$ V.

The details of this circuit are summarized in the analysis table shown in Figure 5-17(b).

To summarize the voltage-divider formula, we can say: The voltage drop across a resistor or group of resistors in a series circuit is equal to the ratio of that resistance (R_X) to the total resistance (R_T), multiplied by the source voltage (V_S).

To show an application of a voltage divider, let us imagine that three voltages of 50, 80, and 100 V were required by an electronic system in order to make it operate. To meet this need, we could use three individual power sources, which would be very expensive, or use one 100 V voltage source connected across three resistors, as shown in Figure 5-18, to divide up the 100 V.

EXAMPLE:

Figure 5-19(a) shows a 24 V voltage source driving a 10 Ω resistor that is located 1000 ft from the battery. If two 1000 ft lengths of AWG No. 13 wire are used to connect the source to the load, what will be the voltage applied across the load?

FIGURE 5-18   Series Circuit Example.

FIGURE 5-19   Series Wire Resistance.

■   *Solution:*

Referring back to Table 2-2, you can see that AWG No. 13 copper cable has a resistance of 2.003 Ω for every 1000 ft. To be more accurate, this means that our circuit should be redrawn as shown in Figure 5-19(b) to show the series resistances of wire 1 and wire 2. Using the voltage-divider formula, we can calculate the voltage drop across wire 1 and wire 2.

$$V_{W1} = \frac{R_{W1}}{R_T} \times V_T$$

$$= \frac{2\ \Omega}{14\ \Omega} \times 24\ V = 3.43\ V$$

Since the voltage drop across wire 2 will also be 3.43 V, the total voltage drop across both wires will be 6.86 V. The remainder, 17.14 V (24 V−6.86 V = 17.14 V), will appear across the load resistor, R_L.

5-4-2   *Variable Voltage Divider*

When discussing variable-value resistors in Chapter 3, we talked about a potentiometer, or variable voltage divider, which consists of a fixed value of resistance between two terminals and a wiper that can be adjusted to vary resistance between its terminal and one of the other two.

As an example, Figure 5-20(a) illustrates a 10 kΩ potentiometer that has been hooked up across a 10 V dc source with a voltmeter between terminals *B* and *C*. If the wiper terminal is positioned midway between *A* and *C*, the voltmeter should read 5 V, and the poten-

FIGURE 5-20   Potentiometer Wiper in Mid-Position.

tiometer will be equivalent to two 5 kΩ resistors in series, as shown in Figure 5-20(b). Kirchhoff's voltage law states that the entire source voltage will be dropped across the resistances in the circuit, and since the resistance values are equal, each will drop half of the source voltage, that is, 5 V.

In Figure 5-21(a), the wiper has been moved down so that the resistance between A and B is equal to 8 kΩ, and the resistance between B and C equals 2 kΩ. This will produce 2 V on the voltmeter, as shown in Figure 5-21(b). The amount of voltage drop is proportional to the resistance, so a larger voltage will be dropped across the larger resistance. Using the voltage-divider formula, you can calculate that the 8 kΩ resistance is 80% of the total resistance and therefore will drop 80% of the voltage:

$$V_{AB} = \frac{R_{AB}}{R_{total}} \times V_S = \frac{8 \text{ k}\Omega}{10 \text{ k}\Omega} \times 10 \text{ V} = 8 \text{ V}$$

The 2 kΩ resistance between B and C is 20% of the total resistance and consequently will drop 20% of the total voltage:

$$V_{BC} = \frac{R_{BC}}{R_{total}} \times V_S = \frac{2 \text{ k}\Omega}{10 \text{ k}\Omega} \times 10 \text{ V} = 2 \text{ V}$$

In Figure 5-22(a), the wiper has been moved up and now 2 kΩ exists between A and B, and 8 kΩ is present between B and C. In this situation, 2 V will be dropped across the 2 kΩ between A and B, and 8 V will be dropped across the 8 kΩ between B and C, as shown in Figure 5-22(b).

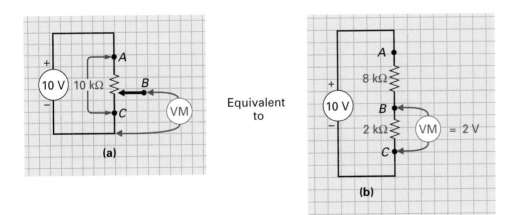

FIGURE 5-21   Potentiometer Wiper in Lower Position.

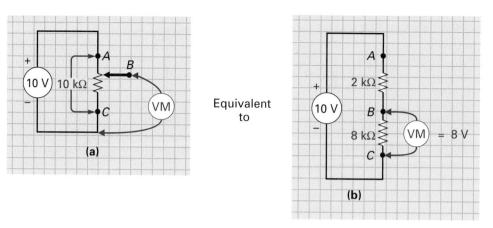

FIGURE 5-22   Potentiometer Wiper in Upper Position.

From this discussion, you can see that the potentiometer can be adjusted to supply different voltages on the wiper. This voltage can be decreased by moving the wiper down to supply a minimum of 0 V as show in Figure 5-23(a), or the wiper can be moved up to supply a maximum of 10 V as shown in Figure 5-23(b). By adjusting the wiper position, the potentiometer can be made to deliver any voltage between its maximum and minimum value, which is why the potentiometer is known as a variable voltage divider.

EXAMPLE:

Figure 5-24 illustrates how a potentiometer can be used to control the output volume of an amplifier that is being driven by a compact disk (CD) player. The preamplifier is producing an output of 2 V, which is developed across a 50 kΩ potentiometer. If the wiper of the potentiometer is in its upper position, the full 2 V from the preamp will be applied into the input of the power amplifier. The power amplifier has a fixed voltage gain (A_V) of 12, and therefore the power amplifier's output is always 12 times larger than the input voltage. An input of 2 V (V_{in}) will therefore produce an output voltage (V_{out}) of 24 V ($V_{out} = V_{in} \times A_V = 2 \text{ V} \times 12 = 24 \text{ V}$).

As the wiper is moved down, less of the 2 V from the preamp will be applied to the power amplifier, and therefore the output of the power amplifier and volume of the music heard will decrease. If the wiper of the potentiometer is adjusted so that a resistance of 20 kΩ exists between the wiper and the lower end of the potentiometer, what will be the input voltage to the power amplifier and output voltage to the speaker?

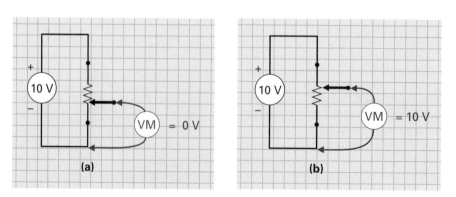

FIGURE 5-23   Minimum and Maximum Settings of a Potentiometer.

FIGURE 5-24 The Potentiometer as a Volume Control.

■ *Solution:*

By using the voltage-divider formula, we can determine the voltage developed across the potentiometer with 20 kΩ of resistance between the wiper (B) and lower end (C).

$$V_{in} = \frac{R_{AB}}{R_{AC}} \times V_{Pre}$$

$$= \frac{20 \text{ k}\Omega}{50 \text{ k}\Omega} \times 2 \text{ V} = 0.8 \text{ V (800 mV)}$$

The voltage at the input of the power amplifier (V_{in}), is applied to the input of the power amplifier, which has an amplification factor or gain of 12, and therefore the output voltage will be

$$V_{out} = V_{in} \times A_V$$

$$= 0.8 \text{ V} \times 12 = 9.6 \text{ V}$$

SELF-TEST EVALUATION POINT FOR SECTION 5-4

Use the following questions to test your understanding of Section 5-4.

1. True or false: A series circuit is also known as a voltage-divider circuit.
2. True or false: The voltage drop across a series resistor is proportional to the value of the resistor.
3. If 6 Ω and 12 Ω resistors are connected across a dc 18 V supply, calculate I_T and the voltage drop across each.
4. State the voltage-divider formula.
5. Which component can be used as a variable voltage divider?
6. Could a rheostat be used in place of a potentiometer?

5-5   POWER IN A SERIES CIRCUIT

As discussed earlier in Chapter 3, power is the rate at which work is done, and work is said to have been done when energy is converted from one energy form to another. Resistors convert electrical energy into heat energy, and the rate at which they dissipate energy is called *power* and is measured in *watts* (joules per second). Resistors all have a resistive value, a tolerance, and a wattage rating. The wattage of a resistor is the amount of heat energy a resistor can safely dissipate per second, and this wattage is directly proportional to the resis-

FIGURE 5-25   **Resistor Wattage Ratings.**

$\frac{1}{8}$ Watt $\frac{1}{4}$ Watt $\frac{1}{2}$ Watt   1 Watt   2 Watts

tor's size. Figure 5-25 reviews the size versus wattage rating of commercially available resistors. As you know, any of the power formulas can be used to calculate wattage. Resistors are manufactured in several different physical sizes, and if, for example, it is calculated that for a certain value of current and voltage a 5 W resistor is needed, and a ½ W resistor is put in its place, the ½ W resistor will burn out, because it is generating heat (5 W) faster than it can dissipate heat (½ W). A 10 W or 25 W resistor or greater could be used to replace a 5 W resistor but anything less than a 5 W resistor will burn out.

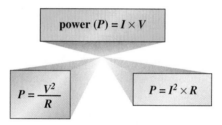

$$\text{power } (P) = I \times V$$

$$P = \frac{V^2}{R} \qquad P = I^2 \times R$$

It is, therefore, necessary that we have some way of calculating which wattage rating is necessary for each specific application.

A question you may be asking is why not just use large-wattage resistors everywhere. The two disadvantages with this are:

1. The larger the wattage, the greater the cost.
2. The larger the wattage, the greater the size and area the resistor occupies within the equipment.

EXAMPLE:

Figure 5-26 shows a 20 V battery driving a 12 V/1 A television set. R_1 is in series with the TV set and is being used to drop 8 V of the 20 V supply, so 12 V will be developed across the television.

a. What is the wattage rating for R_1?

b. What is the series load resistance of the TV set?

c. What is the amount of power being consumed by the TV set?

d. Compile a circuit analysis table for this circuit.

FIGURE 5-26   Series Circuit Example.

	CIRCUIT ANALYSIS TABLE			
	Resistance $R = V/I$	Current $I = V/R$	Voltage $V = I \times R$	Power $P = V \times I$
	$R_1 = 8\,\Omega$	$I_1 = 1$ A	$V_{R1} = 8$ V	$P_1 = 8$ W
	$R_{TV} = 12\,\Omega$	$I_2 = 1$ A	$V_{RTV} = 12$ V	$P_{TV} = 12$ W

$R_T = 20\,\Omega$ $I_T = 1$ A $V_T = V_S = 20$ V $P_T = 20$ W

(b)

FIGURE 5-27 **Series Circuit Example with Values Inserted. (a) Schematic. (b) Circuit Analysis Table.**

■ *Solution:*

a. Everything is known about R_1. Its resistance is 8 Ω, it has 1 A of current flowing through it, and 8 V is being dropped across it. As a result, any one of the three power formulas can be used to calculate the wattage rating of R_1.

$$\text{power } (P) = I \times V = 1 \text{ A} \times 8 \text{ V} = 8 \text{ W}$$

or

$$P = I^2 \times R = 1^2 \times 8 = 8 \text{ W}$$

or

$$P = \frac{V^2}{R} = \frac{8^2}{8} = 8 \text{ W}$$

The nearest commercially available device would be a 10 W resistor. If size is not a consideration, it is ideal to double the wattage needed and use a 16 W resistor.

b. You may recall that any piece of equipment is equivalent to a load resistance. The TV set has 12 V across it and is pulling 1 A of current. Its load resistance can be calculated simply by using Ohm's law and deriving an equivalent circuit, as shown in Figure 5-27(a).

$$R_L \text{ (load resistance)} = \frac{V}{I}$$
$$= \frac{12 \text{ V}}{1 \text{ A}}$$
$$= 12 \text{ }\Omega$$

c. The amount of power being consumed by the TV set is

$$P = V \times I = 12 \text{ V} \times 1 \text{ A} = 12 \text{ W}$$

d. Figure 5-27(b) shows the circuit analysis table for this example.

■ **EXAMPLE:**

Calculate the total amount of power dissipated in the series circuit in Figure 5-28, and insert any calculated values in a circuit analysis table.

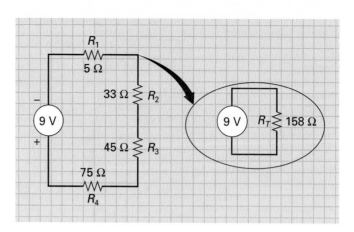

FIGURE 5-28   Series Circuit Example.

■ *Solution:*

The total power dissipated in a series circuit is equal to the sum of all the power dissipated by all the resistors. The easiest way to calculate the total power is to simplify the circuit to one resistance, as shown in Figure 5-28.

$$R_T = R_1 + R_2 + R_3 + R_4$$
$$= 5\ \Omega + 33\ \Omega + 45\ \Omega + 75\ \Omega$$
$$= 158\ \Omega$$

We now have total resistance and total voltage, so we can calculate the total power:

$$P_T = \frac{V_S^2}{R_T} = \frac{9\ V^2}{158\ \Omega} = \frac{81}{158} = 512.7\ \text{milliwatts (mW)}$$

The longer method would have been to first calculate the current through the series current:

$$I = \frac{V_S}{R_T} = \frac{9\ V}{158\ \Omega} = 56.96\ \text{mA} \quad \text{or} \quad 57\ \text{mA}$$

We could then calculate the power dissipated by each separate resistor and add up all the individual values to gain a total power figure. This is illustrated in Figure 5-29(a).

$$P_T = P_1 + P_2 + P_3 + P_4 + \ \cdots$$

total power = addition of all the individual power losses

$$P_T = 16\ \text{mW} + 107\ \text{mW} + 146\ \text{mW} + 243\ \text{mW}$$
$$= 512\ \text{mW}$$

The calculated values for this example are shown in the circuit analysis table in Figure 5-29(b).

5-5-1   *Maximum Power Transfer*

The **maximum power transfer theorem** states that maximum power will be delivered to a load when the resistance of the load (R_L) is equal to the resistance of the source (R_S). The best way to see if this theorem is correct is to apply it to a series of examples and then make a comparison.

CHAPTER 5 / SERIES DC CIRCUITS

$P_1 = I^2 \times R = 57 \text{ mA}^2 \times 5 =$ 16 mW

57 mA

$R_1 = 5\ \Omega$

R_2 $33\ \Omega$ $P_2 = I^2 \times R = 57 \text{ mA}^2 \times 33 =$ 107 mW

9 V

R_3 $45\ \Omega$ $P_3 = I^2 \times R = 57 \text{ mA}^2 \times 45 =$ 146 mW

$R_4 = 75\ \Omega$

$P_T = 512$ mW

$P_4 = I^2 \times R = 57 \text{ mA}^2 \times 75\ \Omega =$ 243 mW

(a)

CIRCUIT ANALYSIS TABLE

Resistance $R = V/I$	Current $I = V/R$	Voltage $V = I \times R$	Power $P = V \times I$
$R_1 = 5\ \Omega$	$I_1 = 57$ mA		$P_1 = 16$ mW
$R_2 = 33\ \Omega$	$I_2 = 57$ mA		$P_2 = 107$ mW
$R_3 = 45\ \Omega$	$I_3 = 57$ mA		$P_3 = 146$ mW
$R_4 = 75\ \Omega$	$I_4 = 57$ mA		$P_4 = 243$ mW

$R_T = 158\ \Omega$ $I_T = 57$ mA $V_T = V_S = 9$ V $P_T = 512$ mW

(b)

FIGURE 5-29 Series Circuit Example with Values Inserted. (a) Schematic. (b) Circuit Analysis Table.

EXAMPLE:

Figure 5-30(a) illustrates a 10 V battery with a 5 Ω internal resistance connected across a load. The load in this case is a light bulb, which has a load resistance of 1 Ω. Calculate the power delivered to this light bulb or load when $R_L = 1\ \Omega$.

Solution:

$$\text{circuit current, } I = \frac{V}{R}$$
$$= \frac{10 \text{ V}}{R_S + R_L}$$
$$= \frac{10 \text{ V}}{6\ \Omega}$$
$$= 1.66 \text{ A}$$

The power supplied to the load is $P = I^2 \times R = 1.66^2 \times 1\ \Omega = 2.8$ W.

EXAMPLE:

Figure 5-30(b) illustrates the same battery and R_S, but in this case connected across a 5 Ω light bulb. Calculate the power delivered to this light bulb now that $R_L = 5\ \Omega$.

Solution:

$$I = \frac{V}{R}$$
$$= \frac{10 \text{ V}}{R_S + R_L}$$
$$= \frac{10 \text{ V}}{10\ \Omega}$$
$$= 1 \text{ A}$$

Power supplied is $P = I^2 \times R = 1^2 \times 5\ \Omega = 5$ W.

(b) Example ②

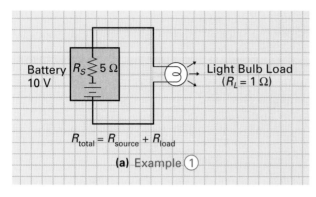

$R_{total} = R_{source} + R_{load}$

(a) Example ①

(c) Example ③

(d) Curve

FIGURE 5-30   Maximum Power Transfer.

■   **EXAMPLE:**

Figure 5-30(c) illustrates the same battery again, but in this case a 10 Ω light bulb is connected in the circuit. Calculate the power delivered to this light bulb now that $R_L = 10$ Ω.

■   *Solution:*

$$I = \frac{V}{R}$$

$$= \frac{10 \text{ V}}{R_S + R_L}$$

$$= \frac{10 \text{ V}}{15 \text{ Ω}}$$

$$= 0.67 \text{ A}$$

Thus $P = I^2 \times R = 0.67^2 \times 10$ Ω $= 4.5$ W.

Optimum Power Transfer

Since the ideal maximum power transfer conditions cannot always be achieved, most designers try to achieve optimum power transfer and have the source resistance and load resistance as close in value as possible.

As can be seen by the graph in Figure 5-30(d), which plots the power supplied to the load against load resistance, maximum power is delivered to the load (5 W) when the load resistance is equal to the source resistance.

The maximum power transfer condition is only used in special cases such as the automobile starter, where the load resistance remains constant and maximum power is needed. In most other cases, where load resistance can vary over a range of values, circuits are designed for a load resistance that will cause the best amount of power to be delivered. This is known as **optimum power transfer.**

Use the following questions to test your understanding of Section 5-5.

1. State the power formula.
2. Calculate the power dissipated by a 12 Ω resistor connected across a 12 V supply.
3. What fixed resistor type should probably be used for Question 2, and what would be a safe wattage rating?
4. What would be the total power dissipated if R_1 dissipates 25 W and R_2 dissipates 3800 mW?

5-6   TROUBLESHOOTING A SERIES CIRCUIT

A resistor will usually burn out and cause an open between its two leads when an excessive current flow occurs. This can normally, but not always, be noticed by a visual check of the resistor, which will appear charred due to the excessive heat. In some cases you will need to use your multimeter (combined ammeter, voltmeter, and ohmmeter) to check the circuit components to determine where a problem exists.

The two basic problems that normally exist in a series circuit are opens and shorts. In most instances a problem is not always as drastic as a short or an open, but may be a variation in a component's value over a long period of time, which will eventually cause a problem.

To summarize, then, we can say that one of three problems can occur to components in a series circuit:

1. A component will open (infinite resistance).
2. A component's value will change over a period of time.
3. A component will short (zero resistance).

The voltmeter is the most useful tool for checking series circuits as it can be used to measure voltage drops by connecting the meter leads across the component or resistor. Let's now analyze a circuit problem and see if we can solve it by logically **troubleshooting** the circuit and isolating the faulty component. To begin with, let us take a look at the effects of an open component.

Troubleshooting
The process of locating and diagnosing malfunctions or breakdowns in equipment by means of systematic checking or analysis.

5-6-1   *Open Component in a Series Circuit*

A component is open when its resistance is the maximum possible (infinity).

▨   **EXAMPLE:**

Figure 5-31(a) illustrates a TV set with a load resistance of 3 Ω. The TV set is off because R_2 has burned out and become an open circuit. How would you determine that the problem is R_2?

▨   *Solution:*

If an open circuit ever occurs in a series circuit, due in this case to R_2 having burned out, there can be no current flow, because series circuits have only one path for current to flow and that path has been broken ($I = 0$ A). Using the voltmeter to check the amount of voltage drop across each resistor, two results will be obtained:

1. The voltage drop across a good resistor will be zero volts.
2. The voltage drop across an open resistor will be equal to the source voltage V_S.

FIGURE 5-31   Troubleshooting an Open in a Series Circuit.

No voltage will be dropped across a good resistor because current is zero, and if $I = 0$, the voltage drop, which is the product of I and R, must be zero ($V = I \times R = 0 \times R = 0$ V). If no voltage is being dropped across the good resistor R_1 and the TV set resistance of 3 Ω, the entire source voltage will appear across the open resistor, R_2, in order that this series circuit comply with Kirchhoff's voltage law: V_S (9 V) $= V_{R1}$ (0 V) $+ V_{R2}$ (9 V) $+ V_L$ (0 V).

To explain this point further, refer to the fluid analogy in Figure 5-31(b). Like R_2, valve 2 has completely blocked any form of flow (water flow $= 0$). Looking at the pressure differences across all three valves, you can see that no pressure difference occurs across valves 1 and 3, but the entire pump pressure is appearing across valve 2, which is the component that has opened the circuit.

5-6-2   *Component Value Variation in a Series Circuit*

Resistors will rarely go completely open unless severely stressed because of excessive current flow. With age, resistors will normally change their resistance value. This occurs slowly and will generally cause a decrease in the resistor's resistance and eventually cause a circuit problem. This lowering of resistance will cause an increase of current, which will cause an increase in the power dissipated. If the wattage of a resistor is exceeded, it can burn out. If they do not burn out but merely blow the circuit fuse due to an increase in current, the problem can be found by measuring the resistance values of each resistor or by measuring how much voltage is dropped across each resistor and comparing these to the calculated expected voltage, based on the parts list supplied by the manufacturer.

5-6-3   *Shorted Component in a Series Circuit*

A component has gone short when its resistance is 0 W. Let's work out a problem using the same example of the three bulbs across the 9 V battery, as shown in Figure 5-32.

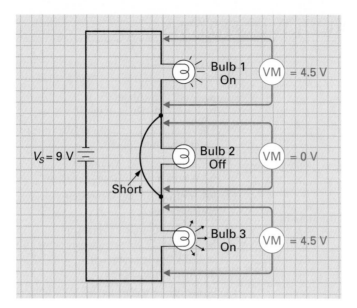

FIGURE 5-32   Troubleshooting a Short in a Series Circuit.

■ **EXAMPLE:**

Bulbs 1 and 3 in Figure 5-32 are on and bulb 2 is off. A piece of wire exists between the two terminals of bulb 2, and the current is taking the lowest resistance path through the wire rather than through the filament resistor of the bulb. Bulb 2 therefore has no current flow through it, and light cannot be generated without current. How would you determine what the problem is?

■ *Solution:*

If bulb 2 was open (burned out), there could be no current flow in the circuit, so bulbs 1 and 2 would be off. Bulb 2 must therefore be shorted. Using the voltmeter to investigate this further, you find that bulb 2 drops 0 V across it because it has no resistance except the very small wire resistance of the bypass, which means that 4.5 V must be dropped across each working bulb (1 and 3) in order to comply with Kirchhoff's voltage law. The loss of bulb 2's resistance causes the overall circuit resistance offered by bulbs 1 and 3 to decrease, which will cause an increase in current and so bulbs 1 and 3 should glow more brightly.

To summarize opens and shorts in series circuits, we can say that:

1. The supply voltage appears across an open component.
2. Zero volts appears across a shorted component.

SELF-TEST EVALUATION POINT FOR SECTION 5-6

Use the following questions to test your understanding of Section 5-6.

1. List the three basic problems that can occur with components in a series circuit.
2. How can an open component be detected in a series circuit?
3. True or false: If a series-connected resistor's value were to decrease, the voltage drop across that same resistor would increase.
4. How could a shorted component be detected in a series circuit?

Multiple-Choice Questions

1. A series circuit:
   a. Is the connecting of components end to end
   b. Provides a single path for current
   c. Functions as a voltage divider
   d. All of the above

2. The total current in a series circuit is equal to:
   a. $I_1 + I_2 + I_3 + \cdots$ c. $I_1 = I_2 = I_3 = \cdots$
   b. $I_1 - I_2$ d. All of the above

3. If R_1 and R_2 are connected in series with a total current of 2 A, what will be the current flowing through R_1 and R_2, respectively?
   a. 1 A, 1 A
   b. 2 A, 1 A
   c. 2 A, 2 A
   d. All of the above could be true on some occasions.

4. The total resistance in a series circuit is equal to:
   a. The total voltage divided by the total current
   b. The sum of all the individual resistor values
   c. $R_1 + R_2 + R_3 + \cdots$
   d. All of the above
   e. None of the above are even remotely true.

5. Which of Kirchhoff's laws applies to series circuits:
   a. His voltage law
   b. His current law
   c. His power law
   d. None of them apply to series circuits, only parallel.

6. The amount of voltage dropped across a resistor is proportional to:
   a. The value of the resistor
   b. The current flow in the circuit
   c. Both (a) and (b)
   d. None of the above

7. If three resistors of 6 kΩ, 4.7 kΩ, and 330 Ω are connected in series with one another, what total resistance will the battery sense?
   a. 11.03 MΩ   c. 6 kΩ
   b. 11.03 Ω   d. 11.03 kΩ

8. The voltage-divider formula states that the voltage drop across a resistor or multiple resistors in a series circuit is equal to the ratio of that _____ to the _____ multiplied by the _____.
   a. Resistance, source voltage, total resistance
   b. Resistance, total resistance, source voltage
   c. Total current, resistance, total voltage
   d. Total voltage, total current, resistance

9. The _____ can be used as a variable voltage divider.
   a. Potentiometer   c. SPDT switch
   b. Fixed resistor   d. None of the above

10. A resistor of larger physical size will be able to dissipate _____ heat than a small resistor.
    a. More   c. About the same
    b. Less   d. None of the above

11. The _____ is the most useful tool when checking series circuits.
    a. Ammeter   c. Voltmeter
    b. Wattmeter   d. Both (a) and (b)

12. When an open component occurs in a series circuit, it can be noticed because:
    a. Zero volts appears across it
    b. The supply voltage appears across it
    c. 1.3 V appears across it
    d. None of the above

13. Power can be calculated by:
    a. The addition of all the individual power figures
    b. The product of the total current and the total voltage
    c. The square of the total voltage divided by the total resistance
    d. All of the above

14. A series circuit is known as a:
    a. Current divider   c. Current subtractor
    b. Voltage divider   d. All of the above

15. In a series circuit only _____ path(s) exists for current flow, while the voltage applied is distributed across all the individual resistors.
    a. Three   c. Four
    b. Several   d. One

Practice Problems

16. If three resistors of 1.7 kΩ, 3.3 kΩ, and 14.4 kΩ are connected in series with one another across a 24 V source as shown in Figure 5-33, calculate:
    a. Total resistance (R_T)
    b. Circuit current
    c. Individual voltage drops
    d. Individual and total power dissipated

17. If 40 Ω and 35 Ω resistors are connected across a 24 V source, what would be the current flow through the resistors, and what resistance would cause half the current to flow?

18. Calculate the total resistance (R_T) of the following series-connected resistors: 2.7 kΩ, 3.4 MΩ, 370 Ω, and 4.6 MΩ.

19. Calculate the value of resistors needed to divide up a 90 V source to produce 45 V and 60 V outputs, with a divider circuit current of 1 A.

20. If $R_1 = 4.7$ kΩ and $R_2 = 6.4$ kΩ and both are connected across a 9 V source, how much voltage will be dropped across R_2?

21. What current would flow through R_1 if it were one-third the ohmic value of R_2 and R_3, and all were connected in series with a total current of 6.5 mA flowing out of V_S?

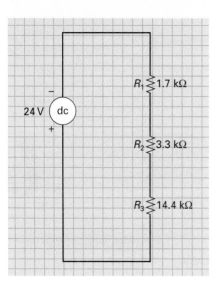

FIGURE 5-33

22. Draw a circuit showing $R_1 = 2.7$ kΩ, $R_2 = 3.3$ kΩ, and $R_3 = 0.027$ MΩ in series with one another across a 20 V source. Calculate:

a. I_T d. P_2 g. V_{R2}
b. P_T e. P_3 h. V_{R3}
c. P_1 f. V_{R1} i. I_{R1}

23. Calculate the current flowing through three light bulbs that are dissipating 120 W, 60 W, and 200 W when they are connected in series across a 120 V source. How is the voltage divided around the series circuit?

24. If three equal-value resistors are series connected across a dc power supply adjusted to supply 10 V, what percentage of the source voltage will appear across R_1?

25. Refer to the following figures and calculate:

a. I (Figure 5-34)
b. R_T and P_T (Figure 5-35)
c. V_S, V_{R1}, V_{R2}, V_{R3}, V_{R4}, P_1, P_2, P_3, and P_4 (Figure 5-36)
d. P_T, I, R_1, R_2, R_3, and R_4 (Figure 5-37)

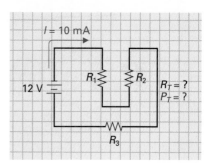

FIGURE 5-35

Troubleshooting Questions

26. If three bulbs are connected across a 9 V battery in series, and the filament in one of the bulbs burned out, causing an open in the bulb, would the other lamps be on? Explain why.

27. Using a voltmeter, how would a short be recognized in a series circuit?

28. If one of three series-connected bulbs is shorted, will the other two bulbs be on? Explain why.

29. When one resistor in a series string is open, explain what would happen to the circuit's:

a. Current
b. Resistance
c. Voltage across the open component
d. Voltage across the other components

30. When one resistor in a series string is shorted, explain what would happen to the circuit's:

a. Current
b. Resistance
c. Voltage across the shorted component
d. Voltage across the other components

FIGURE 5-34

FIGURE 5-36

FIGURE 5-37

Parallel DC Circuits

6

An Apple a Day

One of the early pioneers who laid the foundations of many branches of science was Isaac Newton. He was born in a small farmhouse near Woolsthorpe in Lincolnshire, England, on Christmas Day in 1642. He was an extremely small, premature baby, which worried the midwives who went off to get medicine and didn't expect to find him alive when they came back. Luckily for science, however, he did survive.

Newton's father was an illiterate farmer who died three months before he was born. His mother married the local vicar soon after Newton's birth and left him in the care of his grandmother. This parental absence while he was growing up had a traumatic effect on him and throughout his life affected his relationships with people.

At school in the nearby town of Grantham, Newton showed no interest in classical studies but rather in making working models and studying the world around him. When he was in his early teens, his stepfather died and Newton had to return to the farm to help his mother. Newton proved to be a hopeless farmer; in fact, on one occasion when he was tending sheep he became so engrossed with a stream that he followed it for miles and was missing for hours. Luckily, a schoolteacher recognized Newton's single-minded powers of concentration and convinced his mother to let him return to school, where he performed better and later went off to Cambridge University.

In 1665, in his graduation year, Newton left Cambridge to return home to escape an epidemic of bubonic plague that had spread throughout London. During this time, Newton reflected on his years of seclusion at his mother's cottage and called them the most significant time in his life. It was here on a warm summer's day that Newton saw the apple fall to the ground, leading him to develop his laws of motion and gravitation. It was here that he wondered about the nature of light and later built a prism and proved that white light contains all the colors in a rainbow.

Later, when Newton returned to Cambridge, he demonstrated many of his discoveries but was reluctant to publish the details and did so finally only at the insistence of others. Newton went on to build the first working reflecting astronomical telescope and wrote a paper on optics that was fiercely challenged by the physicist Robert Hooke. Hooke quarreled bitterly with Newton over the years, and there were also heated debates about whether Newton or the German mathematician Gottfried Leibniz invented calculus.

The truth is that many of Newton's discoveries roamed around with him in the English countryside, and even though many of these would, before long, have been put forward by others, it was Newton's genius and skill (and long walks) that tied together all the loose ends.

Introduction

By tracing the path of current, we can determine whether a circuit has series-connected or parallel-connected components. In a series circuit, there is only one path for current, whereas in parallel circuits the current has two or more paths. These paths are known as *branches*. A parallel circuit, by definition, is when two or more components are connected to the same voltage source so that the current can branch out over two or more paths. In a parallel resistive circuit, two or more resistors are connected to the same source, so current splits to travel through each separate branch resistance.

6-1 COMPONENTS IN PARALLEL

Many components, other than resistors, can be connected in parallel, and a **parallel circuit** can easily be identified because current is split into two or more paths. Being able to identify a parallel connection requires some practice, because they can come in many different shapes and sizes. The means for recognizing series circuits is that if you can place your pencil at the negative terminal of the voltage source (battery) and follow the wire connections through components to the positive side of the battery and only have one path to follow, the circuit is connected in series. If, however, you can place your pencil at the negative terminal of the voltage source and follow the wire and at some point have a choice of two or more routes, the circuit is connected with two or more parallel branches. The number of routes determines the number of parallel branches. Figure 6-1 illustrates five examples of parallel resistive circuits.

Parallel Circuit
Also called shunt; circuit having two or more paths for current flow.

□ **EXAMPLE:**

Figure 6-2(a) illustrates four resistors laid out on a table top.

   a. With wire leads, connect all four resistors in parallel on a protoboard, and then connect the circuit to a dc power supply.

   b. Draw the schematic diagram of the parallel-connected circuit.

■ *Solution:*

Figure 6-2(b) shows how to connect the resistors in parallel and the circuit's schematic.

FIGURE 6-1 Parallel Circuits.

(a)

(b)

FIGURE 6-2   Connecting Resistors in Parallel. (a) Problem. (b) Solution.

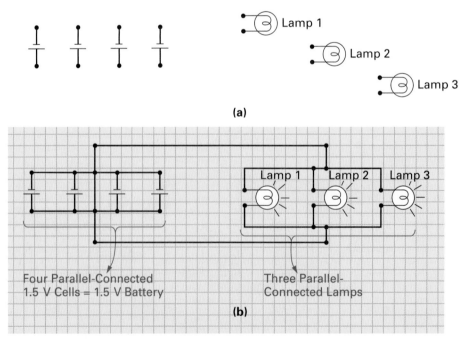

(a)

(b)

FIGURE 6-3   Parallel-Connected Cells and Lamps.

■ **EXAMPLE:**

Figure 6-3(a) shows four 1.5 V cells and three lamps. Using wires, connect all of the cells in parallel to create a 1.5 V source. Then connect all of the three lamps in parallel with one another and finally connect the 1.5 V source across the three parallel-connected lamp load.

■ *Solution:*

In Figure 6-3(b) you can see the final circuit containing a source consisting of four parallel-connected 1.5 V cells, and a load consisting of three parallel-connected lamps. As explained in Chapter 4, when cells are connected in parallel, the total voltage remains the same as for one cell, but the current demand can now be shared.

6-2   VOLTAGE IN A PARALLEL CIRCUIT

Figure 6-4(a) shows a simple circuit with four resistors connected in parallel across the voltage source of a 9 V battery. The current from the negative side of the battery will split between the four different paths or branches, yet the voltage drop across each branch of a parallel circuit is equal to the voltage drop across all the other branches in parallel. This means that if the voltmeter were to measure the voltage across A and B or C and D or E and F or G and H, they would all be the same or, in this example, would all drop 9 V.

It is quite easy to imagine why there will be the same voltage drop across all the resistors, seeing that points A, C, E, and G are all one connection and points B, D, F, and H are all one connection. Measuring the voltage drop with the voltmeter across any of the resistors is the same as measuring the voltage across the battery, as shown in Figure 6-4(b). As long as

FIGURE 6-4   Voltage in a Parallel Circuit.

(a)

(b)

FIGURE 6-5   Parallel-Circuit Voltage Drop.

the voltage source remains constant, the voltage drop will always be common (9 V) across the parallel resistors, no matter what value or how many resistors are connected in parallel. The voltmeter is therefore measuring the voltage between two common points that are directly connected to the battery, and the voltage dropped across all these parallel resistors will be equal to the source voltage.

In Figure 6-5(a) and (b), the same circuit is shown in two different ways, so you can see how the same circuit can look completely different. In both examples, the voltage drop across any of the resistors will always be the same and, as long as the voltage source is not heavily loaded, equal to the source voltage. Just as you can trace the positive side of the battery to all four resistors, you can also trace the negative side to all four resistors.

Mathematically stated, we can say that in a parallel circuit

$$V_{R1} = V_{R2} = V_{R3} = V_{R4} = V_S$$

voltage drop across R_1 = voltage drop across R_2 = voltage drop across R_3 (etc.) = source voltage

To reinforce the concept, let us compare this parallel circuit characteristic to a water analogy, as seen in Figure 6-6. The pressure across valves A and B will always be the same, even if one offers more opposition than the other. This is because the pressure measured across either valve will be the same as checking the pressure difference between piping X and Y. Since the piping at point X and Y run directly back to the pump, the pressure across A and B is the same as the pressure difference across the pump.

FIGURE 6-6   Fluid Analogy of Parallel-Circuit Pressure.

FIGURE 6-7   A Parallel Circuit Example (a) Schematic. (b) Protoboard Circuit.

▢   **EXAMPLE:**

Refer to Figure 6-7 and calculate:

a. Voltage drop across R_1
b. Voltage drop across R_2
c. Voltage drop across R_3

▢   *Solution:*

Since all these resistors are connected in parallel, the voltage across every branch will be the same and equal to the source voltage applied. Therefore,

$$V_{R1} = V_{R2} = V_{R3} = V_S$$
$$7.5\ V = 7.5\ V = 7.5\ V = 7.5\ V$$

SELF-TEST EVALUATION POINT FOR SECTIONS 6-1 AND 6-2

Use the following questions to test your understanding of Sections 6-1 and 6-2.

1. Describe a parallel circuit.
2. True or false: A parallel circuit is also known as a voltage-divider circuit.
3. What would be the voltage drop across R_1 if $V_S = 12$ V and R_1 and R_2 are both in parallel with one another and equal to 24 Ω each?
4. Can Kirchhoff's voltage law be applied to parallel circuits?

6-3   CURRENT IN A PARALLEL CIRCUIT

In addition to providing the voltage law for series circuits, Gustav Kirchhoff (in 1847) was the first to observe and prove that the sum of all the **branch currents** in a parallel circuit ($I_1 + I_2 + I_3$, etc.) was equal to the total current (I_T). In honor of his second discovery, this phenomenon is known as **Kirchhoff's current law,** which states that the sum of all the currents entering a junction is equal to the sum of all the currents leaving that same junction.

Figure 6-8(a) and (b) illustrate two examples of how this law applies. In both examples, the sum of the currents entering a junction is equal to the sum of the currents leaving that same junction. In Figure 6-8(a) the total current arrives at a junction X and splits to produce three branch currents, I_1, I_2, and I_3, which cumulatively equal the total current (I_T) that arrived at the junction X. The same three branch currents combine at junction Y, and the total

Branch Current

A portion of the total current that is present in one path of a parallel circuit.

Kirchhoff's Current Law

The sum of the currents flowing into a point in a circuit is equal to the sum of the currents flowing out of that same point.

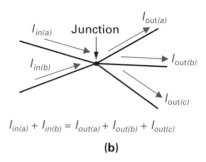

$I_T = I_1 + I_2 + I_3$ or $I_1 + I_2 + I_3 = I_T$

(a)

$I_{in(a)} + I_{in(b)} = I_{out(a)} + I_{out(b)} + I_{out(c)}$

(b)

FIGURE 6-8   Kirchhoff's Current Law.

current (I_T) leaving that junction is equal to the sum of the three branch currents arriving at junction Y.

$$I_T = I_1 + I_2 + I_3 + I_4 = \ldots$$

As another example, in Figure 6-8(b) you can see that there are two branch currents entering a junction [$I_{in(a)}$ and $I_{in(b)}$] and three branch currents leaving that same junction [$I_{out(a)}$, $I_{out(b)}$, and $I_{out(c)}$]. As stated below the illustration, the sum of the input currents will equal the sum of the output currents: $I_{in(a)} + I_{in(b)} = I_{out(a)} + I_{out(b)} + I_{out(c)}$.

◻ **EXAMPLE:**

Refer to Figure 6-9 and calculate the value of I_T.

◼ *Solution:*

By Kirchhoff's current law,

$$I_T = I_1 + I_2 + I_3 + I_4$$
$$? = 2 \text{ mA} + 17 \text{ mA} + 7 \text{ mA} + 37 \text{ mA}$$
$$I_T = 63 \text{ mA}$$

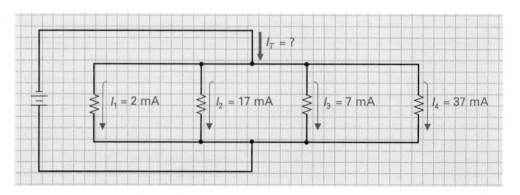

FIGURE 6-9   Calculating Total Current.

EXAMPLE:

Refer to Figure 6-10 and calculate the value of I_1.

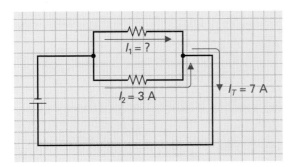

FIGURE 6-10   Calculating Branch Current.

Solution:

By transposing Kirchhoff's current law, we can determine the unknown value (I_1):

$$I_T = I_1 + I_2 \qquad I_T = I_1 + I_2 \qquad (-I_2)$$
$$7\,\text{A} = ? + 3\,\text{A} \quad \text{or} \quad I_T - I_2 = I_1$$
$$I_1 = 4\,\text{A} \qquad I_1 = I_T - I_2 = 7\,\text{A} - 3\,\text{A} = 4\,\text{A}$$

As with series circuits, to find out how much current will flow through a parallel circuit, we need to find out how much opposition or resistance is being connected across the voltage source.

$$I_T = \frac{V_S}{R_T}$$

Total current equals source voltage divided by total resistance.

When we connect resistors in parallel, the total resistance in the circuit will actually decrease. In fact, the total resistance in a parallel circuit will always be less than the value of the smallest resistor in the circuit.

To prove this point, Figure 6-11 shows how two sets of identical resistors (R_1, R_2, and R_3) were used to build both a series and a parallel circuit. The total current flow in the parallel circuit would be larger than the total current in the series circuit, because the parallel circuit has two or more paths for current to flow, while the series circuit only has one.

To explain why the total current will be larger in a parallel circuit, let us take the analogy of a freeway with only one path for traffic to flow. A single-lane freeway is equivalent to a series circuit, and only a small amount of traffic is allowed to flow along this freeway. If the freeway is expanded to accommodate two lanes, a greater amount of traffic can flow along the freeway in the same amount of time. Having more lanes permits a greater total amount of traffic flow. In parallel circuits, more branches will allow a greater total amount of current flow because there is less resistance in more paths than there is with only one path. This concept is summarized in Figure 6-11.

Just as a series circuit is often referred to as a voltage-divider circuit, a parallel circuit is often referred to as a **current-divider** circuit, because the total current arriving at a junction will divide or split into branch currents (Kirchhoff's law), as shown in Figure 6-12.

The current division is inversely proportional to the resistance in the branch, assuming that the voltage across both resistors is constant and equal to the source voltage (V_S). This means that a large branch resistance will cause a small branch current ($I\downarrow = V/R\uparrow$), and a small branch resistance will cause a large branch current ($I\uparrow = V/R\downarrow$).

Current Divider

A parallel network designed to proportionally divide the circuit's total current.

FIGURE 6-11   Series Circuit and Parallel Circuit Current Comparison.

FIGURE 6-12   The Parallel Circuit Current Divider.

■   **EXAMPLE:**

Calculate the following for Figure 6-13(a), and then insert the values in a circuit analysis table.

a. I_1

b. I_2

c. I_T

■   *Solution:*

Since R_1 and R_2 are connected in parallel across the 10 V source, the voltage across both resistors will be 10 V.

a. $I_1 = \dfrac{V_{R1}}{R_1} = \dfrac{10 \text{ V}}{6 \text{ k}\Omega} = 1.6 \text{ mA}$ (smaller branch current through larger branch resistance)

CIRCUIT ANALYSIS TABLE			
Resistance $R = V/I$	Voltage $V = I \times R$	Current $I = V/R$	Power $P = V \times I$
$R_1 = 6\ k\Omega$	$V_{R1} = 10\ V$	$I_1 = 1.6\ mA$	
$R_2 = 3\ k\Omega$	$V_{R2} = 10\ V$	$I_2 = 3.3\ mA$	

$V_T = V_S = 10\ V$ $I_T = 4.9\ mA$

(b)

FIGURE 6-13   Parallel Circuit Example. (a) Schematic. (b) Circuit Analysis Table.

b. $I_2 = \dfrac{V_{R2}}{R_2} = \dfrac{10\ V}{3\ k\Omega} = 3.3\ mA$ (larger branch current through smaller branch resistance)

c. By Kirchhoff's current law,

$$I_T = I_1 + I_2$$
$$= 1.6\ mA + 3.3\ mA$$
$$= 4.9\ mA$$

The circuit analysis table for this example is shown in Figure 6-13(b).

By rearranging Ohm's law, we can arrive at another formula, which is called the *current-divider formula* and can be used to calculate the current through any branch of a multiple-branch parallel circuit.

$$I_x = \frac{R_T}{R_x} \times I_T$$

where I_x = branch current desired
R_T = total resistance
R_x = resistance in branch
I_T = total current

EXAMPLE:

Refer to Figure 6-14 and calculate the following if the total circuit resistance (R_T) is equal to 1 kΩ:

a. $I_1 =$

b. $I_2 =$

c. $I_3 =$

Solution:

Since the source and therefore the voltage across each branch resistor are not known, we will use the current-divider formula to calculate I_1, I_2, and I_3.

a. $I_1 = \dfrac{R_T}{R_1} \times I_T = \dfrac{1\ k\Omega}{2\ k\Omega} \times 10\ mA = 5\ mA$

(smallest branch resistance has largest branch current)

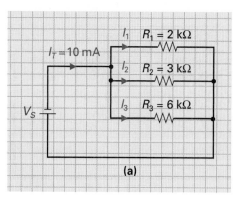

CIRCUIT ANALYSIS TABLE			
Resistance $R = V/I$	Voltage $V = I \times R$	Current $I = V/R$	Power $P = V \times I$
$R_1 = 2\ k\Omega$		$I_1 = 5\ mA$	
$R_2 = 3\ k\Omega$		$I_2 = 3.33\ mA$	
$R_3 = 6\ k\Omega$		$I_3 = 1.67\ mA$	
$R_T = 1\ k\Omega$		$I_T = 10\ mA$	

(b)

FIGURE 6-14   Parallel Circuit Example. (a) Schematic. (b) Circuit Analysis Table.

b. $I_2 = \dfrac{R_T}{R_2} \times I_T = \dfrac{1\ k\Omega}{3\ k\Omega} \times 10\ mA$

   $= 3.33\ mA$

c. $I_3 = \dfrac{R_T}{R_3} \times I_T = \dfrac{1\ k\Omega}{6\ k\Omega} \times 10\ mA$

   $= 1.67\ mA$

   (largest branch resistance has smallest branch current)

To double-check that the values for I_1, I_2, and I_3 are correct, you can apply Kirchhoff's current law, which is

$$I_T = I_1 + I_2 + I_3$$
$$10\ mA = 5\ mA + 3.33\ mA + 1.67\ mA$$
$$= 10\ mA$$

□   **EXAMPLE:**

A common use of parallel circuits is in the residential electrical system. All of the household lights and appliances are wired in parallel, as seen in the typical room wiring circuit in Figure 6-15(a). If it is a cold winter morning, and lamps 1 and 2 are switched ON, together with the space heater and hair dryer, what will the individual branch currents be, and what will be the total current drawn from the source?

■   *Solution:*

Figure 6-15(b) shows the schematic of the pictorial in Figure 6-15(a). Since all resistances are connected in parallel across a 120 V source, the voltage across all devices will be 120 V. Using Ohm's law we can calculate the four branch currents:

$$I_1 = \frac{V_{lamp1}}{R_{lamp1}} = \frac{120\ V}{125\ \Omega} = 960\ mA$$

$$I_2 = \frac{V_{lamp2}}{R_{lamp2}} = \frac{120\ V}{125\ \Omega} = 960\ mA$$

$$I_3 = \frac{V_{hairdryer}}{R_{hairdryer}} = \frac{120\ V}{40\ \Omega} = 3\ A$$

FIGURE 6-15   Parallel Home Electrical System.

$$I_4 = \frac{V_{\text{heater}}}{R_{\text{heater}}} = \frac{120\text{ V}}{12\ \Omega} = 10\text{ A}$$

By Kirchhoff's current law,

$$\begin{aligned}
I_T &= I_1 + I_2 + I_3 + I_4 \\
&= 960\text{ mA} + 960\text{ mA} + 3\text{ A} + 10\text{ A} \\
&= 14.92\text{ A}
\end{aligned}$$

SELF-TEST EVALUATION POINT FOR SECTION 6-3

Use the following questions to test your understanding of Section 6-3.

1. State Kirchhoff's current law.

2. If $I_T = 4$ A and $I_1 = 2.7$ A in a two-resistor parallel circuit, what would be the value of I_2?

3. State the current-divider formula.

4. Calculate I_1 if $R_T = 1$ kΩ, $R_1 = 2$ kΩ, and $V_T = 12$ V.

We now know that parallel circuits will have a larger current flow than a series circuit containing the same resistors due to the smaller total resistance. To calculate exactly how much total current will flow, we need to be able to calculate the total resistance that the parallel circuit presents to the source.

The ability of a circuit to conduct current is a measure of that circuit's conductance, and you will remember from Chapter 1 that conductance (G) is equal to the reciprocal of resistance and is measured in siemens.

$$G = \frac{1}{R} \quad \text{(siemens)}$$

Every resistor in a parallel circuit will have a conductance figure that is equal to the reciprocal of its resistance, and the total conductance (G_T) of the circuit will be equal to the sum of all the individual resistor conductances. Therefore,

$$G_T = G_{R1} + G_{R2} + G_{R3} + \cdots$$

Total conductance is equal to the conductance of R_1 + the conductance of R_2 +
the conductance of R_3 + · · ·

Once you have calculated total conductance, the reciprocal of this figure will give you total resistance. If, for example, we have two resistors in parallel, as shown in Figure 6-16, the conductance for R_1 will equal

$$G_{R1} = \frac{1}{R_1} = \frac{1}{20 \ \Omega} = 0.05 \text{ S}$$

The conductance for R_2 will equal

$$G_{R2} = \frac{1}{R_2} = \frac{1}{40 \ \Omega} = 0.025 \text{ S}$$

The total conductance will therefore equal

$$\begin{aligned} G_{\text{total}} &= G_{R1} + G_{R2} \\ &= 0.05 + 0.025 \\ &= 0.075 \text{ S} \end{aligned}$$

Since total resistance is equal to the reciprocal of total conductance, total resistance for the parallel circuit in Figure 6-16 will be

$$R_{\text{total}} = \frac{1}{G_{\text{total}}} = \frac{1}{0.075 \text{ S}} = 13.3 \ \Omega$$

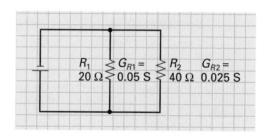

FIGURE 6-16   **Parallel Circuit Conductance and Resistance.**

Combining these three steps (first calculate individual conductances, total conductance and then total resistance) we can arrive at the following *reciprocal formula:*

$$R_{total} = \frac{1}{(1/R_1) + (1/R_2)}$$

This formula states that the conductance of R_1 (G_{R1})
+ conductance of R_2 (G_{R2})
= total conductance (G_T),
and the reciprocal of total conductance is equal to total resistance

In the example for Figure 6-16, this combined general formula for total resistance can be verified by plugging in the example values.

$$R_T = \frac{1}{(1/R_1) + (1/R_2)}$$

$$= \frac{1}{(1/20) + (1/40)}$$

$$= \frac{1}{0.05 + 0.025}$$

$$= \frac{1}{0.075}$$

$$= 13.3 \ \Omega$$

CALCULATOR SEQUENCE

Step	Keypad Entry	Display Response
1.	(Clear Memory)	
2.	[2][0]	20.
3.	[1/x]	5.E-2
4.	[+]	5.E-2
5.	[4][0]	40
6.	[1/x]	2.5E-2
7.	[=]	7.5E-2
8.	[1/x]	13.33333

The *reciprocal formula* for calculating total parallel circuit resistance for any number of resistors is

$$R_T = \frac{1}{(1/R_1) + (1/R_2) + (1/R_3) + (1/R_4) + \cdots}$$

□ **EXAMPLE:**

Referring to Figure 6-17(a), calculate:

  a. Total resistance
  b. Voltage drop across R_2
  c. Voltage drop across R_3

(a)   **(b)**

FIGURE 6-17   Parallel Circuit Example. (a) Schematic. (b) Equivalent Circuit.

■ *Solution:*

Total resistance can be calculated using the reciprocal formula.

$$R_T = \frac{1}{(1/R_1) + (1/R_2) + (1/R_3)}$$

$$= \frac{1}{(1/25\ \Omega) + (1/73\ \Omega) + (1/33\ \Omega)}$$

$$= \frac{1}{0.04 + 0.014 + 0.03}$$

$$= 11.9\ \Omega$$

With parallel resistance circuits, the total resistance is always smaller than the smallest branch resistance. In this example the total opposition of this circuit is equivalent to 11.9 Ω, as shown in Figure 6-17(b).

With parallel resistive circuits, the voltage drop across any branch is equal to the voltage drop across each of the other branches and is equal to the source voltage, in this example 3.9 V.

6-4-1   *Two Resistors in Parallel*

If only two resistors are connected in parallel, a quick and easy formula called the *product-over-sum formula* can be used to calculate total resistance.

$$R_T = \frac{R_1 \times R_2}{R_1 + R_2}$$

$$\text{total resistance} = \frac{\text{product of both resistance values}}{\text{sum of both resistance values}}$$

Using the example shown in Figure 6-18, let us compare the *product-over-sum* formula with the *reciprocal* formula.

(a) PRODUCT-OVER-SUM FORMULA

$$R_T = \frac{R_1 \times R_2}{R_1 + R_2}$$

$$= \frac{3.7\ \text{k}\Omega \times 2.2\ \text{k}\Omega}{3.7\ \text{k}\Omega + 2.2\ \text{k}\Omega}$$

$$= \frac{8.14\ \text{k}\Omega}{5.9\ \text{k}\Omega}$$

$$= 1.38\ \text{k}\Omega$$

(b) RECIPROCAL FORMULA

$$R_T = \frac{1}{(1/R_1) + (1/R_2)}$$

$$= \frac{1}{(1/3.7\ \text{k}\Omega) + (1/2.2\ \text{k}\Omega)}$$

$$= \frac{1}{(270.2 \times 10^{-6}) + (454.5 \times 10^{-6})}$$

$$= \frac{1}{724.7 \times 10^{-6}}$$

$$= 1.38\ \text{k}\Omega$$

FIGURE 6-18   Two Resistors in Parallel.

As you can see from this example, the advantage of the product-over-sum parallel resistance formula (a) is its ease of use. Its disadvantage is that it can only be used for two resistors in parallel. The rule to adopt, therefore, is that if a circuit has two resistors in parallel use the *product-over-sum* formula, and in circuits containing more than two resistors, use the *reciprocal* formula.

6-4-2   *Equal-Value Resistors in Parallel*

If resistors of equal value are connected in parallel, a special case *equal-value formula* can be used to calculate the total resistance.

$$R_T = \frac{\text{value of one resistor } (R)}{\text{number of parallel resistors } (n)}$$

EXAMPLE:

Figure 6-19(a) shows how a stereo music amplifier is connected to drive two 8 Ω speakers, which are connected in parallel with one another. What is the total resistance connected across the amplifier's output terminals?

Speaker 1 (8 Ω)   Speaker 2 (8 Ω)

(a)

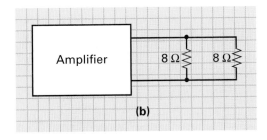

Amplifier   8 Ω   8 Ω

(b)

FIGURE 6-19   **Parallel-Connected Speakers.**

Solution:

Referring to the schematic in Figure 6-19(b), you can see that since both parallel-connected speakers have the same resistance, the total resistance is most easily calculated by using the equal-value formula:

$$R_T = \frac{R}{n} = \frac{8 \,\Omega}{2} = 4 \,\Omega$$

EXAMPLE:

Refer to Figure 6-20 and calculate:

a. Total resistance in part (a)

b. Total resistance in part (b)

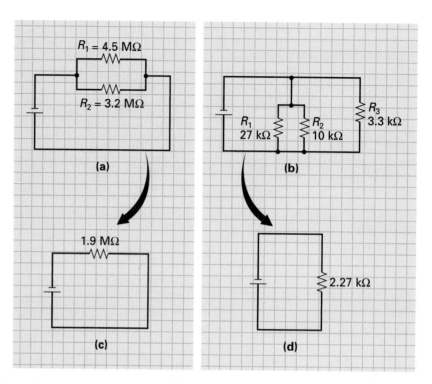

FIGURE 6-20   Parallel Circuit Examples.

Solution:

CALCULATOR SEQUENCE FOR (A)

Step	Keypad Entry	Display Response
1.	4 . 5 E 6	4.5E6
2.	+	
3.	3 . 2 E 6	3.2E6
4.	=	7.7E6
5.	STO (store in memory)	
6.	C/CE	0.
7.	4 . 5 E 6	4.5E6
8.	×	
9.	3 . 2 E 6	3.2E6
10.	=	1.44E13
11.	÷	
12.	RM (Recall memory)	7.7E6
13.	=	1.87E6

a. Figure 6-20(a) has only two resistors in parallel and therefore the product-over-sum resistor formula can be used.

$$R_T = \frac{R_1 \times R_2}{R_1 + R_2}$$

$$= \frac{4.5 \text{ M}\Omega \times 3.2 \text{ M}\Omega}{4.5 \text{ M}\Omega + 3.2 \text{ M}\Omega}$$

$$= \frac{14.4 \text{ M}\Omega}{7.7 \text{ M}\Omega}$$

$$= 1.9 \text{ M}\Omega$$

The equivalent circuit is seen in Figure 6-20(c).

b. Figure 6-20(b) has more than two resistors in parallel, and therefore the reciprocal formula must be used.

$$R_T = \frac{1}{(1/R_1) + (1/R_2) + (1/R_3)}$$

$$= \frac{1}{(1/27 \text{ k}\Omega + (1/10 \text{ k}\Omega + (1/3.3 \text{ k}\Omega)}$$

$$= \frac{1}{440.0 \times 10^{-6}}$$

$$= 2.27 \text{ k}\Omega$$

The equivalent circuit can be seen in Figure 6-20(d).

EXAMPLE:

Find the total resistance of the parallel circuit in Figure 6-21.

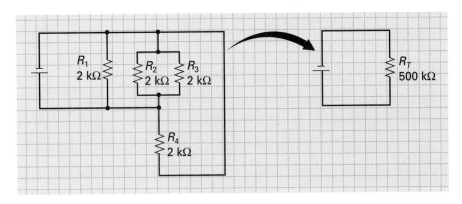

FIGURE 6-21 **Parallel Circuit Example.**

Solution:

Since all four resistors are connected in parallel and are all of the same value, the equal-value resistors in parallel formula can be used to calculate total resistance.

$$R_T = \frac{R}{n} = \frac{2 \text{ k}\Omega}{4} = 500 \ \Omega$$

To summarize what we have learned so far about parallel circuits, we can say that:

1. Components are said to be connected in parallel when the current has to travel two or more paths between the negative and positive sides of the voltage source.

2. The voltage across all the components in parallel is always the same.

3. The total current from the source is equal to the sum of all the branch currents (Kirchhoff's current law).

4. The amount of current flowing through each branch is inversely proportional to the resistance value in that branch.

5. The total resistance of a parallel circuit is always less than the value of the smallest branch resistor.

Use the following questions to test your understanding of Section 6-4.
State what parallel resistance formula should be used in questions 1 through 3 for calculating R_T.

1. Two resistors in parallel.

2. More than two resistors in parallel.

3. Equal-value resistors in parallel.

4. Calculate the total parallel resistance when $R_1 = 2.7$ kΩ, $R_2 = 24$ kΩ, and $R_3 = 1$ MΩ.

6-5 POWER IN A PARALLEL CIRCUIT

As with series circuits, the total power in a parallel resistive circuit is equal to the sum of all the power losses for each of the resistors in parallel.

$$P_T = P_1 + P_2 + P_3 + P_4 + \cdots$$

total power = addition of all the power losses

The formulas for calculating the amount of power dissipated are

$$P = I \times V$$
$$P = \frac{V^2}{R}$$
$$P = I^2 \times R$$

☐ **EXAMPLE:**

Calculate the total amount of power dissipated in Figure 6-22.

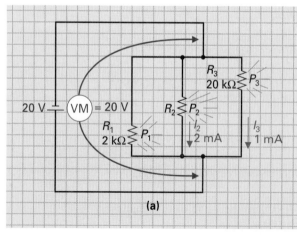

CIRCUIT ANALYSIS TABLE			
Resistance $R = V/I$	Voltage $V = I \times R$	Current $I = V/R$	Power $P = V \times I$
$R_1 = 2$ kΩ	$V_{R1} = 20$ V	$I_1 =$	$P_1 = 200$ mW
$R_2 =$	$V_{R2} = 20$ V	$I_2 = 2$ mA	$P_2 = 40$ mW
$R_3 = 20$ kΩ	$V_{R3} = 20$ V	$I_3 = 1$ mA	$P_3 = 20$ mW
	$V_T = V_S = 20$ V		$P_T = 260$ mW

(a)   **(b)**

FIGURE 6-22 Parallel Circuit Example. (a) Schematic. (b) Circuit Analysis Table.

Solution:

The total power dissipated in a parallel circuit is equal to the sum of all the power dissipated by all the resistors. With P_1, we only know voltage and resistance and therefore we can use the formula

$$P_1 = \frac{V_{R1}^2}{R_1} = \frac{20^2}{2\ k\Omega} = 2.0\ W \quad \text{or} \quad 200\ mW$$

With P_2, we only know current and voltage, and therefore we can use the formula

$$P_2 = I_2 \times V_{R2} = 2\ mA \times 20\ V = 40\ mW$$

With P_3, we know V, I, and R; however, we will use the third power formula:

$$P_3 = I_3^2 \times R_3 = 1\ mA^2 \times 20\ k\Omega = 20\ mW$$

Total power (P_T) equals the sum of all the power or wattage losses for each resistor:

$$
\begin{aligned}
P_T &= P_1 + P_2 + P_3 \\
&= 200\ mW + 40\ mW + 20\ mW \\
&= 260\ mW
\end{aligned}
$$

CALCULATOR SEQUENCE

Step	Keypad Entry	Display Response
1.	[2] [0]	20.
2.	[x²]	400.
3.	[÷]	
4.	[2] [E] [3]	2.E3
5.	[=]	0.2

CALCULATOR SEQUENCE

Step	Keypad Entry	Display Response
1.	[2] [E] [3] [+/−]	2.E-3
2.	[×]	
3.	[2] [0]	20.
4.	[=]	40E-3

CALCULATOR SEQUENCE

Step	Keypad Entry	Display Response
1.	[1] [E] [3] [+/−] [x²]	1.E-6
2.	[×]	1.E-6
3.	[2] [0] [E] [3]	20E3
4.	[=]	20.E-3

EXAMPLE:

In Figure 6-23, there are two ½ W (0.5 W) resistors connected in parallel. Should the wattage rating for each of these resistors be increased or decreased, or can they remain the same?

FIGURE 6-23   Parallel Circuit Example.

Solution:

Since current and voltage are known for both branches, the power formula used in both cases can be $P = I \times V$.

$$
\begin{array}{ll}
P_1 = I \times V & P_2 = I \times V \\
\quad = 20\ mA \times 5\ V & \quad = 200\ mA \times 5\ V \\
\quad = 0.1\ W & \quad = 1\ W
\end{array}
$$

R_1 is dissipating 0.1 W and so a 0.5 W resistor will be fine in this application. On the other hand, R_2 is dissipating 1 W and is only designed to dissipate 0.5 W. R_2 will therefore overheat unless it is replaced with a resistor of the same ohmic value, but with a 1 W or greater rating.

☐ EXAMPLE:

Figure 6-24(a) shows a simplified diagram of an automobile external light system. A 12 V lead–acid battery is used as a source and is connected across eight parallel-connected lamps. The left and right brake lights are controlled by the brake switch, which is attached to the brake pedal. When the light switch is turned on, both the rear taillights and the low-beam headlights are brought into circuit and turned on. The high-beam set of headlights is activated only if the high-beam switch is closed. For the lamp resistances given, calculate the output power of each lamp, when in use.

■ *Solution:*

Figure 6-24(b) shows the schematic diagram of the pictorial in Figure 6-24(a). Since both V and R are known, we can use the V^2/R power formula.

$$\textit{Brake lights: Left lamp wattage } (P) = \frac{V^2}{R} = \frac{12 \, V^2}{4\Omega} = 36 \text{ W}$$

(a)

(b)

FIGURE 6-24 **Parallel Automobile External Light System.**

CHAPTER 6 / PARALLEL DC CIRCUITS

Right lamp wattage is the same as left.

Taillights: Left lamp wattage $(P) = \dfrac{V^2}{R} = \dfrac{12 \ V^2}{6\Omega} = 24$

Right lamp wattage is the same as left.

Low-beam headlights: $(P) = \dfrac{V^2}{R} = \dfrac{12 \ V^2}{3\Omega} = 48 \ W$

Each high-beam headlight is a 48 W lamp.

High-beam headlights: $(P) = \dfrac{V^2}{R} = \dfrac{12 \ V^2}{3\Omega} = 48 \ W$

Each high-beam headlight is a 48 W lamp.

SELF-TEST EVALUATION POINT FOR SECTION 6-5

Use the following questions to test your understanding of Section 6-5.

1. True or false: Total power in a parallel circuit can be obtained by using the same total power formula as for series circuits.

2. If $I_1 = 2$ mA and $V = 24$ V, calculate P_1.

3. If $P_1 = 22$ mW and $P_2 = 6400 \ \mu$W, $PT = $?

4. Is it important to observe the correct wattage ratings of resistors when they are connected in parallel?

6-6   TROUBLESHOOTING A PARALLEL CIRCUIT

In the troubleshooting discussion on series circuits, we mentioned that one of three problems can occur, and these three also apply to parallel circuits:

1. A component will open.
2. A component will short.
3. A component's value will change over a period of time.

As a technician, it is important that you know how to recognize these circuit malfunctions and know how to isolate the cause and make the repair. Since you could be encountering opens, shorts, and component value changes in the parallel circuits you construct in lab, let us step through the troubleshooting procedure you should use.

As an example, Figure 6-25(a) indicates the normal readings that should be obtained from a typical parallel circuit if it is operating correctly. Keep these normal readings in mind, since they will change as circuit problems are introduced in the following section.

6-6-1   *Open Component in a Parallel Circuit*

In Figure 6-25(b), R_1 has opened, so there can be no current flow through the R_1 branch. With one fewer branches for current, the parallel circuit's resistance will increase, causing the total current flow to decrease. The total current flow, which was 7 mA, will actually decrease by the amount that was flowing through the now open branch (1 mA) to a total current flow of 6 mA. If you had constructed this circuit in lab, you would probably not have

Resistance $R = V/I$	Voltage $V = I \times R$	Current $I = V/R$	Power $P = V \times I$
$R_1 = 12$ kΩ	$V_{R1} = 12$ V	$I_1 = 1$ mA	$P_1 = 12$ mW
$R_2 = 3$ kΩ	$V_{R2} = 12$ V	$I_2 = 4$ mA	$P_2 = 48$ mW
$R_3 = 6$ kΩ	$V_{R3} = 12$ V	$I_3 = 2$ mA	$P_3 = 24$ mW
$R_T = 1.7$ kΩ	$V_T = V_S = 12$ V	$I_T = 7$ mA	$P_T = 84$ mW

CIRCUIT ANALYSIS TABLE

(a)

(c)

(d)

FIGURE 6-25   **Troubleshooting an Open in a Parallel Circuit.**

noticed that there was anything wrong with the circuit until you started to make some multimeter measurements. A voltmeter measurement across the parallel circuit, as seen in Figure 6-25(b), would not have revealed any problem since the voltage measured across each resistor will always be the same and equal to the source voltage. A total current measurement, on the other hand, would indicate that there is something wrong since the measured reading does not equal the expected calculated value. Using your ammeter, you could isolate the branch fault by one of the following two methods:

1. If you measure each branch current, you will isolate the problem of R_1, because there will be no current flow through R_1. However, this could take three checks.

2. If you just measure total current (one check), you will notice that total current has decreased by 1 mA, and after making a few calculations you could determine that the only path that had a branch current of 1 mA was through R_1, so R_1 must have gone open.

We can summarize how method 2 could be used to isolate an open in this example circuit:

If R_1 opens, total current decreases by 1 mA to 6 mA ($I_T = I_1 + I_2 + I_3 = 0$ mA + 4 mA + 2 mA = 6 mA).

If R_2 opens, total current decreases by 4 mA to 3 mA ($I_T = I_1 + I_2 + I_3 = 1$ mA + 0 mA + 2 mA = 3 mA).

If R_3 opens, total current decreases by 2 mA to 5 mA ($I_T = I_1 + I_2 + I_3 = 1$ mA + 4 mA + 0 mA = 5 mA).

Method 2 can be used as long as the resistors are unequal. If they are equal, method 1 may be used to locate the 0 current branch.

In most cases, the ohmmeter is ideal for locating open circuits. To make a resistance measurement, you should disconnect power from the circuit, isolate the device from the circuit, and then connect the ohmmeter across the device. Figure 6-25(c) shows why it is necessary for you to isolate the device to be tested from the circuit. With R_1 in circuit, the ohmmeter will be measuring the parallel resistance of R_2 and R_3, even though the ohmmeter is connected across R_1. This will lead to a false reading and confuse the troubleshooting process. Disconnecting the device to be tested from the circuit, as shown in Figure 6-25(d), will give an accurate reading of R_1, then R_2, and then R_3's resistance and lead you to the cause of the problem.

6-6-2   *Shorted Component in a Parallel Circuit*

In Figure 6-26(a), R_2, which has a resistance of 3 kΩ, has been shorted out. The only resistance in this center branch is the resistance of the wire, which would typically be a fraction of an ohm. In this example, let us assume that the resistance in the branch is 1 Ω, and therefore the current in this branch will attempt to increase to:

$$I_2 = \frac{12V}{1\ \Omega} = 12\ A$$

Before the current even gets close to 12 A, the fuse in the dc power supply will blow, as shown in the inset in Figure 6-26(a), and disconnect power from the circuit. This is typically the effect you will see from a circuit short. Replacing the fuse will be a waste of time and money since the fuse will simply blow again once power is connected across the circuit. As before, the tool to use to isolate the cause is the ohmmeter, as shown in Figure 6-26(b). To check the resistance of each branch, disconnect power from the circuit, disconnect the devices to be tested from the circuit to prevent false readings, and then test each branch until the shorted branch is located. Once you have corrected the fault, you will need to change the power supply fuse or reset its circuit breaker.

FIGURE 6-26 **Troubleshooting a Short in a Parallel Circuit.**

6-6-3 *Component Value Variation in a Parallel Circuit*

The resistive values of resistors will change with age. This increase or decrease in resistance will cause a corresponding decrease or increase in branch current and therefore total current. This deviation from the desired value can also be checked with the ohmmeter. Be careful to take into account the resistor's tolerance because this can make you believe that the resistor's value has changed, when it is, in fact, within the tolerance rating.

1. An open component will cause no current flow within that branch. The total current will decrease, and the voltage across the component will be the same as the source voltage.

2. A shorted component will cause maximum current through that branch. The total current will increase and the voltage source fuse will normally blow.

3. A change in resistance value will cause a corresponding opposite change in branch current and total current.

SELF-TEST EVALUATION POINT FOR SECTION 6-6

Use the following questions to test your understanding of Section 6-6.
How would problems 1 through 3 be recognized in a parallel circuit?

1. An open component

2. A shorted component

3. A component's value variation

4. True or false: An ammeter is typically used to troubleshoot series circuits, whereas a voltmeter is typically used to troubleshoot parallel circuits.

REVIEW QUESTIONS

Multiple-Choice Questions

1. A parallel circuit has _____ path(s) for current to flow.
   a. One c. Only three
   b. Two or more d. None of the above

2. If a source voltage of 12 V is applied across four resistors of equal value in parallel, the voltage drops across each resistor would be equal to:
   a. 12 V c. 4 V
   b. 3 V d. 48 V

3. What would be the voltage drop across two 25 Ω resistors in parallel if the source voltage were equal to 9 V?
   a. 50 V c. 12 V
   b. 25 V d. None of the above

4. If a four-branch parallel circuit has 15 mA flowing through each branch, the total current into the parallel circuit will be equal to:
   a. 15 mA c. 30 mA
   b. 60 mA d. 45 mA

5. If the total three-branch parallel circuit current is equal to 500 mA, and 207 mA is flowing through one branch and 153 mA through another, what would be the current flow through the third branch?
   a. 707 mA c. 140 mA
   b. 653 mA d. None of the above

6. A large branch resistance will cause a _____ branch current.
   a. Large c. Medium
   b. Small d. None of the above are true

7. What would be the conductance of a 1 kΩ resistor?
   a. 10 mS c. 2 kΩ
   b. 1 mS d. All of the above

8. If only two resistors are connected in parallel, the total resistance equals:
   a. The sum of the resistance values
   b. Three times the value of one resistor
   c. The product over the sum
   d. All of the above

9. If resistors of equal value are connected in parallel, the total resistance can be calculated by:
   a. One resistor value divided by the number of parallel resistors
   b. The sum of the resistor values
   c. The number of parallel resistors divided by one resistor value
   d. All of the above could be true.

10. The total power in a parallel circuit is equal to the:
    a. Product of total current and total voltage
    b. Reciprocal of the individual power losses
    c. Sum of the individual power losses
    d. Both (a) and (b)
    e. Both (a) and (c)

Practice Problems

11. Calculate the total resistance of four 30 kΩ resistors in parallel.

12. Find the total resistance for each of the following parallel circuits:

    a. 330 Ω and 560 Ω
    b. 47 kΩ, 33 kΩ, and 22 kΩ
    c. 2.2 MΩ, 3 kΩ, and 220 Ω

13. If 10 V is connected across three 25 Ω resistors in parallel, what will be the total and individual branch currents?

14. If a four-branch parallel circuit has branch currents equal to 25 mA, 37 mA, 220 mA, and 0.2 A, what is the total circuit current?

15. If three resistors of equal value are connected across a 14 V supply and the total resistance is equal to 700 Ω, what is the value of each branch current?

16. If three 75 W light bulbs are connected in parallel across a 110 V supply, what is the value of each branch current? What is the branch current through the other two light bulbs if one burns out?

17. If 33 kΩ and 22 kΩ resistors are connected in parallel across a 20 V source, calculate:

    a. Total resistance   **d.** Total power dissipated
    b. Total current   **e.** Individual resistor power
    c. Branch currents   dissipated

18. If four parallel-connected resistors are each dissipating 75 mW, what is the total power being dissipated?

19. Calculate the branch currents through the following parallel resistor circuits when they are connected across a 10 V supply:

    a. 22 kΩ and 33 kΩ   **b.** 220 Ω, 330 Ω, and 470 Ω

20. If 30 Ω and 40 Ω resistors are connected in parallel, which resistor will generate the greatest amount of heat?

21. Calculate the total conductance and resistance of the following parallel circuits:

    a. Three 5 Ω resistors   **c.** 1 MΩ, 500 MΩ, 3.3 MΩ
    b. Two 200 Ω resistors   **d.** 5 Ω, 3 Ω, 2 Ω

22. Connect the three resistors in Figure 6-27 in parallel across a 12 V battery and then calculate the following:

    a. V_{R1}, V_{R2}, V_{R3} **d.** P_T
    b. I_1, I_2, I_3 **e.** P_1, P_2, P_3
    c. I_T **f.** G_{R1}, G_{R2}, G_{R3}

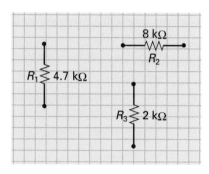

FIGURE 6-27   Connect in Parallel Across a 12-volt Source.

23. Calculate R_T in Figure 6-28 (a), (b), (c), and (d).

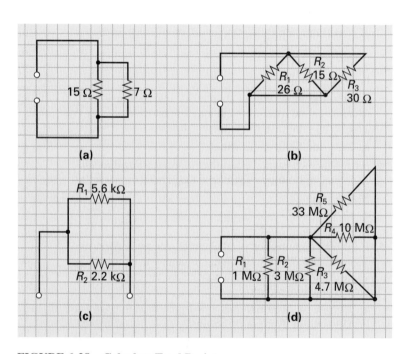

FIGURE 6-28   Calculate Total Resistance.

24. Calculate the branch currents through four 60 W bulbs connected in parallel across 110 V. How much is the total current, and what would happen to the total current if one of the bulbs were to burn out? What change would occur in the remaining branch currents?

25. Calculate the following in Figure 6-29:

    a. I_2 **c.** V_S, I_1, I_2
    b. I_T **d.** R_2, I_1, I_2, P_T

Troubleshooting Questions

26. An open component in a parallel circuit will cause _____ _____ current flow within that branch, which will cause the total current to _____.

    a. Maximum, increase   **c.** Maximum, decrease
    b. Zero, decrease   **d.** Zero, increase

27. A shorted component in a parallel circuit will cause _____ current through a branch, and consequently the total current will _____.

    a. Maximum, increase   **c.** Maximum, decrease
    b. Zero, decrease   **d.** Zero, increase

28. If a 10 kΩ and two 20 kΩ resistors are connected in parallel across a 20 V supply, and the total current measured is 2 mA, determine whether a problem exists in the circuit and, if it does, isolate the problem.

29. What situation would occur and how would we recognize the problem if one of the 20 kΩ resistors in Question 28 were to short?

30. With age, the resistance of a resistor will _____, resulting in a corresponding but opposite change in _____

    _____.

    a. Increase, branch current
    b. Change, source voltage
    c. Decrease, source resistance
    d. Change, branch current

FIGURE 6-29   **Calculate the Unknown.**

Series–Parallel DC Circuits

7

The Christie Bridge Circuit

Who invented the Wheatstone bridge circuit? It was obviously Sir Charles Wheatstone. Or was it?

The Wheatstone bridge was actually invented by S.H. Christie of the Royal Military Academy at Woolwich, England. He described the circuit in detail in the *Philosophical Transactions* paper dated February 28, 1833. Christie's name, however, was unknown and his invention was ignored.

Ten years later, Sir Charles Wheatstone called attention to Christie's circuit. Sir Charles was very well known, and from that point on, and even to this day, the circuit is known as a Wheatstone bridge. Later, Werner Siemens would modify Christie's circuit and invent the variable-resistance arm bridge circuit, which would also be called a Wheatstone bridge.

No one has given full credit to the real inventors of these bridge circuits, until now!

The Christie Bridge   The Siemens Bridge

Introduction

Very rarely are we lucky enough to run across straightforward series or parallel circuits. In general, all electronic equipment is composed of many components that are interconnected to form a combination of series and parallel circuits. In this chapter, we will be combining our knowledge of the series and parallel circuits discussed in the previous two chapters.

7-1 SERIES- AND PARALLEL-CONNECTED COMPONENTS

Figure 7-1(a) through (f) shows six examples of **series–parallel resistive circuits.** The most important point to learn is how to distinguish between the resistors that are connected in series and the resistors that are connected in parallel, which will take a little practice.

One thing that you may not have noticed when examining Figure 7-1 is that:

Circuit 7-1(a) is equivalent to 7-1(b)

Circuit 7-1(c) is equivalent to 7-1(d)

Circuit 7-1(e) is equivalent to 7-1(f)

When analyzing these series–parallel circuits, always remember that current flow determines whether the resistor is connected in series or parallel. Begin at the negative side of the battery and apply these two rules:

1. If the total current has only one path to follow through a component, that component is connected in series.

2. If the total current has two or more paths to follow through two or more components, those components are connected in parallel.

Series–Parallel Circuit
Network or circuit that contains components that are connected in both series and parallel.

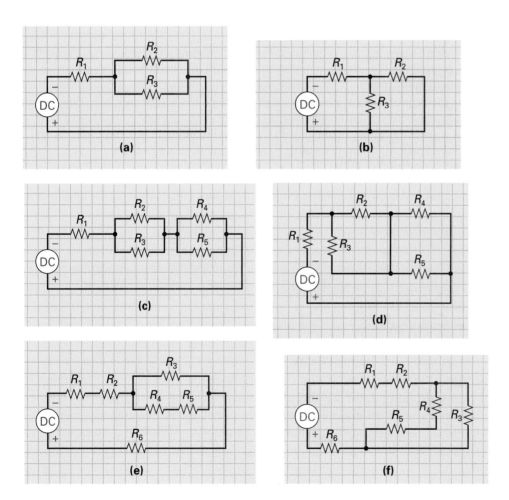

FIGURE 7-1   Series–Parallel Resistive Circuits.

Referring again to Figure 7-1, you can see that series or parallel resistor networks are easier to identify in parts (a), (c), and (e) than in parts (b), (d), and (f). Redrawing the circuit so that the components are arranged from left to right or from top to bottom is your first line of attack in your quest to identify series– and parallel–connected components.

EXAMPLE:

Refer to Figure 7-2 and identify which resistors are connected in series and which are in parallel.

FIGURE 7-2   Series–Parallel Circuit Example.

Solution:

First, let's redraw the circuit so that the components are aligned either from left to right as shown in Figure 7-3(a), or from top to bottom as shown in Figure 7-3(b). Placing your pencil at the negative terminal of the battery on whichever figure you prefer, either Figure 7-3(a) or (b), trace the current paths through the circuit toward the positive side of the battery, as illustrated in Figure 7-4.

The total current arrives first at R_1. There is only one path for current to flow, which is through R_1, and therefore R_1 is connected in series. The total current proceeds on past R_1 and arrives at a junction where current divides and travels through two branches, R_2 and R_3. Since current had to split into two paths, R_2 and R_3 are therefore connected in parallel. After the parallel connection of R_2 and R_3, total current combines and travels to the positive side of the battery.

In this example, therefore, R_1 is in series with the parallel combination of R_2 and R_3.

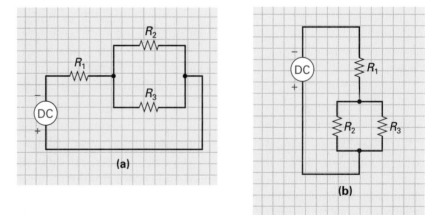

FIGURE 7-3   Redrawn Series–Parallel Circuit. (a) Left to Right. (b) Top to Bottom.

(a)

(b)

FIGURE 7-4   **Tracing Current through a Series–Parallel Circuit.**

▢ EXAMPLE:

Refer to Figure 7-5 and identify which resistors are connected in series and which are connected in parallel.

▢ *Solution:*

Figure 7-6 illustrates the simplified, redrawn schematic of Figure 7-5. Total current leaves the negative terminal of the battery, and all of this current has to travel through R_1, which is therefore a series-connected resistor. Total current will split at junction A, and consequently, R_3 and R_4 with R_2 make up a parallel combination. The current that flows through R_3 (I_2) will also flow through R_4 and therefore R_3 is in series with R_4. I_1 and I_2 branch currents combine at junction B to produce total current, which has only one path to follow through the series resistor R_5, and finally, to the positive side of the battery.

   In this example, therefore, R_3 and R_4 are in series with one another and both are in parallel with R_2, and this combination is in series with R_1 and R_5.

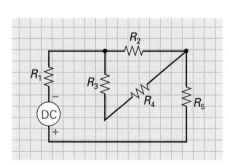

FIGURE 7-5   **Series–Parallel Circuit Example.**

FIGURE 7-6   **Redrawn Series–Parallel Circuit Example.**

FIGURE 7-7   Total Series–Parallel Circuit Resistance.

7-2   TOTAL RESISTANCE IN A SERIES–PARALLEL CIRCUIT

No matter how complex or involved the series–parallel circuit, there is a simple three-step method to simplify the circuit to a single equivalent total resistance. Figure 7-7 illustrates an example of a series–parallel circuit. Once you have analyzed and determined the series–parallel relationship, we can proceed to solve for total resistance.

The three-step method is:

Step A: Determine the equivalent resistances of all branch series-connected resistors.

Step B: Determine the equivalent resistances of all parallel-connected combinations.

Step C: Determine the equivalent resistance of the remaining series-connected resistances.

Let's now put this procedure to work with the example circuit in Figure 7-7.

STEP A   Solve for all branch series-connected resistors. In our example, this applies only to R_3 and R_4, and since this is a series connection, we have to use the series resistance formula.

$$R_{3,4} = R_3 + R_4 = 8 + 2 = 10\,\Omega \qquad \text{(series resistance formula)}$$

With R_3 and R_4 solved, the circuit now appears as indicated in Figure 7-8.

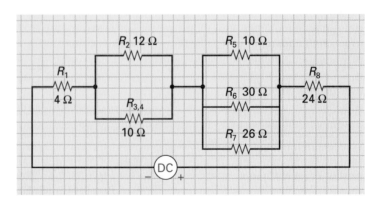

FIGURE 7-8   After Completing Step A.

FIGURE 7-9   After Completing Step B.

FIGURE 7-10   After Completing Step C.

STEP B   Solve for all parallel combinations. In this example, they are the two parallel combinations of (a) R_2 and $R_{3,4}$ and (b) R_5 and R_6 and R_7. Since these are parallel connections, use the parallel resistance formulas.

$$R_{2,3,4} = \frac{R_2 \times R_{3,4}}{R_2 + R_{3,4}} = \frac{12 \times 10}{12 + 10} = 5.5\ \Omega \qquad \text{(product-over-sum formula)}$$

$$R_{5,6,7} = \frac{1}{(1/R_5) + (1/R_6) + (1/R_7)} = 5.8\ \Omega \qquad \text{(reciprocal formula)}$$

With $R_{2,3,4}$ and $R_{5,6,7}$ solved, the circuit now appears as illustrated in Figure 7-9.

STEP C   Solve for the remaining series resistances. There are now four remaining series resistances, which can be reduced to one equivalent resistance (R_{eq}) or total resistance (R_T). As seen in Figure 7-10, by using the series resistance formula, the total equivalent resistance for this example circuit will be

$$\begin{aligned} R_{eq} &= R_1 + R_{2,3,4} + R_{5,6,7} + R_8 \\ &= 4\ \Omega + 5.5\ \Omega + 5.8\ \Omega + 24\ \Omega \\ &= 39.3\ \Omega \end{aligned}$$

EXAMPLE:

Find the total resistance of the circuit in Figure 7-11.

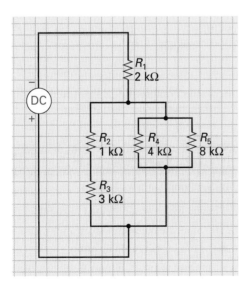

FIGURE 7-11   Calculate Total Resistance.

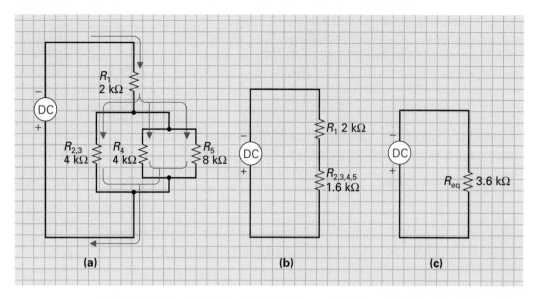

FIGURE 7-12 **Calculating Total Resistance. (a) Step A. (b) Step B. (c) Step C.**

■ *Solution:*

STEP A Solve for all branch series-connected resistors. This applies to R_2 and R_3 (series connection):

$$R_{2,3} = R_2 + R_3 = 1 \text{ k}\Omega + 3 \text{ k}\Omega = 4 \text{ k}\Omega$$

The resulting circuit, after completing step A, is illustrated in Figure 7-12(a).

STEP B Solve for all parallel combinations. Looking at Figure 7-12(a), which shows the circuit resulting from step A, you can see that current branches into three paths, so the parallel reciprocal formula must be used for this step.

$$R_{2,3,4,5} = \frac{1}{(1/R_{2,3}) + (1/R_4) + (1/R_5)}$$

$$= \frac{1}{(1/4 \text{ k}\Omega) + (1/4 \text{ k}\Omega) + (1/8 \text{ k}\Omega)} = 1.6 \text{ k}\Omega$$

The resulting circuit, after completing step B, is illustrated in Figure 7-12(b).

STEP C Solve for the remaining series resistances. Looking at Figure 7-12(b), which shows the circuit resulting from step B, you can see that there are two remaining series resistances. The equivalent resistance (R_{eq}) is therefore equal to

$$R_{eq} = R_1 + R_{2,3,4,5} = 2 \text{ k}\Omega + 1.6 \text{ k}\Omega = 3.6 \text{ k}\Omega$$

The total equivalent resistance, after completing all three steps, is illustrated in Figure 7-12(c).

SELF-TEST EVALUATION POINT FOR SECTIONS 7-1 AND 7-2

Use the following questions to test your understanding of Sections 7-1 and 7-2.

1. How can we determine which resistors are connected in series and which are connected in parallel in a series–parallel circuit?

2. Calculate the total resistance if two series-connected 12 kΩ resistors are connected in parallel with a 6 kΩ resistor.

3. State the three-step procedure used to determine total resistance in a circuit made up of both series and parallel resistors.

4. Sketch the following series–parallel resistor network made up of three resistors. R_1 and R_2 are in series with each other and are connected in parallel with R_3. If $R_1 = 470 \ \Omega$, $R_2 = 330 \ \Omega$, and $R_3 = 270 \ \Omega$, what is R_T?

FIGURE 7-13 Series–Parallel Circuit Example.

7-3 VOLTAGE DIVISION IN A SERIES–PARALLEL CIRCUIT

There is a simple three-step procedure for finding the voltage drop across each part of the series–parallel circuit. Figure 7-13 illustrates an example of a series–parallel circuit to which we will apply the three-step method for determining voltage drop.

STEP 1 Determine the circuit's total resistance. This is achieved by following the three-step method used previously for calculating total resistance.

Step A:

$$R_{3,4} = 4 + 8 = 12 \ \Omega$$

Step B:

$$R_{2,3,4} = \frac{1}{(1/R_2) + (1/R_{3,4})} = 6 \ \Omega$$

$$R_{5,6,7} = \frac{1}{(1/R_5) + (1/R_6) + (1/R_7)} = 12 \ \Omega$$

Figure 7-14 illustrates the equivalent circuit up to this point. We end up with one series resistor (R_1) and two series equivalent resistors ($R_{2,3,4}$ and $R_{5,6,7}$). R_T is therefore equal to 28 Ω.

STEP 2 Determine the circuit's total current. This step is achieved simply by utilizing Ohm's law.

$$I_T = \frac{V_T}{R_T} = \frac{84 \ \text{V}}{28 \ \Omega} = 3 \ \text{A}$$

STEP 3 Determine the voltage across each series resistor and each parallel combination (series equivalent resistor) in Figure 7-14. Since these are all in series, the same current (I_T) will flow through all three.

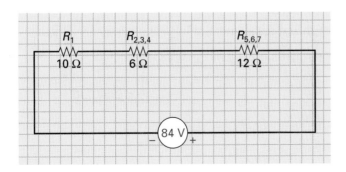

FIGURE 7-14 After Completing Step 1.

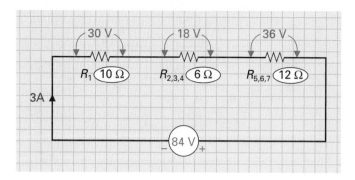

FIGURE 7-15 After Completing Steps 2 and 3.

$$V_{R1} = I_T \times R_1 = 3\,\text{A} \times 10\,\Omega = 30\,\text{V}$$
$$V_{R2,3,4} = I_T \times R_{2,3,4} = 3\,\text{A} \times 6\,\Omega = 18\,\text{V}$$
$$V_{R5,6,7} = I_T \times R_{5,6,7} = 3\,\text{A} \times 12\,\Omega = 36\,\text{V}$$

The voltage drops across the series resistor (R_1) and series equivalent resistors ($R_{2,3,4}$ and $R_{5,6,7}$) are illustrated in Figure 7-15.

Kirchhoff's voltage law states that the sum of all the voltage drops is equal to the source voltage applied. This law can be used to confirm that our calculations are all correct:

$$V_T = V_{R1} + V_{R2,3,4} + V_{R5,6,7}$$
$$= 30\,\text{V} + 18\,\text{V} + 36\,\text{V}$$
$$= 84\,\text{V}$$

To summarize, refer to Figure 7-16, which shows these voltage drops inserted into our original circuit. As you can see from this illustration:

30 V is dropped across R_1.

18 V is dropped across R_2.

18 V is dropped across both R_3 and R_4.

36 V is dropped across R_5.

36 V is dropped across R_6.

36 V is dropped across R_7.

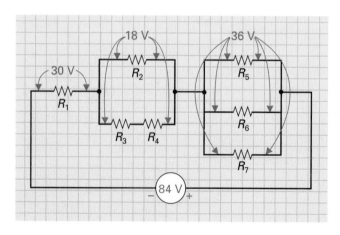

FIGURE 7-16 Detail of Step 3.

CHAPTER 7 / SERIES-PARALLEL DC CIRCUITS

Use the following questions to test your understanding of Section 7-3.

1. State the three-step procedure used to calculate the voltage drop across each part of a series–parallel circuit.
2. Referring to Figure 7-13, double the values of all the resistors. Would the voltage drops calculated previously change, and if so, what would they be?

7-4   BRANCH CURRENTS IN A SERIES–PARALLEL CIRCUIT

In the preceding example, step 2 calculated the total current flowing in a series–parallel circuit. The next step is to find out exactly how much current is flowing through each parallel branch. This will be called step 4. Figure 7-17 shows the previously calculated data inserted in the appropriate places in our example circuit.

STEP 4   Total current (I_T) will exist at points A, B, C, and D. Between A and B, current has only one path to flow, which is through R_1. R_1 is therefore a series resistor, so $I_1 = I_T = 3$ A. Between points B and C, current has two paths: through R_2 (12 Ω) and through R_3 and R_4 (12 Ω).

$$I_2 = \frac{V_{R2}}{R_2} = \frac{18 \text{ V}}{12 \text{ Ω}} = 1.5 \text{ A}$$

$$I_{3,4} = \frac{V_{R3,4}}{R_{3,4}} = \frac{18 \text{ V}}{12 \text{ Ω}} = 1.5 \text{ A}$$

Not surprisingly, the total current of 3 A is split equally due to both branches having equal resistance.

The two 1.5 A branch currents will combine at point C to produce once again the total current of 3 A. Between points C and D, current has three paths to flow through, R_5, R_6, and R_7.

$$I_5 = \frac{V_{R5}}{R_5} = \frac{36 \text{ V}}{60 \text{ Ω}} = 0.6 \text{ A}$$

$$I_6 = \frac{V_{R6}}{R_6} = \frac{36 \text{ V}}{24 \text{ Ω}} = 1.5 \text{ A}$$

$$I_7 = \frac{V_{R7}}{R_7} = \frac{36 \text{ V}}{40 \text{ Ω}} = 0.9 \text{ A}$$

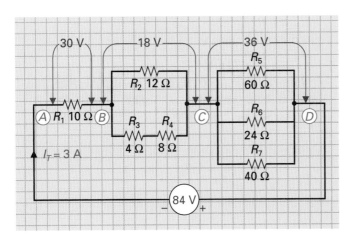

FIGURE 7-17   Series–Parallel Circuit Example with Previously Calculated Data.

All three branch currents will combine at point D to produce the total current of 3 A ($I_T = I_5 + I_6 + I_7 = 0.6 + 1.5 + 0.9 = 3$ A), proving Kirchhoff's current law.

7-5   POWER IN A SERIES–PARALLEL CIRCUIT

Whether resistors are connected in series or in parallel, the total power in a series–parallel circuit is equal to the sum of all the individual power losses.

$$P_T = P_1 + P_2 + P_3 + P_4 + \cdots$$

total power = addition of all power losses

The formulas for calculating the amount of power lost by each resistor are

$$P = \frac{V^2}{R} \qquad\qquad P = I \times V \qquad\qquad P = I^2 \times R$$

Let us calculate the power dissipated by each resistor. This final calculation will be called step 5.

STEP 5   Since resistance, voltage, and current are known, either of the three formulas for power can be used to determine power.

$$P_1 = \frac{V_{R1}^2}{R_1} = \frac{30 \text{ V}^2}{10 \text{ }\Omega} = 90 \text{ W}$$

$$P_2 = \frac{V_{R2}^2}{R_2} = \frac{18 \text{ V}^2}{12 \text{ }\Omega} = 27 \text{ W}$$

$$P_3 = I_{R3}^2 \times R_3 = 1.5 \text{ A}^2 \times 4 \text{ }\Omega = 9 \text{ W}$$

$$P_4 = I_{R4}^2 \times R_4 = 1.5 \text{ A}^2 \times 8 \text{ }\Omega = 18 \text{ W}$$

$$P_5 = \frac{V_{R5}^2}{R_5} = \frac{36 \text{ V}^2}{60 \text{ }\Omega} = 21.6 \text{ W}$$

$$P_6 = \frac{V_{R6}^2}{R_6} = \frac{36 \text{ V}^2}{24 \text{ }\Omega} = 54 \text{ W}$$

$$P_7 = \frac{V_{R7}^2}{R_7} = \frac{36 \text{ V}^2}{40 \text{ }\Omega} = 32.4 \text{ W}$$

$$\begin{aligned} P_T &= P_1 + P_2 + P_3 + P_4 + P_5 + P_6 + P_7 \\ &= 90 + 27 + 9 + 18 + 21.6 + 54 + 32.4 \\ &= 252 \text{ W} \end{aligned}$$

or

$$P_T = \frac{V_T^2}{R_T} = \frac{84 \text{ V}^2}{28 \text{ }\Omega}$$
$$= 252 \text{ W}$$

The total power dissipated in this example circuit is 252 W. All the information can now be inserted in a final diagram for the example, as shown in Figure 7-18.

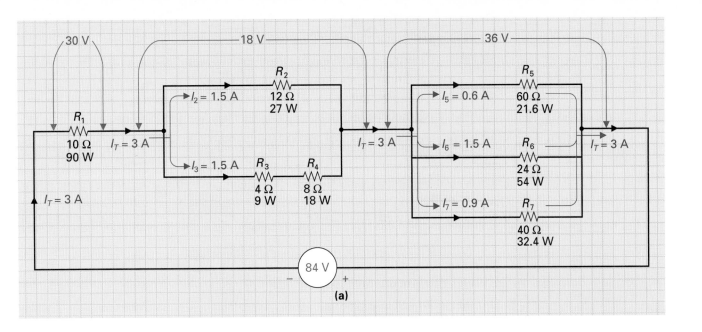

(a)

CIRCUIT ANALYSIS TABLE			
Resistance $R = V/I$	Voltage $V = I \times R$	Current $I = V/R$	Power $P = V \times I$
$R_1 = 10\ \Omega$	$V_{R1} = 30\ V$	$I_1 = 3\ A$	$P_1 = 90\ W$
$R_2 = 12\ \Omega$	$V_{R2} = 18\ V$	$I_2 = 1.5\ A$	$P_2 = 27\ W$
$R_3 = 4\ \Omega$	$V_{R3} = 6\ V$	$I_3 = 1.5\ A$	$P_3 = 9\ W$
$R_4 = 8\ \Omega$	$V_{R4} = 12\ V$	$I_4 = 1.5\ A$	$P_4 = 18\ W$
$R_5 = 60\ \Omega$	$V_{R5} = 36\ V$	$I_5 = 0.6\ A$	$P_5 = 21.6\ W$
$R_6 = 24\ \Omega$	$V_{R6} = 36\ V$	$I_6 = 1.5\ A$	$P_6 = 54\ W$
$R_7 = 40\ \Omega$	$V_{R7} = 36\ V$	$I_7 = 0.9\ A$	$P_7 = 32.4\ W$
$R_T = 28\ \Omega$	$V_T = V_S = 84V$	$I_T = 3\ A$	$P_T = 252\ W$

(b)

**FIGURE 7-18   Series–Parallel Circuit Example with All Information Inserted.
(a) Schematic. (b) Circuit Analysis Table.**

7-6   FIVE-STEP METHOD FOR SERIES–PARALLEL CIRCUIT ANALYSIS

Let's now combine and summarize all the steps for calculating resistance, voltage, current, and power in a series–parallel circuit by solving another problem. Before we begin, however, let us review the five-step procedure.

Solving for Resistance, Voltage, Current, and Power in a Series–Parallel Circuit

STEP 1   Determine the circuit's total resistance.

Step A   Solve for series-connected resistors in all parallel combinations.

Step B   Solve for all parallel combinations.

Step C   Solve for remaining series resistances.

☐ **EXAMPLE:**

Referring to Figure 7-19, calculate:

a. Total resistance

b. Total current

c. Voltage drop across all resistors

d. Current through each resistor

e. Total power dissipated by the circuit

■ *Solution:*

This problem has asked us to calculate everything about the series–parallel circuit shown in Figure 7-19, and is an ideal application for our five-step series–parallel circuit analysis procedure.

STEP 1 Determine the circuit's total resistance.

Step A: There are no series resistors within parallel combinations.

Step B: There are two-resistor (R_2, R_3) and three-resistor (R_5, R_6, R_7) parallel combinations in this circuit.

$$R_{2,3} = \frac{1}{(1/R_2) + (1/R_3)} = 222.2 \ \Omega$$

$$R_{5,6,7} = \frac{1}{(1/R_5) + (1/R_6) + (1/R_7)} = 500 \ \Omega$$

FIGURE 7-19 Apply the Five-Step Procedure to This Series–Parallel Circuit Example.

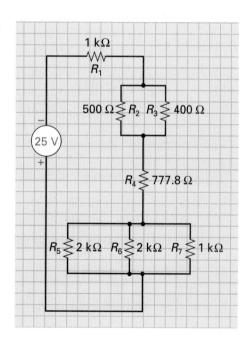

CHAPTER 7 / SERIES-PARALLEL DC CIRCUITS

FIGURE 7-20 **Circuit Resulting after Step 1B.**

FIGURE 7-21 **Circuit Resulting after Step 1C.**

Figure 7-20 illustrates the circuit resulting after step B.

Step C: Solve for the remaining four resistances to gain the circuit's total resistance (R_T) or equivalent resistance (R_{eq}).

$$R_{eq} = R_1 + R_{2,3} + R_4 + R_{5,6,7}$$
$$= 1000 \ \Omega + 222.2 \ \Omega + 777.8 \ \Omega + 500 \ \Omega$$
$$= 2500 \ \Omega \quad \text{or} \quad 2.5 \ k\Omega$$

Figure 7-21 illustrates the circuit resulting after step C.

STEP 2 Determine the circuit's total current.

$$I_T = \frac{V_S}{R_T} = \frac{25 \ V}{2.5 \ k\Omega} = 10 \ mA$$

STEP 3 Determine the voltage across each series resistor and each series equivalent resistor. To achieve this, we utilize the diagram obtained after completing step B (Figure 7-20):

$$V_{R1} = I_T \times R_1 = 10 \ mA \times 1 \ k\Omega = 10 \ V$$
$$V_{R2,3} = I_T \times R_{2,3} = 10 \ mA \times 222.2 \ \Omega = 2.222 \ V$$
$$V_{R4} = I_T \times R_4 = 10 \ mA \times 777.8 \ \Omega = 7.778 \ V$$
$$V_{R5,6,7} = I_T \times R_{5,6,7} = 10 \ mA \times 500 \ \Omega = 5 \ V$$

Figure 7-22 illustrates the results after step 3.

STEP 4 Determine the value of current through each parallel resistor (Figure 7-23). R_1 and R_4 are series-connected resistors, and therefore their current will equal 10 mA.

$$I_1 = 10 \ mA$$
$$I_4 = 10 \ mA$$

The current through the parallel resistors is calculated by Ohm's law.

FIGURE 7-22 Circuit Resulting after Step 3.

FIGURE 7-23 Series–Parallel Circuit Example with Step 1, 2, and 3 Data Inserted.

$$I_2 = \frac{V_{R2}}{R_2} = \frac{2.222 \text{ V}}{500 \text{ }\Omega} = 4.4 \text{ mA}$$

$$I_3 = \frac{V_{R3}}{R_3} = \frac{2.222 \text{ V}}{400 \text{ }\Omega} = 5.6 \text{ mA}$$

$$\left.\begin{array}{l} I_T = I_2 + I_3 \\ 10 \text{ mA} = 4.4 \text{ mA} + 5.6 \text{ mA} \end{array}\right\} \text{Kirchhoff's current law}$$

$$I_5 = \frac{V_{R5}}{R_5} = \frac{5 \text{ V}}{2 \text{ k}\Omega} = 2.5 \text{ mA}$$

$$I_6 = \frac{V_{R6}}{R_6} = \frac{5 \text{ V}}{2 \text{ k}\Omega} = 2.5 \text{ mA}$$

$$I_7 = \frac{V_{R7}}{R_7} = \frac{5 \text{ V}}{1 \text{ k}\Omega} = 5 \text{ mA}$$

$$\left.\begin{array}{l} I_T = I_5 + I_6 + I_7 \\ 10 \text{ mA} = 2.5 \text{ mA} + 2.5 \text{ mA} + 5 \text{ mA} \end{array}\right\} \text{Kirchhoff's current law}$$

STEP 5   Determine the total power dissipated by the circuit.

$$P_T = P_1 + P_2 + P_3 + P_4 + P_5 + P_6 + P_7$$

or

$$P_T = \frac{V_T^{\ 2}}{R_T}$$

Each resistor's power figure can be calculated and the sum would be the total power dissipated by the circuit. Since the problem does not ask for the power dissipated by each individual resistor, but for the total power dissipated, it will be easier to use the formula:

$$P_T = \frac{V_T{}^2}{R_T}$$
$$= \frac{25\ \text{V}^2}{2.5\ \text{k}\Omega}$$
$$= 0.25\ \text{W}$$

SELF-TEST EVALUATION POINT FOR 7-4, 7-5, AND 7-6

Use the following questions to test your understanding of Sections 7-4, 7-5, and 7-6.

1. State the five-step method used for series–parallel circuit analysis.
2. Design your own five-resistor series–parallel circuit, assign resistor values and a source voltage, and then apply the five-step analysis method.

7-7 SERIES–PARALLEL CIRCUITS

7-7-1 Loading of Voltage-Divider Circuits

Loading
The adding of a load to a source.

The straightforward voltage divider was discussed in Chapter 5, but at that point we did not explore some changes that will occur if a load resistance is connected to the voltage divider's output. Figure 7-24 shows a voltage divider, and as you can see, the advantage of a voltage-divider circuit is that it can be used to produce several different voltages from one main voltage source by the use of a few chosen resistor values.

In our discussion on load resistance, we discussed how every circuit or piece of equipment offers a certain amount of resistance, and this resistance represents how much a circuit or piece of equipment will load down the source supply.

Figure 7-25 shows an example voltage-divider circuit that is being used to develop a 10 V source from a 20 V dc supply. Figure 7-25(a) illustrates this circuit in the unloaded condition, and by making a few calculations you can analyze this circuit condition.

FIGURE 7-24   Voltage-Divider Circuit.

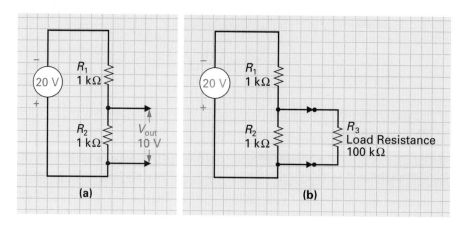

FIGURE 7-25 **Voltage-Divider Circuit. (a) Unloaded Output Voltage. (b) Loaded Output Voltage.**

STEP 1	$R_T = R_1 + R_2 = 1 \text{ k}\Omega + 1 \text{ k}\Omega = 2 \text{ k}\Omega$
STEP 2	$I_T = \dfrac{V_T}{R_T} = \dfrac{20 \text{ V}}{2 \text{ k}\Omega} = 10 \text{ mA}$

The current that flows through a voltage divider, without a load connected, is called the **bleeder current.** In this example the bleeder current is equal to 10 mA. It is called the bleeder current because it is continually drawing or bleeding this current from the voltage source.

Bleeder Current

Current drawn continuously from a voltage source. A bleeder resistor is generally added to lessen the effect of load changes or provide a voltage drop across a resistor.

STEP 3 $V_{R1} = V_{R2}$ (as resistors are the same value)
$$V_{R1} = 10 \text{ V}$$
$$V_{R2} = 10 \text{ V}$$

In Figure 7-25(b) we have connected a piece of equipment represented as a resistance (R_3) across the 10 V supply. This automatically turns the previous series circuit of R_1 and R_2 into a series–parallel circuit made up of R_1, R_2, and the 100 kΩ load resistance. By making a few more calculations, we can discover the changes that have occurred by connecting this load resistance.

STEP 1   Total resistance (R_T)

Step B:
$$R_{2,3} = \frac{R_2 \times R_3}{R_2 + R_3} = \frac{1 \text{ k}\Omega \times 100 \text{ k}\Omega}{1 \text{ k}\Omega + 100 \text{ k}\Omega} = 990.1 \text{ }\Omega$$

Step C:
$$R_{1,2,3} = R_1 + R_{2,3}$$
$$= 1 \text{ k}\Omega + 990.1$$
$$= 1.99 \text{ k}\Omega$$

STEP 2   Total current (I_T)

$$I_T = \frac{20 \text{ V}}{1.99 \text{ k}\Omega} = 10.05 \text{ mA}$$

STEP 3 $V_{R1} = I_T \times R_1 = 10.05 \text{ mA} \times 1 \text{ k}\Omega = 10.05 \text{ V}$
$V_{R2,3} = I_T \times R_{2,3} = 10.05 \text{ mA} \times 990.1 \text{ }\Omega = 9.95 \text{ V}$

STEP 4 $I_1 = I_T = 10.05 \text{ mA}$

$$I_2 = \frac{V_{R2}}{R_2} = \frac{9.95 \text{ V}}{1 \text{ k}\Omega} = 9.95 \text{ mA}$$

$$I_3 = \frac{V_{R3}}{R_3} = \frac{9.95 \text{ V}}{100 \text{ k}\Omega} = 99.5 \text{ }\mu\text{A}$$

$$I_2 + I_3 = I_T$$
$$9.95 \text{ mA} + 99.5 \text{ μA} = 10.05 \text{ mA}$$
$$\Big\} \text{ Kirchhoff's current law}$$

As you can see, the load resistance is pulling 99.5 μA from the source, and this pulls the voltage down to 9.95 V from the required 10 V that was desired and is normally present in the unloaded condition.

When designing a voltage divider, design engineers need to calculate how much current a particular load will pull and then alter the voltage-divider resistor values to offset the loading effect when the load is connected.

7-7-2   *The Wheatstone Bridge*

In 1850, Charles Wheatstone developed a circuit to measure resistance. This circuit, which is still widely used today, is called the **Wheatstone bridge** and is illustrated in Figure 7-26(a). In Figure 7-26(b), the same circuit has been redrawn so that the series and parallel resistor connections are easier to see.

Wheatstone Bridge
A four-arm, generally resistive, bridge that is used to measure resistance.

Balanced Bridge

Figure 7-27 illustrates an example circuit in which four resistors are connected together to form a series–parallel arrangement. Let us now use the five-step procedure to find out exactly what resistance, current, voltage, and power values exist throughout the circuit.

STEP 1   Total resistance (R_T)

Step A:
$$R_{1,3} = R_1 + R_3 = 10 + 20 = 30$$
$$R_{2,4} = R_2 + R_4 = 10 + 20 = 30$$

Step B:
$$R_T: (R_{1,2,3,4}) = \frac{R_{1,3} \times R_{2,4}}{R_{1,3} + R_{2,4}}$$
$$= \frac{30 \times 30}{30 + 30} = 15 \text{ }\Omega$$

Total resistance $= 15 \text{ }\Omega$

STEP 2   Total current (I_T)

$$I_T = \frac{V_T}{R_T} = \frac{30 \text{ V}}{15 \text{ }\Omega} = 2 \text{ A}$$

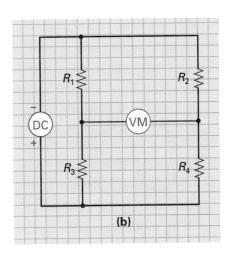

(a)   (b)

FIGURE 7-26   Wheatstone Bridge. (a) Actual Circuit. (b) Redrawn Simplified Circuit.

FIGURE 7-27   Wheatstone Bridge Circuit Example.

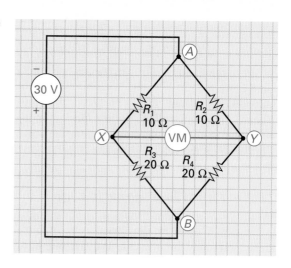

STEP 3   Since $R_{1,3}$ is in parallel with $R_{2,4}$, 30 V will appear across both $R_{1,3}$ and $R_{2,4}$.

$$V_T = V_{R1,3} = V_{R2,4} = 30 \text{ V}$$

voltage-divider formula $$V_{R1} = \frac{R_1}{R_{1,3}} \times V_T$$

$$= \frac{10}{30} \times 30 = 10 \text{ V}$$

$$V_{R3} = \frac{R_3}{R_{1,3}} \times V_T = 20 \text{ V}$$

$$V_{R2} = \frac{R_2}{R_{2,4}} \times V_T = 10 \text{ V}$$

$$V_{R4} = \frac{R_4}{R_{2,4}} \times V_T = 20 \text{ V}$$

STEP 4 $$I_{1,3} = \frac{V_{R1,3}}{R_{1,3}} = \frac{30 \text{ V}}{30 \text{ }\Omega} = 1 \text{ A}$$

$$I_{2,4} = \frac{V_{R2,4}}{R_{2,4}} = \frac{30 \text{ V}}{30 \text{ }\Omega} = 1 \text{ A}$$

$$\left. \begin{array}{l} I_{1,3} + I_{2,4} = I_T \\ 1 \text{ A} + 1 \text{ A} = 2 \text{ A} \end{array} \right\} \text{Kirchhoff's current law}$$

STEP 5   Total power dissipated (P_T)

$$P_T = I_T{}^2 \times R_T$$
$$= 2 \text{ A}^2 \times 15 \text{ }\Omega$$
$$= 4 \text{ A}^2 \times 15 \text{ }\Omega$$
$$= 60 \text{ W}$$

Figure 7-28 shows all of the step results inserted in the Wheatstone bridge example schematic. The Wheatstone bridge is said to be in the balanced condition when the voltage at point X equals the voltage at point Y ($V_{R3} = V_{R4}$, 20 V = 20 V). This same voltage exists across R_3 and R_4, so the voltmeter, which is measuring the voltage difference between X and Y, will indicate 0 V potential difference, and the circuit is said to be a *balanced bridge*.

Unbalanced Bridge

In Figure 7-29 we have replaced R_3 with a variable resistor and set it to 10 Ω. The R_2 and R_4 resistor combination will not change its voltage drop. However, R_1 and R_3, which are

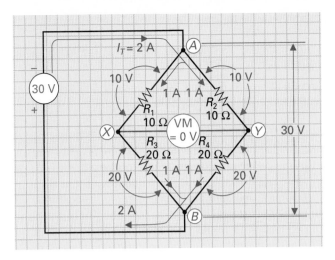

FIGURE 7-28 Balanced Wheatstone Bridge.

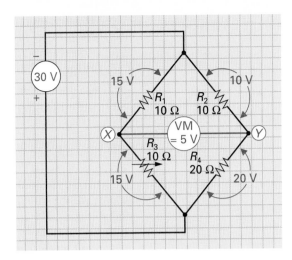

FIGURE 7-29 Unbalanced Wheatstone Bridge.

now equal, will each split the 30 V supply, producing 15 V across R_3. The voltmeter will indicate the difference in potential (5 V) from the voltage across R_3 at point X (15 V) and across R_4 at point Y (20 V). The voltmeter is actually measuring the imbalance in the circuit, which is why this circuit in this condition is known as an *unbalanced bridge.*

Determining Unknown Resistance

Figure 7-30 shows how a Wheatstone bridge circuit can be used to find the value of an unknown resistor (R_{un}). The variable-value resistor (R_{va}) is a calibrated resistor, which means that its resistance has been checked against a known, accurate resistance and its value can be adjusted and read from a calibrated dial.

FIGURE 7-30 Using a Wheatstone Bridge to Determine an Unknown Resistance.
(a) Schematic. (b) Pictorial.

The procedure to follow to find the value of the unknown resistor is as follows:

1. Adjust the variable-value resistor until the voltmeter indicates that the Wheatstone bridge is balanced (voltmeter indicates 0 V).

2. Read the value of the variable-value resistor. As long as $R_1 = R_2$, the variable resistance value will be the same as the unknown resistance value.

$$R_{va} = R_{un}$$

Since R_1 and R_2 are equal to one another, the voltage will be split across the two resistors, producing 10 V at point Y. The variable-value resistor must therefore be adjusted so that it equals the unknown resistance, and therefore the same situation will occur, in that the 20 V source will be split, producing 10 V at point X, indicating a balanced zero-volt condition on the voltmeter. For example, if the unknown resistance is equal to 5 Ω, then only when the variable-value resistor is adjusted and equal to 5 Ω would 10 V appear at point X and allow the circuit meter to read zero volts, indicating a balance. The variable-value resistor resistance could be read (5 Ω) and the unknown resistor resistance would be known (5 Ω).

□ **EXAMPLE:**

What is the unknown resistance in Figure 7-31?

FIGURE 7-31 **Wheatstone Bridge Circuit Example.**

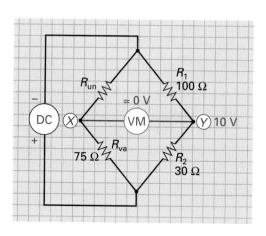

■ *Solution:*

The bridge is in a balanced condition as the voltmeter is reading a 0 V difference between points X and Y. In the previous section we discovered that if $R_1 = R_2$, then:

$$R_{va} = R_{un}$$

In this case, R_1 does not equal R_2, so a variation in the formula must be applied to take into account the ratio of R_1 and R_2.

$$R_{un} = R_{va} \times \frac{R_1}{R_2}$$

$$= 75\ \Omega \times \frac{100}{30}$$

$$= 75\ \Omega \times 3.33 = 250\ \Omega$$

Since R_1 is 3.33 times greater than R_2, then R_{un} must be 3.33 times greater than R_{va} if the Wheatstone bridge is in the balanced condition.

CALCULATOR SEQUENCE

Step	Keypad Entry	Display Response
1.	1 0 0	100
2.	÷	
3.	3 0	30
4.	=	3.333
5.	×	
6.	7 5	75
7.	=	250

Use the following questions to test your understanding of Section 7-7.

1. What is meant by loading of a voltage-divider circuit?
2. Sketch a Wheatstone bridge circuit and list an application of this circuit.

7-8 TROUBLESHOOTING SERIES–PARALLEL CIRCUITS

Troubleshooting is defined as the process of locating and diagnosing malfunctions or breakdowns in equipment by means of systematic checking or analysis. As discussed in previous resistive-circuit troubleshooting procedures, there are basically only three problems that can occur:

1. A component will open. This usually occurs if a resistor burns out or a wire or switch contact breaks.
2. A component will short. This usually occurs if a conductor, such as solder, wire, or some other conducting material, is dropped or left in the equipment, making or connecting two points that should not be connected.
3. There is a variation in a component's value. This occurs with age in resistors over a long period of time and can eventually cause a malfunction of the equipment.

Using the example circuit in Figure 7-32, we will step through a few problems, beginning with an open component. Throughout the troubleshooting, we will use the voltmeter whenever possible, as it can measure voltage by just connecting the leads across the

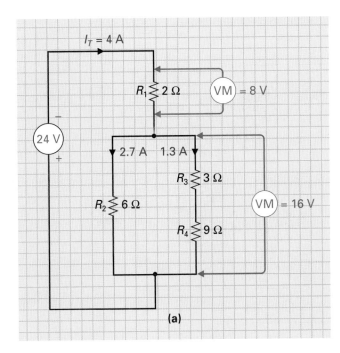

CIRCUIT ANALYSIS TABLE

Resistance $R = V/I$	Voltage $V = I \times R$	Current $I = V/R$	Power $P = V \times I$
$R_1 = 2\ \Omega$	$V_{R1} = 8\ \text{V}$	$I_1 = 4\ \text{A}$	$P_1 = 32\ \text{W}$
$R_2 = 6\ \Omega$	$V_{R2} = 16\ \text{V}$	$I_2 = 2.7\ \text{A}$	$P_2 = 43.2\ \text{W}$
$R_3 = 3\ \Omega$	$V_{R3} = 4\ \text{V}$	$I_3 = 1.3\ \text{A}$	$P_3 = 5.2\ \text{W}$
$R_4 = 9\ \Omega$	$V_{R4} = 12\ \text{V}$	$I_4 = 1.3\ \text{A}$	$P_4 = 15.6\ \text{W}$

$R_T = 6\ \Omega$ $V_T = V_S = 24\text{V}$ $I_T = 4\ \text{A}$ $P_T = 96\ \text{W}$

(b)

FIGURE 7-32 Series–Parallel Circuit Example. (a) Schematic. (b) Circuit Analysis Table.

component, rather than the ammeter, which has to be placed in the circuit, in which case the circuit path has to be opened. In some situations, using an ammeter can be difficult.

To begin, let's calculate the voltage drops and branch current obtained when the circuit is operating normally.

STEP 1 (A) $\quad R_{3,4} = R_3 + R_4 = 3\,\Omega + 9\,\Omega = 12\,\Omega$

(B) $\quad R_{2,3,4} = \dfrac{R_2 \times R_{3,4}}{R_2 + R_{3,4}} = \dfrac{6\,\Omega \times 12\,\Omega}{6\Omega + 12\,\Omega} = 4\,\Omega$

(C) $\quad R_{1,2,3,4} = R_T = R_1 + R_{2,3,4}$
$$= 2\,\Omega + 4\,\Omega = 6\,\Omega$$

STEP 2 $\quad I_T = \dfrac{V_T}{R_T} = \dfrac{24\text{ V}}{6\,\Omega} = 4\text{ A}$

STEP 3 $\quad V_{R1} = I_{R1} \times R_1 = 4\text{ A} \times 2\,\Omega = 8\text{ V}$
$$V_{R2,3,4} = I_{R2,3,4} \times R_{2,3,4} = 4\text{ A} \times 4\,\Omega = 16\text{ V}$$
(Kirchhoff's voltage law)

STEP 4 $\quad I_1 = 4\text{ A} \quad$ (series resistor)

$$I_2 = \dfrac{V_{R2}}{R_2} = \dfrac{16\text{ V}}{6\,\Omega} = 2.7\text{ A}$$

$$I_{3,4} = \dfrac{V_{R3,4}}{R_{3,4}} = \dfrac{16\text{ V}}{12\,\Omega} = 1.3\text{ A}$$

All these results have been inserted in the schematic in Figure 7-32(a) and in the circuit analysis table in Figure 7-32(b).

7-8-1 Open Component

R_1 Open (Figure 7-33)

With R_1 open, there cannot be any current flow through the circuit as there is not a path from one side of the power supply to the other. This fault can be recognized easily because approximately all of the applied 24 V will be measured across the open resistor (R_1), and 0 V will appear across all the other resistors.

FIGURE 7-33   Open Series-Connected Resistor in a Series–Parallel Circuit.

R_3 Open (Figure 7-34)

With R_3 open, there will be no current through the branch made up of R_3 and R_4. The current path will be through R_1 and R_2, and therefore the total resistance will now increase ($R_T\uparrow$) from 6 Ω to

$$R_T = R_1 + R_2$$
$$= 2\,\Omega + 6\,\Omega = 8\,\Omega$$

This 8 Ω is an increase in circuit resistance from the normal resistance, which was 6 Ω, which implies that an open has occurred to increase resistance. The total current will decrease ($I_T\downarrow$) from 4 A to

$$I_T = \frac{V_T}{R_T} = \frac{24\ \text{V}}{8\ \Omega} = 3\ \text{A}$$

The voltage drop across the resistors will be

$$V_{R1} = I_T \times R_1 = 3\ \text{A} \times 2\ \Omega = 6\ \text{V}$$
$$V_{R2} = I_T \times R_2 = 3\ \text{A} \times 6\ \Omega = 18\ \text{V}$$

If one of the parallel branches is opened, the overall circuit resistance will always increase. This increase in the total resistance will cause an increase in the voltage dropped across the parallel branch (the greater the resistance, the greater the voltage drop), which enables the technician to localize the fault area and also to determine that the fault is an open.

The voltage measured with a voltmeter will be

$$V_{R1} = 6\ \text{V}$$
$$V_{R2} = 18\ \text{V}$$
$$V_{R3} = 18\ \text{V (open)}$$
$$V_{R4} = 0\ \text{V}$$

This identifies the problem as R_3 being open, as it drops the entire parallel circuit voltage (18 V) across itself, whereas normally the voltage would be dropped proportionally across R_3 and R_4, which are in series with one another.

FIGURE 7-34   Open Parallel-Connected Resistor in a Series–Parallel Circuit.

FIGURE 7-35 **Shorted Series-Connected Resistor in a Series–Parallel Circuit.**

7-8-2 *Shorted Component*

R_1 Shorted (Figure 7-35)

With R_1 shorted, the total circuit resistance will decrease ($R_T\downarrow$), causing an increase in circuit current ($I_T\uparrow$). This increase in current will cause an increase in the voltage dropped across the parallel branch. However, the fault can be located once you measure the voltage across R_1, which will read 0 V, indicating that this resistor has almost no resistance, as it has no voltage drop across it.

R_3 Shorted (Figure 7-36)

With R_3 shorted, there will be a decrease in the circuit's total resistance from 6 Ω to

$$R_{2,3,4} = \frac{R_2 \times R_{3,4}}{R_2 + R_{3,4}} = \frac{6 \times 9}{6 + 9} = 3.6$$

$$R_{1,2,3,4} = R_1 + R_{2,3,4}$$
$$= 2\ \Omega + 3.6\ \Omega$$
$$= 5.6\ \Omega$$

This decrease in total resistance ($R_T\downarrow$) will cause an increase in total current ($I_T\uparrow$), which implies that a short has occurred to decrease resistance. The total current will now increase from 4 A to

$$I_T = \frac{V_T}{R_T} = \frac{24\ \text{V}}{5.6\ \Omega} = 4.3\ \text{A}$$

The voltage drops across the resistors will be

$$V_{R1} = I_T \times R_1 = 4.3\ \text{A} \times 2\ \Omega = 8.6\ \text{V}$$
$$V_{R2,3,4} = I_T \times R_{2,3,4} = 4.3\ \text{A} \times 3.6\ \Omega = 15.4\ \text{V}$$

When one of the parallel branch resistors is shorted, the overall circuit resistance will always decrease. This decrease in total resistance will cause a decrease in the voltage dropped across the parallel branch (the smaller the resistance, the smaller the voltage drop), and this enables the technician to localize the faulty area and also to determine that the fault is a short.

FIGURE 7-36   Shorted Parallel-Connected Resistor in a Series–Parallel Circuit.

The voltage measured with the voltmeter (VM) will be

$$V_{R1} = 8.6 \text{ V}$$
$$V_{R2} = 15.4 \text{ V}$$
$$V_{R3} = 0 \text{ V (short)}$$
$$V_{R4} = 15.4 \text{ V}$$

This identifies the problem as R_3 being a short, as 0 V is being dropped across it.

In summary, you may have noticed that (Figure 7-37):

1. An open component causes total resistance to increase ($R_T\uparrow$) and, therefore, total current to decrease ($I_T\downarrow$), and the open component, if in series, has the supply voltage across it, and if in a parallel branch, has the parallel branch voltage across it.

FIGURE 7-37   Summary of Symptoms for Opens and Shorts.

FIGURE 7-38   **Parallel-Connected Resistor Value Decrease in Series–Parallel Circuit.**

2. A shorted component causes total resistance to decrease ($R_T\downarrow$) and, therefore, total current to increase ($I_T\uparrow$), and the shorted component, if in series or parallel branches, will have 0 V across it.

7-8-3   *Resistor Value Variation*

R_2 Resistance Decreases (Figure 7-38)

If the resistance of R_2 decreases, the total circuit resistance (R_T) will decrease and the total circuit current (I_T) will increase. The result of this problem and the way in which the fault can be located is that when the voltage drop across R_1 and the parallel branch is tested, there will be an increased voltage drop across R_1 due to the increased current flow, and a decrease in the voltage drop across the parallel branch due to a decrease in the parallel branch resistance.

With open and shorted components, the large voltage (open) or small voltage (short) drop across a component enables the technician to identify the faulty component. A variation in a component's value, however, will vary the circuit's behavior. With this example, the symptoms could have been caused by a combination of variations. So once the area of the problem has been localized, the next troubleshooting step is to disconnect power and then remove each of the resistors in the suspected faulty area and verify that their resistance values are correct by measuring their resistances with an ohmmeter. In this problem we had an increase in voltage across R_1 and a decrease in voltage across the parallel branch when measuring with a voltmeter, which was caused by R_2 decreasing. The same swing in voltage readings could also have been obtained by an increase in the resistance of R_1.

R_2 Resistance Increases (Figure 7-39)

If the resistance of R_2 increases, the total circuit resistance (R_T) will increase and the total circuit current (I_T) will decrease. The voltage drop across R_1 will decrease and the voltage across the branch will increase in value. Once again, these measured voltage changes could be caused by the resistance of R_2 increasing or the resistance of R_1 decreasing.

To reinforce your understanding let's work out a few examples of troubleshooting other series–parallel circuits.

FIGURE 7-39 **Parallel-Connected Resistor Value Increase in Series–Parallel Circuit.**

▦ EXAMPLE:

If bulb 1 in Figure 7-40 goes open, what effect will it have, and how will the fault be recognized?

▦ *Solution:*

A visual inspection of the circuit shows all bulbs off, as there is no current path from one side of the battery to the other, since bulb 1 is connected in series and has opened the only path. Since all bulbs are off, the faulty bulb cannot be visually isolated. However, you can easily localize the faulty bulb by using one of two methods:

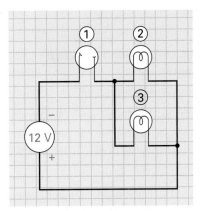

1. Use the voltmeter and check the voltage across each bulb. Bulb 1 would have 12 V across it, while 2 and 3 would have 0 V across them. This isolates the faulty component to bulb 1, since all the supply voltage is being measured across it.

2. By analyzing the circuit diagram, you can see that only one bulb can open and cause all the bulbs to go out, and that is bulb 1. If bulb 2 opens, 1 and 3

FIGURE 7-40 **Series-Connected Open Bulb in Series–Parallel Circuit.**

would still be on, and if bulb 3 opens, 1 and 2 would still remain on. With power off, the ohmmeter could verify this open.

▦ EXAMPLE:

One resistor in Figure 7-41 has shorted. From the voltmeter reading shown, determine which one.

▦ *Solution:*

If the supply voltage is being measured across the parallel branch of R_2 and R_3, there cannot be any other resistance in circuit, so R_1 must have shorted. The next step would be to locate the

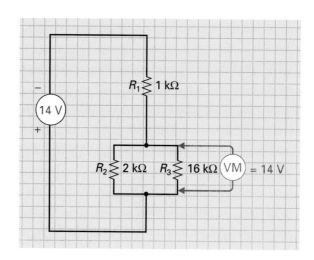

FIGURE 7-41 **Find the Shorted Resistor.**

component, R_1, and determine what has caused it to short. If we were not told that a resistor had shorted, the same symptom could have been caused if R_2 and R_3 were both open, and therefore the open parallel branch would allow no current to flow and maximum supply voltage would appear across it. The individual component resistance, when checked, will isolate the problem.

▢ **EXAMPLE:**

Determine if there is an open or short in Figure 7-42. If so, isolate it by the two voltage readings that are shown in the circuit diagram.

▧ *Solution:*

Performing a few calculations, you should come up with a normal total circuit resistance of 14 kΩ and a total circuit current of 1 mA. This should cause 12 V across R_1 and 2 V across the parallel branch under no-fault conditions. The decrease in the voltage drop across the series resistor R_1 leads you to believe that there has been a decrease in total circuit current, which must have been caused by a total resistance increase, which points to an open component (assuming that only an open or short can occur and not a component value variation).

If R_1 was open, all the 14 V would have been measured across R_1, which did not occur.

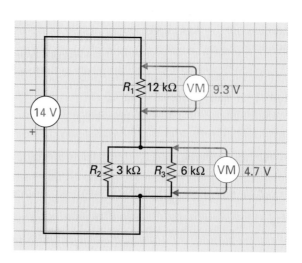

FIGURE 7-42 **Does a Problem Exist?**

CHAPTER 7 / SERIES-PARALLEL DC CIRCUITS

If R_3 opened,

$$\text{total resistance} = 15 \text{ k}\Omega$$

$$\text{total current} = 0.93 \text{ mA} \left(\frac{V_T}{R_T} = \frac{14 \text{ V}}{15 \text{ k}\Omega} \right)$$

$$V_{R1} = I_T \times R_1 = 11.16 \text{ V}$$

Since the voltage dropped across R_1 was 9.3 V, R_3 is not the open.

If R_2 opened,

$$\text{total resistance} = 18 \text{ k}\Omega$$

$$\text{total current} = 0.78 \text{ mA} \left(\frac{V}{R} = \frac{14 \text{ V}}{18 \text{ k}\Omega} \right)$$

$$V_{R1} = I_T \times R_1 = 9.36 \text{ V}$$

This circuit's problem is resistor R_2, which has opened.

SELF-TEST EVALUATION POINT FOR SECTION 7-8

Use the following questions to test your understanding of Section 7-8.

Describe how to isolate the following problems in a series–parallel circuit.

1. An open component
2. A shorted component
3. A resistor value variation

REVIEW QUESTIONS

Multiple-Choice Questions

1. A series–parallel circuit is a combination of:
   a. Components connected end to end
   b. Series (one current path) circuits
   c. Both series and parallel circuits
   d. Parallel (two or more current path) circuits

2. Total resistance in a series–parallel circuit is calculated by applying the _____ resistance formula to series-connected resistors and the _____ resistance formula to resistors connected in parallel.
   a. Series, parallel   c. Series, series
   b. Parallel, series   d. Parallel, parallel

3. Total current in a series–parallel circuit is determined by dividing the total _____ by the total _____.
   a. Power, current   c. Current, resistance
   b. Voltage, resistance   d. Voltage, power

4. Branch current within series–parallel circuits can be calculated by:
   a. Ohm's law
   b. The current-divider formula
   c. Kirchhoff's current law
   d. All of the above

5. A _____ ground has the negative side of the source voltage connected to ground, while a _____ ground has the positive side of the source voltage connected to ground.
   a. Positive, negative   c. Positive, earth
   b. Chassis, earth   d. Negative, positive

6. The output voltage will always _____ when a load or voltmeter is connected across a voltage divider.
   a. Decrease   c. Increase
   b. Remain the same   d. All of the above could be considered true.

7. A Wheatstone bridge was originally designed to measure:
   a. An unknown voltage   c. An unknown power
   b. An unknown current   d. An unknown resistance

8. A balanced bridge has an output voltage:
   a. Equal to the supply voltage
   b. Equal to half the supply voltage
   c. Of 0 V
   d. Of 5 V

9. In a series–parallel resistive circuit, an open series-connected resistor will cause _____ current, whereas an open parallel-connected resistor will result in a total current _____.

a. An increase in, decrease   **c.** Zero, decrease

b. A decrease in, increase   **d.** None of the above

10. In a series–parallel resistive circuit, a shorted series-connected resistor will cause _____ current, whereas a shorted parallel-connected resistor will result in a total current _____.

   a. An increase in, increase   **c.** An increase in, decrease

   b. A decrease in, decrease   **d.** A decrease in, increase

Practice Problems

11. R_3 and R_4 are in series with one another and are both in parallel with R_5. This parallel combination is in series with two series-connected resistors, R_1 and R_2. $R_1 = 2.5\ k\Omega$, $R_2 = 10\ k\Omega$, $R_3 = 7.5\ k\Omega$, $R_4 = 2.5\ k\Omega$, $R_5 = 2.5\ M\Omega$, and $V_S = 100\ V$. For these values, calculate:

   a. Total resistance
   b. Total current
   c. Voltage across series resistors and parallel combinations
   d. Current through each resistor
   e. Total and individual power figures

12. Referring to the example in Question 11, calculate the voltage at every point of the circuit with respect to ground.

13. A 10 V source is connected across a series–parallel circuit made up of R_1 in parallel with a branch made up of R_2 in series with a parallel combination of R_3 and R_4. $R_1 = 100\ \Omega$, $R_2 = 100\ \Omega$, $R_3 = 200\ \Omega$, and $R_4 = 300\ \Omega$. For these values, apply the five-step procedure, and also determine the voltage at every point of the circuit with respect to ground.

Troubleshooting Questions

14. Referring to the example circuit in Figure 7-21, describe the effects you would get if a resistor were to short, and if a resistor were to open.

15. Design a simple five-resistor series–parallel circuit and insert a source voltage and resistance values. Apply the five-step series–parallel circuit procedure, and then theoretically open and short all the resistors and calculate what effect would occur and how you would recognize the problem.

16. Carbon composition resistors tend to increase in resistance with age, while most other types generally decrease in resistance. What effects would resistance changes have on their respective voltage drops?

Alternating Current (AC)

8

The Laser

In 1898, H. G. Wells's famous book, *The War of the Worlds,* had Martian invaders with laserlike death rays blasting bricks, incinerating trees, and piercing iron as if it were paper. In 1917, Albert Einstein stated that, under certain conditions, atoms or molecules could absorb light and then be stimulated to release this borrowed energy. In 1954, Charles H. Townes, a professor at Columbia University, conceived and constructed with his students the first "maser" (acronym for "microwave amplification by stimulated emission of radiation"). In 1958, Townes and Arthur L. Shawlow wrote a paper showing how stimulated emission could be used to amplify light waves as well as microwaves, and the race was on to develop the first "laser." In 1960, Theodore H. Maiman, a scientist at Hughes Aircraft Company, directed a beam of light from a flash lamp into a rod of synthetic crystal, which responded with a burst of crimson light so bright that it outshone the sun.

An avalanche of new lasers emerged, some as large as football fields, while others were no bigger than a pinhead. They can be made to produce invisible infrared or ultraviolet light or any visible color in the rainbow, and the high-power lasers can vaporize any material a million times faster and more intensely than a nuclear blast, while the low-power lasers are safe to use in children's toys.

At present, the laser is being used by the FBI to detect fingerprints that are 40 years old, in defense programs, in compact disk players, in underground fiber optic communication to transmit hundreds of telephone conversations, to weld car bodies, to drill holes in baby-bottle nipples, to create three-dimensional images called holograms, and as a surgeon's scalpel in the operating room. Not a bad beginning for a device that when first developed was called "a solution looking for a problem."

Introduction

In this chapter you will be introduced to alternating current (ac). While direct current (dc) can be used in many instances to deliver power or represent information, there are certain instances in which ac is preferred. For example, ac is easier to generate and is more efficiently transmitted as a source of electrical power for the home or business. Audio (speech and music) and video (picture) information are generally always represented in electronic equipment as an alternating current or alternating voltage signal.

In this chapter, we will begin by describing the difference between dc and ac, and then examine where ac is used. Following this we will discuss all the characteristics of ac waveform shapes, and finally, the differences between electricity and electronics.

8-1   THE DIFFERENCE BETWEEN DC AND AC

One of the best ways to describe anything new is to begin by redescribing something known and then discuss the unknown. The known topic in this case is **direct current.** Direct current (dc) is the flow of electrons in one DIRECTion and one direction only. Dc voltage is non-varying and normally obtained from a battery or power supply unit, as seen in Figure 8-1(a). The only variation in voltage from a battery occurs due to the battery's discharge, but even then, the current will still flow in only one direction, as seen in Figure 8-1(b). A dc voltage of 9 or 6 V could be illustrated graphically as shown in Figure 8-1(c). Whether 9 or 6 V, the voltage can be seen to be constant or the same at any time.

Alternating current (ac) flows first in one direction and then in the opposite direction. This reversing current is produced by an alternating voltage source, as shown in Figure 8-2(a), which reaches a maximum in one direction (positive), decreases to zero, and then reverses itself and reaches a maximum in the opposite direction (negative). This is graphically illustrated in Figure 8-2(b). During the time of the positive voltage alternation, the polarity of the voltage will be as shown in Figure 8-2(c), so current will flow from

Direct Current

Current flow in only one direction.

Alternating Current

Electric current that rises from zero to a maximum in one direction, falls to zero, and then rises to a maximum in the opposite direction, and then repeats another cycle, the positive and negative alternations being equal.

FIGURE 8-1   Direct Current. (a) DC Sources. (b) DC Flow. (c) Graphic Representation of DC.

FIGURE 8-2 **Alternating Current.**

negative to positive in a counterclockwise direction. During the time of the negative voltage alternation, the polarity of the voltage will reverse, as shown in Figure 8-2(d), causing current to flow once again from negative to positive, but in this case, in the opposite clockwise direction.

Use the following questions to test your understanding of Section 8-1.

1. Give the full names of the following abbreviations: (a) ac; (b) dc.

2. The polarity of a/an _____ voltage source will continually reverse, and therefore so will the circuit current.

3. The polarity of a/an _____ voltage source will remain constant, and therefore current will flow in only one direction.

4. State briefly the difference between ac and dc.

8-2   WHY ALTERNATING CURRENT?

The question that may be troubling you at this point is: If we have been managing fine for the past chapters with dc, why do we need ac?

There are two main applications for ac:

1. *Power transfer:* to supply electrical power for lighting, heating, cooling, appliances, and machinery in both home and industry

2. *Information transfer:* to communicate or carry information, such as radio music and television pictures, between two points

To begin with, let us discuss the first of these applications, power transfer.

8-2-1   *Power Transfer*

There are three advantages that ac has over dc from a power point of view, and these are:

1. Flashlights, radios, and portable televisions all use batteries (dc) as a source of power. In these applications where a small current is required, batteries will last a good length of time before there is a need to recharge or replace them. Many appliances and most industrial equipment need a large supply of current, and in this situation a generator would have to be used to generate this large amount of current. Generators operate in the opposite way to motors, in that a **generator** converts a mechanical rotation input into an electrical output. Generators can be used to generate either dc or ac, but ac generators can be larger, less complex internally, and cheaper to operate, and this is the first reason why we use ac instead of dc for supplying power.

2. From a power point of view, ac is always used by electric companies when transporting power over long distances to supply both the home and industry with electrical energy. Recalling the power formula, you will remember that power is proportional to either current or voltage squared ($P \propto I^2$ or $P \propto V^2$), which means that to supply power to the home or industry, we would supply either a large current or voltage. As you can see in Figure 8-3, between the electric power plant and home or industry are power lines carrying the power. The amount of power lost (heat) in these power lines can be calculated by using the formula $P = I^2 \times R$, where I is the current flowing through the line and R is the resistance of the power lines. This means that the larger the current, the greater the amount of power lost in the lines in the form of heat and therefore the less the amount of power supplied to the home or industry. For this reason, power companies transport electric energy at a very high voltage between 200,000 and 600,000 V. Since the voltage is high, the current can be low and provide the same amount of power to the consumer ($P = V\uparrow \times I\downarrow$). Yet, by keeping the current low, the amount of heat loss generated in the power lines is minimal.

    Now that we have discovered why it is more efficient over a long distance to transport high voltages than high current, what does this have to do with ac? An ac voltage can easily and efficiently be transformed up or down to a higher or lower voltage by utilizing a device known as a **transformer,** and even though dc voltages can be stepped up and down, the method is inefficient and more complex.

3. Nearly all electronic circuits and equipment are powered by dc voltages, which means that once the ac power arrives at the home or industry, in most cases it will have to be converted into dc power to operate electronic equipment. It is a relatively simple process to convert ac to dc, but conversion from dc to ac is a complex and comparatively inefficient process.

Figure 8-3 illustrates ac power distribution from the electric power plant to the home and industry. The ac power distribution system begins at the electric power plant, which has the powerful large generators driven by turbines to generate large ac voltages. The turbines can be driven by either falling water (hydroelectric), or from steam, which is produced with intense heat by burning either coal, gas, or oil or from a nuclear reactor (thermoelectric). The turbine supplies the mechanical energy to the generator, to be transformed into ac electrical energy.

The generator generates an ac voltage of approximately 22,000 V, which is stepped up by transformers to approximately 500,000 V. This voltage is applied to the long-distance

Generator
Device used to convert a mechanical energy input into an electrical energy output.

Transformer
Device consisting of two or more coils that are used to couple electric energy from one circuit to another, yet maintain electrical isolation between the two.

Hydroelectric
Falling water turns turbines.

Electric Power Plant

Step-up Transformers

Long-Distance Transmission Lines: Connect cities and towns to power plant.

22,000 V AC

500,000 V AC

Thermoelectric
Heat is used to generate steam, which turns turbines.

TURBINES

ELECTRIC GENERATOR

Short-Distance Transmission Lines: Distribute power throughout cities and towns.

500,000 V AC

Step-down Transformers

66,000 V AC

Industrial Customer: AC is used to drive motors, presses, machinery, and so on.

Step-down Transformers

4,800 V AC

Utility Pole: Step-down transformer mounted on pole.

Commercial Customer: AC is converted to DC internally by copiers, computers, typewriters, and so on.

240/120 V AC

Residential Customer: Arriving AC is used for heating, cooling, cooking, and lighting while appliances internally convert AC to DC.

FIGURE 8-3   AC Power Distribution.

transmission lines, which connect the power plant to the city or town. At each city or town, the voltage is tapped off the long-distance transmission lines and stepped down to approximately 66,000 V and is distributed to large-scale industrial customers. The 66,000 V is stepped down again to approximately 4800 V and distributed throughout the city or town by short-distance transmission lines. This 4800 V is used by small-scale industrial customers and residential customers who receive the ac power via step-down transformers on utility poles, which step down the 4800 V to 240 V and 120 V.

A large amount of equipment and devices within industry and the home will run directly from the ac power, such as heating, lighting, and cooling. Some equipment that runs on dc, such as televisions and computers, will accept the 120 V ac and internally convert it to the dc voltages required. Figure 8-4 illustrates a TV set that is operating from the 120 V ac from the wall outlet and internally converting it to dc so that it can power the circuits necessary to produce both audio (sound) and video (picture) information. This unit that converts ac to dc is called a **rectifier.**

Rectifier

Device that achieves rectification.

FIGURE 8-4   Rectification (Converting AC to DC) by Appliances.

8-2-2   *Information Transfer*

Information, by definition, is the property of a signal or message that conveys something meaningful to the recipient. **Communication,** which is the transfer of information between two points, began with speech and progressed to handwritten words in letters and printed words in newspapers and books. To achieve greater distances of communication, face-to-face communications evolved into telephone and radio communications.

A simple communication system can be seen in Figure 8-5(a).

The voice information or sound wave produced by the sender is a variation in air pressure, and travels at the speed of sound, as detailed in Figure 8-5. Sound waves or sounds are normally generated by a vibrating reed or plucked string in the case of musical instruments. In this example the sender's vocal cords vibrate backward and forward, producing a rarefaction or decreased air pressure, where few air molecules exist, and a compression or increased air pressure, where many air molecules exist. Like the ripples produced by a stone falling in a pond, the sound waves produced by the sender are constantly expanding and traveling outward.

The microphone is in fact a **transducer** (energy converter), because it converts the sound wave (which is a form of mechanical energy) into electrical energy in the form of voltage and current, which varies in the same manner as the sound wave and therefore contains the sender's message or information.

The **electrical wave,** shown in Figure 8-5(c), is a variation in voltage or current and can only exist in a wire conductor or circuit. This electrical signal travels at the speed of light.

The speaker, like the microphone, is also an electroacoustical transducer that converts the electrical energy input into a mechanical sound-wave output. These sound waves strike

Communication

Transmission of information between two points.

Transducer

Any device that converts energy from one form to another.

Electrical Wave

Traveling wave propagated in a conductive medium that is a variation in voltage or current and travels at slightly less than the speed of light.

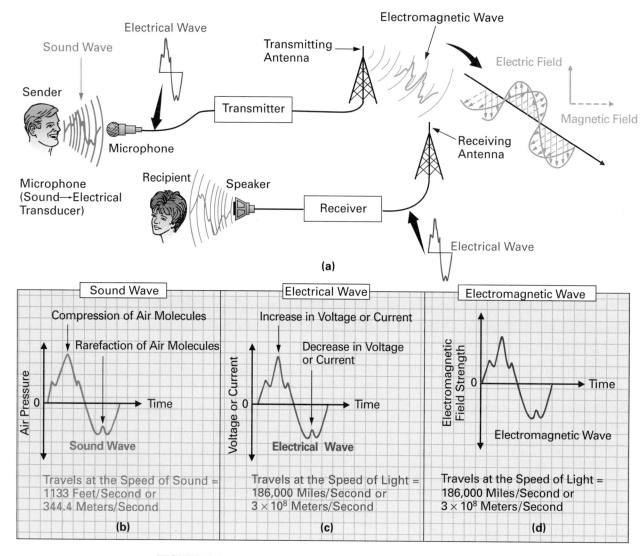

FIGURE 8-5   Information Transfer.

the outer eardrum, causing the ear diaphragm to vibrate, and these mechanical vibrations actuate nerve endings in the ear, which convert the mechanical vibrations into electrochemical impulses that are sent to the brain. The brain decodes this information by comparing these impulses with a library of previous sounds and so provides the sensation of hearing.

To communicate between two distant points, a wire must be connected between the microphone and speaker. However, if an electrical wave is applied to an antenna, the electrical wave is converted into a radio or electromagnetic wave, as shown in the inset in Figure 8-5(a), and communication is established without the need of a connecting wire—hence the term **wireless communication.** Antennas are designed to radiate and receive electromagnetic waves, which vary in field strength, as shown in Figure 8-5(d), and can exist in either air or space. These radio waves, as they are also known, travel at the speed of light and allow us to achieve great distances of communication.

More specifically, radio waves are composed of two basic components. The electrical voltage applied to the antenna is converted into an electric field and the electrical current into a magnetic field. This **electromagnetic** (electric–magnetic) **wave** is used to carry a variety of information, such as speech, radio broadcasts, television signals, and so on.

In summary, the sound wave is a variation in air pressure, the electrical wave is a variation of voltage or current, and the electromagnetic wave is a variation of electric and magnetic field strength.

Wireless Communication

Term describing radio communication that requires no wires between the two communicating points.

Electromagnetic (Radio) Wave

Wave that consists of both an electric and magnetic variation, and travels at the speed of light.

■ **EXAMPLE:**

How long will it take the sound wave produced by a rifle shot to travel 9630.5 feet?

■ *Solution:*

This problem makes use of the following formula:

$$\text{Distance} = \text{velocity} \times \text{time}$$

or

D = distance
v = velocity
t = time

If someone travels at 20 mph for 2 hours, the person will travel 40 miles ($D = v \times t = 20$ mph $\times$ 2 hours = 40 miles). In this problem, the distance (9630.5) and the sound wave's velocity (1130 ft/sec) are known, and so by rearranging the formula, we can find time:

$$\text{time} = \frac{\text{distance}}{\text{velocity}} = \frac{9630.5 \text{ ft}}{1130 \text{ ft/s}} = 8.5 \text{ s}$$

■ **EXAMPLE:**

How long will it take an electromagnetic (radio) wave to reach a receiving antenna that is 2000 miles away from the transmitting antenna?

■ *Solution:*

In this problem, both distance (2000 miles) and velocity (186,000 miles/s) are known, and time has to be calculated:

$$\text{time} = \frac{\text{distance}}{\text{velocity}} = \frac{2000 \text{ miles}}{186,000 \text{ miles/s}}$$
$$= 1.075 \times 10^{-2} \quad \text{or} \quad 10.8 \text{ ms}$$

8-2-3   *Electrical Equipment or Electronic Equipment?*

In the beginning of this chapter, it was stated that ac is basically used in two applications: (1) power transfer and (2) information transfer. These two uses for ac help define the difference between electricity and electronics. Electronic equipment manages the flow of information, while electrical equipment manages the flow of power. In summary:

EQUIPMENT	MANAGES
Electrical	Power (large values of V and I)
Electronic	Information (small values of V and I)

To use an example, we can say that a dc power supply is a piece of electrical equipment since it is designed to manage the flow of power. A TV set, however, is an electronic system since its electronic circuits are designed to manage the flow of audio (sound) and video (picture) information.

Since most electronic systems include a dc power supply, we can also say that the electrical circuits manage the flow of power, and this power supply enables the electronic circuits to manage the flow of information.

TIME LINE

Studying the experiments of Maxwell and Heinrich Hertz, Guglielmo Marconi (1874-1937) invented a practical system of telegraphy communication. In an evolutionary process, Marconi extended his distance of communication from 1 1/2 miles in 1896 to 6000 miles in 1902. In September 1899, Marconi equipped two U.S. ships with equipment and used them in the Atlantic Ocean to transmit to America the progress of the America's Cup yacht race.

Use the following questions to test your understanding of Section 8-2-2 and 8-2-3.

1. The _____ wave is a variation in air pressure, the _____ wave is a variation in field strength, and the _____ wave is a variation of voltage or current.

2. Sound waves travel at the speed of sound, which is _____, while electrical and electromagnetic waves travel at the speed of light, which is _____.

3. A human ear is designed to receive _____ waves, an antenna is designed to transmit or receive _____ waves, and an electronic circuit is designed to pass only _____ waves.

4. _____ equipment manages the flow of information, and these ac waveforms normally have small values of current and voltage.

5. _____ equipment manages the flow of power, and these ac wave-forms normally have large values of current and voltage.

8-3   AC WAVE SHAPES

In all fields of electronics, whether medical, industrial, consumer, or data processing, different types of information are being conveyed between two points, and electronic equipment is managing the flow of this information.

Let's now discuss the basic types of ac wave shapes. The way in which a wave varies in magnitude with respect to time describes its wave shape. All ac waves can be classified into one of six groups, and these are illustrated in Figure 8-6.

8-3-1   *The Sine Wave*

Sine Wave

Wave whose amplitude is the sine of a linear function of time. It is drawn on a graph that plots amplitude against time or radial degrees relative to the angular rotation of an alternator.

The **sine wave** is the most common type of waveform. It is the natural output of a generator that converts a mechanical input, in the form of a rotating shaft, into an electrical output in the form of a sine wave. In fact, for one cycle of the input shaft, the generator will produce one sinusoidal ac voltage waveform, as shown in Figure 8-7. When the input shaft of the generator is at 0°, the ac output is 0 V. As the shaft is rotated through 360°, the ac output voltage will rise to a maximum positive voltage at 90°, fall back to 0 V at 180°, and then reach a maximum negative voltage at 270°, and finally return to 0 V at 360°. If this ac voltage is applied across a closed circuit, it produces a current that continually reverses or alternates in each direction.

Figure 8-8 illustrates the sine wave, with all the characteristic information inserted, which at first glance looks a bit ominous. Let's analyze and discuss each piece of information individually, beginning with the sine wave's amplitude.

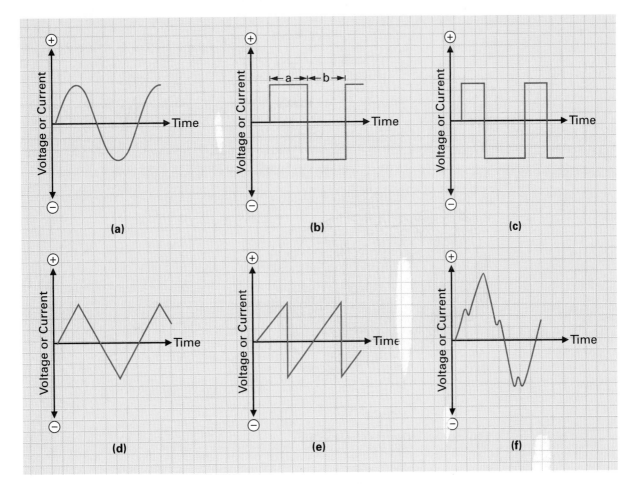

FIGURE 8-6 AC Wave Shapes. (a) Sine Wave. (b) Square Wave. (c) Pulse Wave. (d) Triangular Wave. (e) Sawtooth Wave. (f) Irregular Wave.

Amplitude

Figure 8-9 plots direction and amplitude against time. The **amplitude** or magnitude of a wave is often represented by a **vector** arrow, also illustrated in Figure 8-9. The vector's length indicates the magnitude of the current or voltage, while the arrow's point is used to show the direction, or polarity.

Peak Value

The peak of an ac wave occurs on both the positive and negative alternation, but is only at the peak (maximum) for an instant. Figure 8-10(a) on p. 20? illustrates an ac current waveform rising to a positive peak of 10 A, falling to zero, and then reaching a negative pe of 10 A in the reverse direction. Figure 8-10(b) shows an ac voltage waveform reaching p itive and negative peaks of 9 V.

Peak-to-Peak Value

The **peak-to-peak value** of a sine wave is the value of voltage or current between the positive and negative maximum values of a waveform. For example, the peak-to-peak value of the current waveform in Figure 8-19(a) is equal to $I_{p\text{-}p} = 2 \times I_p = 20$ A. In Figure 8-10(b), it would be equal to $V_{p\text{-}p} = 2 \times V_p = 18$ V.

$$p\text{-}p = 2 \times peak$$

Amplitude

Magnitude or size an alternation varies from zero.

Vector

Quantity that has both magnitude and direction. They are normally represented as a line, the length of which indicates magnitude and the orientation of which, due to the arrowhead on one end, indicates direction.

Peak Value

Maximum or highest-amplitude level.

Peak-to-Peak Value

Difference between the maximum positive and maximum negative values.

FIGURE 8-7   **Degrees of a Sine Wave.**

RMS or Effective Value

Both the positive and negative alternation of a sine wave can accomplish the same amount of work, but the ac waveform is only at its maximum value for an instant in time, spending most of its time between peak currents. Our examples in Figure 8-10(a) and (b), therefore, cannot supply the same amount of power as a dc value of 10 A or 9 V.

The effective value of a sine wave is equal to 0.707 of the peak value.

$$\text{rms} = 0.707 \times \text{peak}$$

This **root-mean-square (rms) value** of 0.707 can always be used to tell us how effective an ac sine wave will be. For example, a 10 A dc source would be 10 A effective because it is continually at its peak value and always delivering power to the circuit to which it is connected, while a 10 A ac source would only be 7.07 A effective, as seen in Figure 8-11, because it is at 10 A for only a short period of time. As another example, a 10 V ac sine-wave alternation would be as effective or supply the same amount of power to a circuit as a 7.07 V dc source.

Unless otherwise stated, ac values of voltage or current are always given in rms. The peak value can be calculated by transposing the original rms formula of rms = peak × 0.707, and ending up with

$$\text{peak} = \frac{\text{rms}}{0.707}$$

Since 1/0.707 = 1.414, the peak can also be calculated by

CHAPTER 8 / ALTERNATING CURRENT (AC)

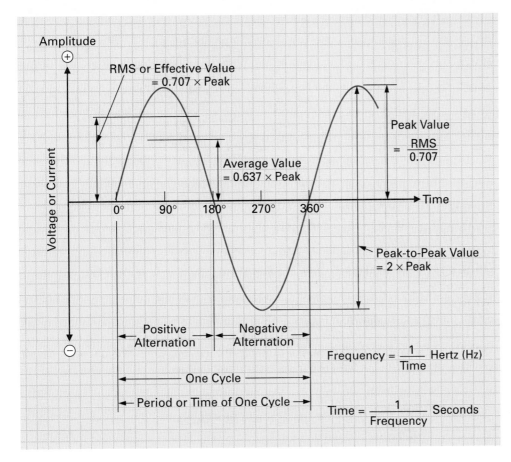

FIGURE 8-8 Sine Wave.

$$\text{peak} = \text{rms} \times 1.414$$

Average Value

The **average value** of the positive or negative alternation is found by taking either the positive or negative alternation, and listing the amplitude or vector length of current or volt-

Average Value

Mean value found when the area of a wave above a line is equal to the area of the wave below the line.

FIGURE 8-9 Sine-Wave Amplitude.

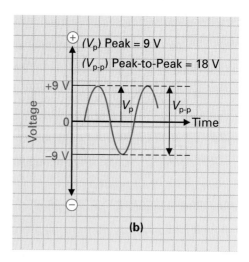

FIGURE 8-10   Peak and Peak-to-Peak of a Sine Wave.

ages at 1° intervals, as shown in Figure 8-12(a). The sum of all these values is then divided by the total number of values (averaging), which for all sine waves will calculate out to be 0.637 of the peak voltage or current. For example, the average value of a sine-wave alternation with a peak of 10 V, as seen in Figure 8-12(b), is equal to

$$\text{average} = 0.637 \times \text{peak}$$

$$= 0.637 \times 10 \text{ V}$$
$$= 6.37 \text{ V}$$

EXAMPLE:

Calculate V_p, $V_\text{p-p}$, V_rms, and V_avg of a 16 V peak sine wave.

■ *Solution:*

$$V_\text{p} = 16 \text{ V}$$
$$V_\text{p-p} = 2 \times V_\text{p} = 2 \times 16 \text{ V} = 32 \text{ V}$$
$$V_\text{rms} = 0.707 \times V_\text{p} = 0.707 \times 16 \text{ V} = 11.3 \text{ V}$$
$$V_\text{avg} = 0.637 \times V_\text{p} = 0.637 \times 16 \text{ V} = 10.2 \text{ V}$$

FIGURE 8-11   Effective Equivalent.

FIGURE 8-12 **Average Value of a Sine-Wave Alternation = 0.637 × Peak.**

□ **EXAMPLE:**

Calculate V_p, V_{p-p}, and V_{avg} of a 120 V (rms) ac main supply.

■ *Solution:*

$$V_p = \text{rms} \times 1.414 = 120 \text{ V} \times 1.414 = 169.68 \text{ V}$$
$$V_{p-p} = 2 \times V_p = 2 \times 169.68 \text{ V} = 339.36 \text{ V}$$
$$V_{avg} = 0.637 \times V_p = 0.637 \times 169.68 \text{ V} = 108.09 \text{ V}$$

The 120 V (rms) that is delivered to every home and business has a peak of 169.68 V. This ac value will deliver the same power as 120 V dc.

Frequency and Period

As shown in Figure 8-13, the **period** (*t*) is the time required for one complete cycle (positive and negative alternation) of the sinusoidal current or voltage waveform. A *cycle,* by definition, is the change of an alternating wave from zero to a positive peak, to zero, then to a negative peak, and finally, back to zero.

Period

Time taken to complete one complete cycle of a periodic or repeating waveform.

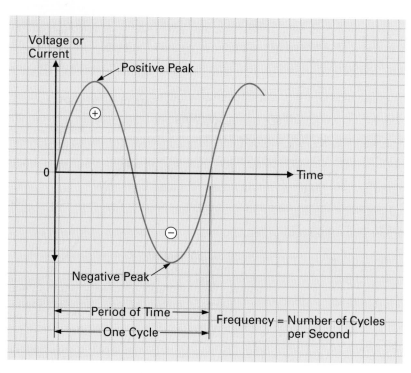

FIGURE 8-13   Frequency and Period.

Frequency is the number of repetitions of a periodic wave in a unit of time. It is symbolized by f and is given the unit hertz (cycles per second), in honor of a German physicist, Heinrich Hertz.

Sinusoidal waves can take a long or a short amount of time to complete one cycle. This time is related to frequency in that period and is equal to the reciprocal of frequency, and vice versa.

$$f\,(\text{hertz}) = \frac{1}{t}$$

$$t\,(\text{seconds}) = \frac{1}{f}$$

where t = period, f = frequency

For example, the ac voltage of 120 V (rms) arrives at the household electrical outlet alternating at a frequency of 60 hertz (Hz). This means that 60 cycles arrive at the household electrical outlet in 1 second. If 60 cycles occur in 1 second, as seen in Figure 8-14(a), it is actually taking $\frac{1}{60}$ of a second for one of the 60 cycles to complete its cycle, which calculates out to be

$$\tfrac{1}{60} \text{ of 1 second} = \frac{1}{60 \text{ cycles}} \times 1 \text{ second} = 16.67 \text{ milliseconds (ms)}$$

So the time or period of one cycle can be calculated by using the formula period $(t) = 1/f = 1/60$ Hz = 16.67 ms, as shown in Figure 8-14(b).

If the period or time of a cycle is known, the frequency can be calculated. For example:

$$\text{frequency } (f) = \frac{1}{\text{period}} = 1/16.67 \text{ ms} = 60 \text{ Hz}$$

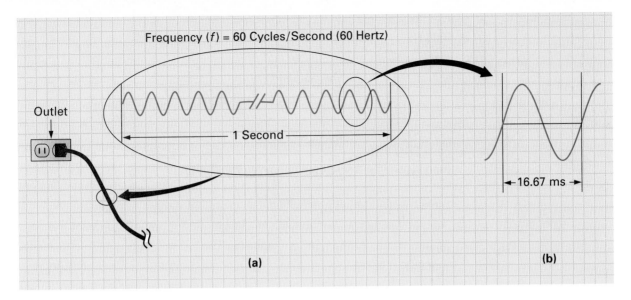

Figure 8-14 **120 V, 60 Hz AC Supply.**

As illustrated in Figure 8-3, all homes in the United States receive at their wall outlets an ac voltage of 120 V rms at a frequency of 60 Hz. This frequency was chosen for convenience, as a lower frequency would require larger transformers, and if the frequency were too low, the slow switching (alternating) current through light bulbs would cause the lights to flicker. A higher frequency than 60 Hz was found to cause an increase in the amount of heat generated in the core of all power distribution transformers due to eddy currents and hysteresis losses. A frequency of 60 Hz was chosen in the United States; however, other countries, such as England and most of Europe, use an ac power line frequency of 50 Hz (240 V).

EXAMPLE:

If a sine wave has a period of 400 μs, what is its frequency?

Solution:

$$\text{frequency } (f) = \frac{1}{\text{time } (t)} = \frac{1}{400 \ \mu s} = 2.5 \text{ kHz or 2500 cycles per second}$$

EXAMPLE:

If it takes a sine wave 25 ms to complete two cycles, how many of the cycles will be received in 1 s?

Solution:

If the period of two cycles is 25 ms, one cycle period will equal 12.5 ms. The number of cycles per second or frequency will equal

$$f = \frac{1}{t} = \frac{1}{12.5 \text{ ms}} = 80 \text{ Hz} \quad \text{or} \quad 80 \text{ cycles/second}$$

EXAMPLE:

Calculate the period of the following:

a. 100 MHz

b. 40 cycles every 5 seconds

c. 4.2 kilocycles/second

d. 500 kHz

■ *Solution:*

$$f = \frac{1}{t} \qquad \text{therefore, } t = \frac{1}{f}$$

a. $t = \dfrac{1}{100 \text{ MHz}} = 10$ nanoseconds (ns)

b. 40 cycles/5 s = 8 cycles/second (8 Hz)

$$t = \frac{1}{8 \text{ Hz}} = 125 \text{ ms}$$

c. $t = \dfrac{1}{4.2 \text{ kHz}} = 238 \text{ }\mu s$

d. $t = \dfrac{1}{500 \text{ kHz}} = 2 \text{ }\mu s$

Wavelength

Wavelength

Distance between two points of corresponding phase and is equal to waveform velocity or speed divided by frequency.

Wavelength, as its name states, is the physical length of one complete cycle and is generally measured in meters. The wavelength (λ, lambda) of a complete cycle is dependent on the frequency and velocity of the transmission:

$$\lambda = \frac{\text{velocity}}{\text{frequency}}$$

Electromagnetic waves. Radio waves travel at the speed of light in air or a vacuum, which is 3×10^8 meters/second or 3×10^{10} cm/second.

$$\lambda \text{ (m)} = \frac{3 \times 10^8 \text{ m/s}}{f \text{ (Hz)}} \qquad \text{or} \qquad \lambda \text{ (cm)} = \frac{3 \times 10^{10} \text{ cm/s}}{\text{frequency (Hz)}}$$

(There are 100 centimeters [cm] in 1 meter [m], therefore cm = 10^{-2}, and m = 10^0 or 1.) Subsequently, the higher the frequency, the shorter the wavelength, which is why a short-wave radio receiver is designed to receive high frequencies ($\lambda\downarrow = 3 \times 10^8/f\uparrow$).

■ **EXAMPLE:**

Calculate the wavelength of the electromagnetic waves illustrated in Figure 8-15.

■ *Solution:*

a. $\lambda = \dfrac{3 \times 10^8}{f \text{ (Hz)}} \text{ m/s} = \dfrac{3 \times 10^8}{10 \text{ kHz}} = 30,000 \text{ m} \quad \text{or} \quad 30 \text{ km}$

b. $\lambda = \dfrac{3 \times 10^{10}}{f \text{ (Hz)}} \text{ cm/s} = \dfrac{3 \times 10^{10}}{2182 \text{ kHz}} = 13,748.9 \text{ cm} \quad \text{or} \quad 137.489 \text{ m}$

c. $\lambda = \dfrac{3 \times 10^{10}}{f \text{ (Hz)}} \text{ cm/s} = \dfrac{3 \times 10^{10}}{4.0 \text{ GHz}} = \dfrac{3 \times 10^{10}}{4 \times 10^9} = 7.5 \text{ cm} \quad \text{or} \quad 0.075 \text{ m}$

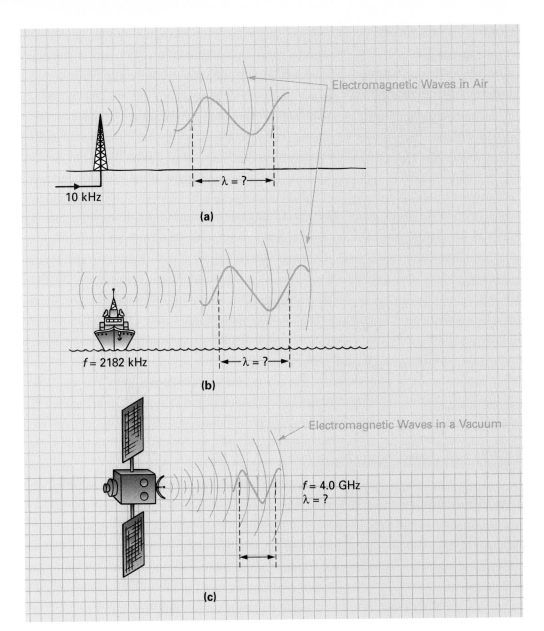

FIGURE 8-15 **Electromagnetic Wavelength Examples.**

Sound waves. Sound waves travel at a slower speed than electromagnetic waves, as their mechanical vibrations depend on air molecules, which offer resistance to the traveling wave. For sound waves, the wavelength formula will be equal to

$$\lambda \text{ (m)} = \frac{344.4 \text{ m/s}}{f \text{ (Hz)}}$$

EXAMPLE:

Calculate the wavelength of the sound waves illustrated in Figure 8-16.

Solution:

a. $\lambda \text{ (m)} = \dfrac{344.4 \text{ m/s}}{f \text{ (Hz)}} = \dfrac{344.4}{35 \text{ kHz}} = 9.8 \times 10^{-3} \text{ m} = 9.8$

(a)

Frequency Range = 300 Hz to 3 kHz
Wavelength Range = ? to ?

(b)

FIGURE 8-16   Sound Wavelength Examples.

b. 300 Hz: λ (m) $= \dfrac{344.4 \text{ m/s}}{300 \text{ Hz}} = 1.15$ m

3000 Hz: λ (m) $= \dfrac{344.4 \text{ m/s}}{3000 \text{ Hz}} = 0.115$ m   or   11.5 cm

Phase

Angular relationship between two waves, normally between current and voltage in an ac circuit.

Phase Relationships

The **phase** of a sine wave is always relative to another sine wave of the same frequency. Figure 8-17(a) illustrates two sine waves that are in phase with one another, while

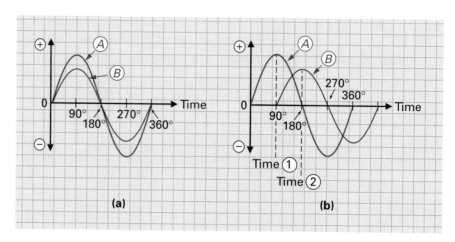

FIGURE 8-17   Phase Relationship. (a) In Phase. (b) Out of Phase.

Figure 8-17(b) shows two sine waves that are out of phase with one another. Sine wave *A* is our reference, since the positive-going zero crossing is at 0°, its positive peak is at 90°, its negative-going zero crossing is at 180°, its negative peak is at 270°, and the cycle completes at 360°. In Figure 8-17(a), sine wave *B* is in phase with *A* since its peaks and zero crossings occur at the same time as sine wave *A*'s.

In Figure 8-17(b), sine wave *B* has been shifted to the right by 90° with respect to the reference sine wave *A*. This **phase shift** or **phase angle** of 90° means that sine wave *A leads B* by 90°, or *B lags A* by 90°. Sine wave *A* is said to lead *B* as its positive peak, for example, occurs first at time 1, while the positive peak of *B* occurs later at time 2.

EXAMPLE:

What are the phase relationships between the two waveforms illustrated in Figure 8-18(a) and (b)?

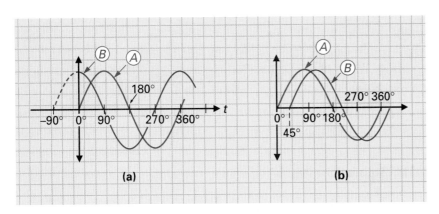

FIGURE 8-18   Phase Relationship Examples.

Solution:

a. The phase shift or angle is 90°. Sine wave *B* leads sine wave *A* by 90°, or *A* lags *B* by 90°.

b. The phase shift or angle is 45°. Sine wave *A* leads sine wave *B* by 45°, or *B* lags *A* by 45°.

8-3-2   *The Square Wave*

The **square wave** is a periodic (repeating) wave that alternates from a positive peak value to a negative peak value, and vice versa, for equal lengths of time.

In Figure 8-19 you can see an example of a square wave that is at a frequency of 1 kHz and has a peak of 10 V. If the frequency of a wave is known, its period or time of one cycle can be calculated by using the formula $t = 1/f = 1/1 \text{ kHz} = 1 \text{ ms}$ or $\frac{1}{1000}$ of a second. One complete cycle will take 1 ms to complete, so the positive and negative alternations will each last for 0.5 ms.

If the peak of the square wave is equal to 10 V, the peak-to-peak value of this square wave will equal $V_{p\text{-}p} = 2 \times V_p = 20 \text{ V}$.

To summarize this example, the square wave alternates from a positive peak value of +10 V to a negative peak value of −10 V for equal time lengths (half-cycles) of 0.5 ms.

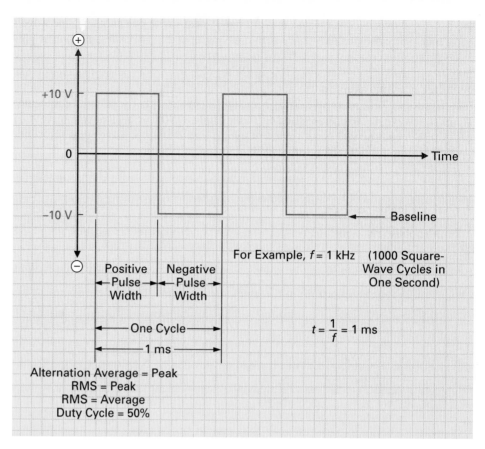

FIGURE 8-19 Square Wave.

Duty Cycle

Duty cycle is an important relationship, which has to be considered when discussing square waveforms. The **duty cycle** is the ratio of a pulse width (positive or negative pulse or cycle) to the overall period or time of the wave and is normally given as a percentage.

$$\text{duty cycle } (\%) = \frac{\text{pulse width } (P_w)}{\text{period } (t)} \times 100\%$$

The duty cycle of the example square wave in Figure 8-19 will equal

$$
\begin{aligned}
\text{duty cycle } (\%) &= \frac{\text{pulse width } (P_w)}{\text{period } (t)} \times 100\% \\
&= \frac{0.5 \text{ ms}}{1 \text{ ms}} \times 100\% \\
&= 50\%
\end{aligned}
$$

Since a square wave always has a positive and a negative alternation that are equal in time, the duty cycle of all square waves is equal to 50%, which actually means that the positive cycle lasts for 50% of the time of one cycle.

Average

The average or mean value of a square wave can be calculated by using the formula

CHAPTER 8 / ALTERNATING CURRENT (AC)

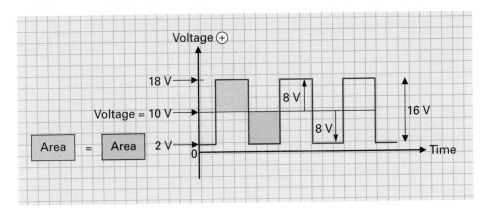

FIGURE 8-20 **2 to 18 V Square Wave.**

$$V \text{ or } I \text{ average} = \text{baseline} + (\text{duty cycle} \times \text{peak to peak})$$

The average of the complete square-wave cycle in Figure 8-19 should calculate out to be zero, as the amount above the line equals the amount below. If we apply the formula to this example, you can see that

$$
\begin{aligned}
V_{avg} &= \text{baseline} + (\text{duty cycle} \times \text{peak to peak}) \\
&= -10\,\text{V} + (0.5 \times 20\,\text{V}) \\
&= -10\,\text{V} + 10\,\text{V} \\
&= 0\,\text{V}
\end{aligned}
$$

However, a square wave does not always alternate about 0. For example, Figure 8-20 illustrates a 16 $V_{p\text{-}p}$ square wave that rests on a baseline of 2 V. The average value of this square wave is equal to

$$
\begin{aligned}
V_{avg} &= \text{baseline} + (\text{duty cycle} \times \text{peak to peak}) \\
&= 2\,\text{V} \times (0.5 \times 16\,\text{V}) \\
&= (+2) + (+8\,\text{V}) \\
&= 10\,\text{V}
\end{aligned}
$$

☐ **EXAMPLE:**

Calculate the duty cycle and V_{avg} of a square wave of 0 to 5 V.

▨ *Solution:*

The duty cycle of a square wave is always 0.5 or 50%. The average is:

$$
\begin{aligned}
V_{avg} &= \text{baseline} + (\text{duty cycle} \times V_{p\text{-}p}) \\
&= 0\,\text{V} + (0.5 \times 5\,\text{V}) \\
&= 0\,\text{V} + 2.5\,\text{V} = 2.5\,\text{V}
\end{aligned}
$$

Up to this point, we have seen the ideal square wave, which has instantaneous transition from the negative to the positive values, and vice versa, as shown in Figure 8-21(a). In fact, the transitions from negative to positive (positive or leading edge) and from positive to negative (negative or trailing edge) are not as ideal as shown here. It takes a small amount of time for the wave to increase to its positive value (the **rise time**), and an equal amount of time for a wave to decrease to its negative value (the **fall time**). Rise time (T_R), by definition, is the time it takes for an edge to rise from 10% to 90% of its full amplitude, while fall time (T_F) is the time it takes for an edge to fall from 90% to 10% of its full amplitude, as shown in Figure 8-21(b).

Rise Time

Time it takes a positive edge of a pulse to rise from 10% to 90% of its peak value.

Fall Time

Time it takes a negative edge of a pulse to fall from 90% to 10% of its peak value.

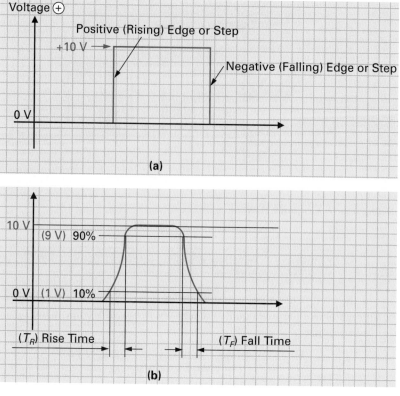

FIGURE 8-21 Square Wave's Rise and Fall Times. (a) Ideal. (b) Actual.

With a waveform such as that in Figure 8-21(b), it is difficult, unless a standard is used, to know exactly what points to use when measuring the width of either the positive or negative alternation. The standard width is always measured between the two 50% amplitude points, as shown in Figure 8-22.

8-3-3   The Rectangular or Pulse Wave

The **rectangular wave** is similar to the square wave in many respects, in that it is a periodic wave that alternately switches between one of two fixed values. The difference in the rectangular wave is that it does not remain at the two peak values for equal lengths of time, as shown in the examples in Figure 8-23(a) and (b).

FIGURE 8-22   Pulse Width of a Square Wave.

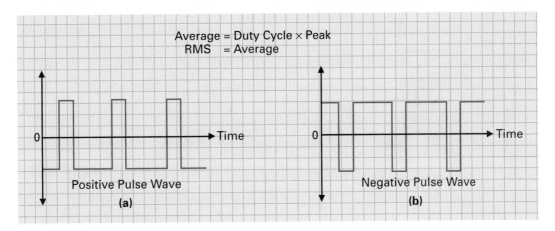

FIGURE 8-23   Rectangular or Pulse Wave.

In Figure 8-23(a), the rectangular wave remains at its negative level for a longer period than its positive, while the rectangular wave in Figure 8-23(b) stays at its positive value for the longer period of time and is only momentarily at its negative value.

PRF, PRT, and Pulse Length

When discussing a rectangular wave, a few terms change. Instead of stating the cycles per second as frequency, it is called **pulse repetition frequency** (PRF), which is far more descriptive. The reciprocal of frequency is time, and with rectangular pulse waveforms the reciprocal of the PRF is **pulse repetition time** (PRT). With rectangular or pulse waves, therefore, frequency is equivalent to PRF, time to PRT, and the only difference is the name.

Let us look at the example in Figure 8-24 of a 5 V rectangular wave at a frequency of 1 kHz and a pulse width of 1 μs, and practice with these new terms. With a pulse repetition frequency of 1 kHz, the time between the leading edges of pulses (PRT) will be 1/1 kHz = 1 ms. **Pulse width** (P_w), **pulse duration** (P_d), or **pulse length** (P_l) are all terms that describe the length of time for which the pulse lasts, and in this example it is equal to 1 μs, which means that 999 μs exists between the end of one pulse and the beginning of the next.

Pulse Repetition Frequency

The number of times per second that a pulse is transmitted.

Pulse Repetition Time

The time interval between the start of two consecutive pulses.

Pulse Width, Pulse Length, or Pulse Duration

The time interval between the leading edge and trailing edge of a pulse at which the amplitude reaches 50% of the peak pulse amplitude.

FIGURE 8-24   PRF and PRT of a Pulse Wave.

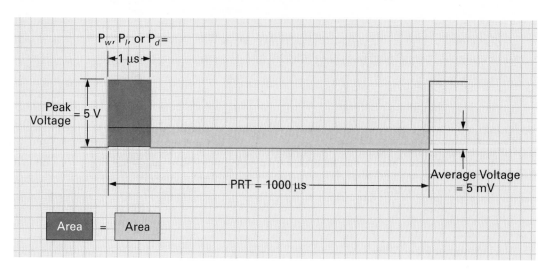

FIGURE 8-25 **Average of a Pulse Wave.**

Duty Cycle

The duty cycle is calculated in exactly the same way as for the square wave and is a ratio of the pulse width to the overall time (PRT). In our example in Figure 8-24 the duty cycle will be equal to

$$\text{duty cycle } (\%) = \frac{\text{pulse width } (P_w)}{\text{PRT}} \times 100\%$$

$$= \frac{1 \ \mu s}{1000 \ \mu s} \times 100\%$$

$$= \text{duty cycle figure of } 0.001 \times 100\%$$

$$= 0.001 \times 100\%$$

$$= 0.1\%$$

The result tells us that the positive pulse lasts for 0.1% of the total time (PRT).

Average

The average or mean value of this waveform is calculated by using the same square-wave formula. The average of the pulse wave in Figure 8-24 will be

$$V \text{ or } I \text{ average} = \text{baseline} + (\text{duty cycle} \times \text{peak to peak})$$
$$V_{\text{avg}} = 0 \text{ V} + (0.001 \times 5 \text{ V})$$
$$= 0 \text{ V} + (5 \text{ mV})$$
$$= 5 \text{ mV}$$

Figure 8-25 illustrates the average value of this rectangular waveform. If the voltage and width of the positive pulse are taken and spread out over the entire PRT, they will have a mean level equal, in this example, to 5 mV.

■ **EXAMPLE:**

Calculate the duty cycle and average voltage of the following radar pulse waveform:

$$\text{peak voltage, } V_p = 20 \text{ kV}$$
$$\text{pulse length, } P_l = 1 \ \mu s$$
$$\text{baseline voltage} = 0 \text{ V}$$
$$\text{PRF} = 3300 \text{ pulses per second (pps)}$$

Solution:

$$\text{duty cycle} = \frac{\text{pulse length } (P_l)}{\text{PRT}} \times 100\%$$

$$= \frac{1 \ \mu s}{303 \ \mu s} \times 100\% \left(\text{PRT} = \frac{1}{\text{PRF}} = \frac{1}{3300} = 303 \ \mu s \right)$$

$$= (3.3 \times 10^{-3}) \times 100\%$$

$$= 0.33\%$$

$$V_{avg} = \text{baseline} + (\text{duty cycle} \times V_{p\text{-}p})$$

$$= 0 \ V + [(3.3 \times 10^{-3}) \times 20 \times 10^3]$$

$$= 66 \ V$$

8-3-4   *The Triangular Wave*

A **triangular wave** consists of a positive and negative ramp of equal values, as shown in Figure 8-26. Both the positive and negative ramps have a linear increase and decrease, respectively. Linear, by definition, is the relationship between two quantities that exists when a change in a second quantity is directly proportional to a change in the first quantity. The two quantities in this case are voltage or current and time. As shown in Figure 8-26, if the increment of change of voltage ΔV (pronounced "delta vee") is changing at the same rate as the increment of time, Δt ("delta tee"), the ramp is said to be **linear.**

8-3-5   *The Sawtooth Wave*

On an oscilloscope display (time-domain presentation), the **sawtooth wave** is very similar to a triangular wave, in that a sawtooth wave has a linear ramp. However, unlike the triangular wave, which reverses and has an equal but opposite ramp back to its starting level, the sawtooth "flies" back to its starting point immediately and then repeats the previous ramp, as seen in Figure 8-27, which shows both a positive and a negative ramp sawtooth.

Triangular Wave
A repeating wave that has equal positive and negative ramps that have linear rates of change with time.

Linear
Relationship between input and output in which the output varies in direct proportion to the input.

Sawtooth Wave
Repeating waveform that rises from zero to a maximum value linearly and then falls to zero and repeats.

FIGURE 8-26   Triangular Wave.

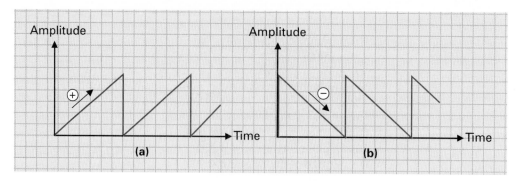

FIGURE 8-27   Sawtooth Wave. (a) Positive Ramp. (b) Negative Ramp.

Use the following questions to test your understanding of Section 8-3.

1. Sketch the following waveforms:

   a. Sine wave

   b. Square wave

   c. Rectangular wave

   d. Triangular wave

   e. Sawtooth wave

2. What are the wavelength formulas for sound waves and electromagnetic waves, and why are they different?

8-4   MEASURING AND GENERATING AC SIGNALS

As a technician or engineer, you are going to be required to diagnose and repair a problem in the shortest amount of time possible. To aid in the efficiency of this fault finding and repair process, you can make use of certain pieces of test equipment. Humans have five kinds of sensory systems: touch, taste, sight, sound, and smell. Four can be used for electronic troubleshooting: sight, sound, touch, and smell.

Electronic test equipment can be used either to *sense* a circuit's condition or to *generate* a signal to see the response of the component or circuit to that signal.

8-4-1   *The AC Meter*

Up to this point, we have seen how a multimeter, like the one shown in Figure 8-28, can be used to measure direct current and voltage. Most multimeters can be used to measure either dc or ac. When the technician or engineer wishes to measure ac, the ac current or voltage is converted to dc internally by a circuit known as a rectifier, as shown in Figure 8-29, before passing on to the meter.

The dc produced by the rectification process is in fact pulsating, as shown in Figure 8-30, so the current through the meter is a series of pulses rising from zero to maximum (peak) and from maximum back to zero. Frequencies below 10 Hz (lower frequency limit) will cause the digits on a digital multimeter to continually increase in value and then decrease in value as the meter follows the pulsating dc. This makes it difficult to read the

Ohms(Ω), Amps (A), Volts (V) Indication

Display Window

Liquid Crystal Display (LCD)

LSI CMOS Analog/ Digital Processor

High Impact Case

Audible Beeper Signals Continuity with 3500–Hz Tone.

Voltage Ranges

Rotary Switch Provides Function Selection.

Metal Oxide Varistor (MOV) Protects Against High Transient Voltages.

Resistance Ranges

Current Ranges Protected by Fuse

Nine-volt Battery Clip

Silicon Seal Makes Meter Water, Dirt, and Grime Resistant

FIGURE 8-28   The Digital Multimeter.

meter. This effect will also occur with an analog multimeter which, from 10 Hz to approximately 2 kHz, will not be able to follow the fluctuation, causing the needle to remain in a position equal to the average value of the pulsating dc from the rectifier (0.637 of peak). Most meters are normally calibrated internally to indicate rms values (0.707 of peak) rather than average values, because this effective value is most commonly used when expressing ac voltage or current. The upper frequency limit of the ac meter is approximately 2 to 8 kHz. Beyond this limit the meter becomes progressively inaccurate due to the reactance of the capacitance in the rectifier. This reactance, which will be discussed later, will result in inaccurate indications due to the change in opposition at different ac input frequencies.

Rectifier Converts AC to DC

AC Input

FIGURE 8-29   AC Meter Uses a Rectifier to Produce DC.

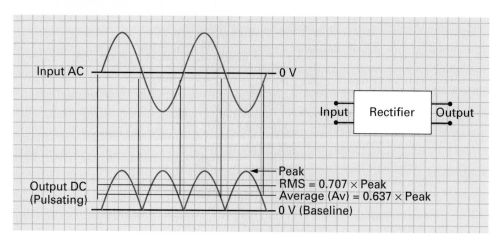

FIGURE 8-30   Rectifier.

Use the following questions to test your understanding of Section 8-4-1.

1. Can a multimeter be used to measure an AC voltage?
2. What is a rectifier?

8-4-2   *The Oscilloscope*

Oscilloscope

Instrument used to view signal amplitude, frequency, and shape at different points throughout a circuit.

Figure 8-31 illustrates a typical **oscilloscope** (sometimes abbreviated to *scope*), which is used primarily to display the shape and spacing of electrical signals. The oscilloscope displays the actual sine, square, rectangular, triangular, or sawtooth wave shape that is occurring at any point in a circuit. This display is made on a cathode-ray tube (CRT), which is also used in television sets and computers to display video information. From the display on the CRT, we can measure or calculate time, frequency, and amplitude characteristics such as rms, average, peak, and peak-to-peak.

(a)

(b)

FIGURE 8-31   Typical Oscilloscope. (a) Oscilloscope. (b) Oscilloscope Probe.

Controls

Oscilloscopes come with a wide variety of features and functions, but the basic operational features are almost identical. Figure 8-32 illustrates the front panel of a typical oscilloscope. Some of these controls are difficult to understand without practice and experience, so practical experimentation is essential if you hope to gain a clear understanding of how to operate an oscilloscope.

General controls (see Figure 8-32)
Intensity control: Controls the brightness of the trace, which is the pattern produced on the screen of a CRT.
Focus control: Used to focus the trace.
Power OFF/ON: Switch will turn on oscilloscope while indicator shows when oscilloscope is turned on.

Some oscilloscopes have the ability to display more than one pattern or trace on the CRT screen, as seen by the examples in Figure 8-33. A dual-trace oscilloscope can produce two traces or patterns on the CRT screen at the same time, whereas a single-trace oscilloscope can trace out only one pattern on the screen. The dual-trace oscilloscope is very useful, as it allows us to make comparisons between the phase, amplitude, shape, and timing of two signals from two separate test points. One signal or waveform is applied to the channel *A* input of the oscilloscope, while the other waveform is applied to the channel *B* input.

Channel selection (see Figure 8-32)
Mode switch: This switch allows us to select which channel input should be displayed on the CRT screen.
CHA: The input arriving at channel *A*'s jack is displayed on the screen as a single trace.
CHB: The input arriving at channel *B*'s jack is displayed on the screen as a single trace.
Dual: The inputs arriving at jacks *A* and *B* are both displayed on the screen, as a dual trace.

FIGURE 8-32   Oscilloscope Controls.

(a)

(b)

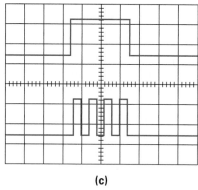

(c)

FIGURE 8-33   Sample of Dual-Trace Oscilloscope Displays for Comparison.

(a) **Calibration** Output   This output connection provides a point where a fixed 1 V peak-to-peak square-wave signal can be obtained at a frequency of 1 kHz. This signal is normally fed into either channel A or B's input to test probes and the oscilloscope operation.

(b) Channel *A* and *B* Horizontal Controls

↔ *Position control:* This control will move the position of the one (single-trace) or two (dual-trace) waveforms horizontally (left or right) on the CRT screen.

Sweep time/cm switch: The oscilloscope contains circuits that produce a beam of light that is swept continually from the left to the right of the CRT screen. When no input signal is applied, this sweep will produce a straight horizontal line in the center of the screen. When an input signal is present, this horizontal sweep is influenced by the input signal, which moves it up and down to produce a pattern on the CRT screen the same as the input pattern (sine, square, sawtooth, and so on). This sweep time/cm switch selects the speed of the sweep from left to right, and it can be either fast (0.2 microseconds per centimeter; 0.2 μs/cm) or slow (0.5 second per centimeter; 0.5 s/cm). A low-frequency input signal (long cycle time or period) will require a long time setting (0.5 s/cm) so that the sweep can capture and display one or more cycles of the input. A number of settings are available, with lower time settings displaying fewer cycles and higher time settings showing more cycles of an input.

(c) **Triggering** Controls   These provide the internal timing control between the sweep across the screen and the input waveform.

Triggering level control: This determines the point where the sweep starts.

Slope switch (+): Sweep is triggered on positive-going slope.

(−): Sweep is triggered on negative-going slope.

Source switch, CHA: The input arriving into channel *A* jack triggers the sweep.

CHB: The input arriving into channel *B* jack triggers the sweep.

EXT: The signal arriving at the external trigger jack is used to trigger the sweep.

(d) Channel *A* and *B* Vertical Controls   The *A* and *B* channel controls are identical.

Volts/cm switch: This switch sets the number of volts to be displayed by each major division on the vertical scale of the screen.

⇕*Position control:* Moves the trace up or down for easy measurement or viewing.

AC-DC-GND *switch:* In the AC position, a capacitor on the input will pass the ac component entering the input jack, but block any dc components.
In the GND position, the input is grounded (0 V) so that the operator can establish a reference.
In the DC position, both ac and dc components are allowed to pass on to and be displayed on the screen.

Measurements

The oscilloscope is probably the most versatile of test equipment, as it can be used to test:

DC voltage
AC voltage
Waveform duration
Waveform frequency
Waveform shape

(a) Voltage Measurement   The screen is divided into eight vertical and ten horizontal divisions, as shown in Figure 8-34. This 8 × 10 cm grid is called the *graticule.* Every vertical division has a value depending on the setting of the volts/cm control. For example, if the volts/cm control is set to 5 V/div or 5 V/cm, the waveform shown in Figure 8-35(a), which rises up four major divisions, will have a peak positive alternation value of 20 V (4 div × 5 V/div = 20 V).

As another example, look at the positive alternation in Figure 8-35(b). The positive alternation rises up three major divisions and then extends another three subdivisions, which are each equal to 1 V because five subdivisions exist within one major division, and one major division is, in this example, equal to 5 V. The positive alternation shown in Figure 8-35(b) therefore has a peak of three major divisions (3 × 5 V/cm = 15 V), plus three subdivisions (3 × 1 V = 3 V), which equals 18 volts peak.

FIGURE 8-34   **Oscilloscope Grid.**

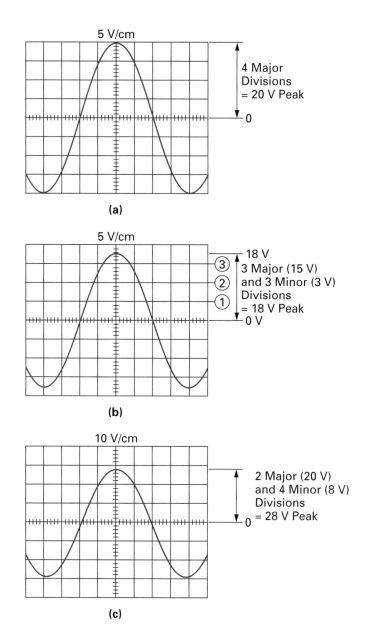

FIGURE 8-35 Measuring AC Peak Voltage.

In Figure 8-35(c), we have selected the 10 volt/cm position, which means that each major division is equal to 10 V and each subdivision is equal to 2 V. In this example, the waveform peak will be equal to two major divisions ($2 \times 10 = 20$ V), plus four subdivisions (4×2 V $= 8$ V), which is equal to 28 V. Once the peak value of a sine wave is known, the peak to peak, average, and rms can be calculated mathematically.

When measuring a dc voltage with the oscilloscope, the volts/cm is applied in the same way, as shown in Figure 8-36. A positive dc voltage in this situation will cause deflection toward the top of the screen, whereas a negative voltage will cause deflection toward the bottom of the screen.

To determine the dc voltage, count the number of major divisions and to this add the number of minor divisions. In Figure 8-36, a major division equals 1 V/cm and, therefore, a minor division equals 0.2 V/cm, so the dc voltage being measured is interpreted as $+2.6$ V.

(b) Time and Frequency Measurement   The frequency of an alternating wave, such as that seen in Figure 8-37(a), is inversely proportional to the amount of time it takes to

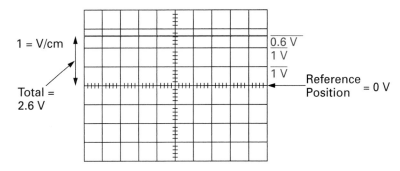

FIGURE 8-36 Measuring DC Voltage.

complete one cycle ($f = 1/t$). Consequently, if time can be measured, frequency can be determined.

The time/cm control relates to the horizontal line on the oscilloscope graticule and is used to determine the period of a cycle so that frequency can be calculated. For example, in Figure 8-37(b), a cycle lasts five major horizontal divisions, and since the 20 μs/division setting has been selected, the period of the cycle will equal 5 × 20 μs/division = 100 μs. If the period is equal to 100 μs, the frequency of the waveform will be equal to $f = 1/t = 1/100$ μs = 10 kHz.

□ EXAMPLE:

A complete sine-wave cycle occupies four horizontal divisions and four vertical divisions from peak to peak. If the oscilloscope is set on the 20 ms/cm and 500 mV/cm, calculate:

  a. V_{p-p}
  b. t
  c. f
  d. V_p
  e. V_{avg} of peak
  f. V_{rms}

■ *Solution:*

4 horizontal divisions × 20 ms/div = 80 ms
4 vertical divisions × 500 mV/div = 2 V

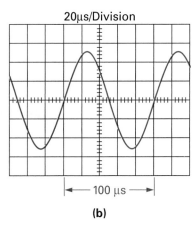

FIGURE 8-37 Time and Frequency Measurement.

a. $V_{p\text{-}p} = 2\ \text{V}$

b. $t = 80\ \text{ms}$

c. $f = \dfrac{1}{t} = 12.5\ \text{Hz}$

d. $V_p = 0.5 \times V_{p\text{-}p} = 1\ \text{V}$

e. $V_{avg} = 0.637 \times V_p = 0.637\ \text{V}$

f. $V_{rms} = 0.707 \times V_p = 0.707\ \text{V}$

SELF-TEST EVALUATION POINT FOR SECTION 8-4-2

Use the following questions to test your understanding of Section 8-4-2.

1. Name the device within the oscilloscope on which the waveforms can be seen.

2. On the 20 µs/div time setting, a full cycle occupies four major divisions. What is the waveform's frequency and period?

3. On the 2 V/cm voltage setting, the waveform swings up and down two major divisions (a total of 4 cm of vertical swing). Calculate the waveform's peak and peak-to-peak voltage.

4. What is the advantage(s) of the dual-trace oscilloscope?

8-4-3   *The Function Generator and Frequency Counter*

Function Generator

Signal generator that can function as a sine, square, rectangular, triangular, or sawtooth waveform generator.

Frequency Counter

Meter used to measure the frequency or cycles per second of a periodic wave.

In many cases you will wish to generate a waveform of a certain shape and frequency, and then apply this waveform to your newly constructed circuit. One of the most versatile waveform generators is the **function generator,** so called because it can function as a sine wave, square wave, rectangular wave, triangular wave, or sawtooth wave generator. Figure 8-38 shows a photograph of a typical function generator.

It is very important that anyone involved in the design, manufacture, and servicing of electronic equipment be able to accurately measure the frequency of a periodic wave. Without the ability to measure frequency accurately, there could be no communications, home entertainment, or a great number of other systems. A **frequency counter,** like the one shown in Figure 8-39, can be used to analyze the frequency of any periodic wave applied to its input jack and provide a readout on the display of its frequency.

FIGURE 8-38   Function Generator.

The overrange indicator flashes when input frequency is beyond the selected range of the counter.

Input Jack

Power ON/OFF Switch

FIGURE 8-39   Frequency Counter.

REVIEW QUESTIONS

Multiple-Choice Questions

1. A current that rises from zero to maximum positive, decreases to zero, and then reverses to reach a maximum in the opposite direction (negative) is known as:
   a. Alternating current
   b. Pulsating direct current
   c. Steady direct current
   d. All of the above

2. The advantage(s) of ac over dc from a power distribution point of view is/are:
   a. Generators can supply more power than batteries
   b. Ac can be transformed to a high or low voltage easily, minimizing power loss
   c. Ac can easily be converted into dc
   d. All of the above
   e. Only (a) and (c)

3. The most common type of alternating wave shape is the:
   a. Square wave   c. Rectangular wave
   b. Sine wave   d. Triangular wave

4. The peak-to-peak value of a sine wave is equal to:
   a. Twice the rms value
   b. 0.707 times the rms value
   c. Twice the peak value
   d. 1.14 × the average value

5. The rms value of a sine wave is also known as the:
   a. Effective value   c. Peak value
   b. Average value   d. All of the above

6. The peak value of a 115 V (rms) sine wave is:
   a. 115 V   c. 162.7 V
   b. 230 V   d. Two of the above could be true

7. The mathematical average value of a sine wave cycle is:
   a. $0.637 \times$ peak   c. $1.414 \times$ rms
   b. $0.707 \times$ peak   d. Zero

8. The frequency of a sine wave is equal to the reciprocal of _____.
   a. The period   d. Both (a) and (b)
   b. One cycle time   e. None of the above
   c. One alternation

9. What is the period of a 1 MHz sine wave?
   a. 1 ms
   b. One millionth of a second
   c. 10 ms
   d. 100 μs

10. What is the frequency of a sine wave that has a cycle time of 1 ms?

    a. 1 MHz   c. 200 m
    b. 1 kHz   d. 10 kHz

11. The pulse width (P_w) is the time between the _____ points on the positive and negative edges of a pulse.

    a. 10%   c. 50%
    b. 90%   d. All of the above

12. The duty cycle is the ratio of _____ to period.

    a. Peak   c. Pulse length
    b. Average power d. Both (a) and (c)

13. With a pulse waveform, PRF can be calculated by taking the reciprocal of:

    a. The duty cycle   c. P_d
    b. PRT   d. P_l

14. The sound wave exists in _____ and travels at approximately _____.

    a. Space, 1130 ft/s   c. Air, 3×10^6 m/s
    b. Wires, 186,282.397 miles/s   d. None of the above

15. The electrical and electromagnetic waves travel at a speed of:

    a. 186,000 miles/s   c. 162,000 nautical miles/s
    b. 3×10^8 meters/s   d. All of the above

Practice Problems

16. Calculate the periods of the following sine-wave frequencies:

    a. 27 kHZ   d. 365 Hz
    b. 3.4 MHz e. 60 Hz
    c. 25 Hz   f. 200 kHz

17. Calculate the frequency for each of the following values of time:

    a. 16 ms   d. 0.05 s
    b. 1 s   e. 200 µs
    c. 15 µs   f. 350 ms

18. A 22 V peak sine wave will have the following values:

    a. Rms voltage =   c. Peak-to-peak voltage =
    b. Average voltage =

19. A 40 mA rms sine wave will have the following values:

    a. Peak current =   c. Average current =
    b. Peak-to-peak current =

20. What is the duty cycle of a 10 V peak square wave at a frequency of 1 kHz?

21. If one cycle of a sine wave occupies 4 cm on the oscilloscope horizontal grid and 5 cm from peak to peak on the vertical grid, calculate frequency, period, rms, average, and peak for the following control settings:

    a. 0.5 V/cm, 20 µs/cm   c. 50 mV/cm, 0.2 µs/cm
    b. 10 V/cm, 10 ms/cm

22. Assuming the same graticule and switch settings of the oscilloscope in Figure 12-9, what would be the lowest setting of the volts/cm and time/division switches to fully view a 6 V rms, 350 kHz sine wave?

23. If the volts/cm switch is positioned to 10 V/cm and the waveform extends 3.5 divisions from peak to peak, what is the peak-to-peak value of this wave?

24. If a square wave occupies 5.5 horizontal cm on the 1 µs/cm position, what is its frequency?

25. Which settings would you use on an autoset scopemeter to fully view an 8 V rms, 20 kHz triangular wave?

Capacitance and Capacitors

BACK TO THE FUTURE

Born in England in 1791, Charles Babbage became very well known for both his mathematical genius and eccentric personality. Babbage's ultimate pursuit was that of mathematical accuracy. He delighted in spotting errors in everything from log tables (used by astronomers, mathematicians, and navigators) to poetry. In fact, he once wrote to poet Alfred Lord Tennyson, pointing out an inaccuracy in his line "Every moment dies a man—every moment one is born." Babbage explained to Tennyson that since the world population was actually increasing and not, as he indicated, remaining constant, the line should be rewritten to read "Every moment dies a man—every moment one and one-sixteenth is born."

In 1822, Babbage described in a paper and built a model of what he called "a difference engine," which could be used to calculate mathematical tables. The Royal Society of Scientists described his machine as "highly deserving of public encouragement," and a year later the government awarded Babbage £1500 for his project. Babbage originally estimated that the project should take 3 years; however, the design had its complications, and after 10 years of frustrating labor, in which the government grants increased to £17,000, Babbage was still no closer to completion. Finally, the money stopped and Babbage reluctantly decided to let his brainchild go.

In 1833, Babbage developed an idea for a much more practical machine, which he named "the analytical engine." It was to be a more general machine that could be used to solve a variety of problems, depending on instructions supplied by the operator. It would include two units, called a "mill" and a "store," both of which would be made of cogs and wheels. The store, which was equivalent to a modern-day computer memory, could hold up to 100 forty-digit numbers. The mill, which was equivalent to a modern computer's arithmetic and logic unit (ALU), could perform both arithmetic and logic operations on variables or numbers retrieved from the store, and the result could be stored in the store and then acted upon again or printed out. The program of instructions directing these operations would be fed into the analytical engine in the form of punched cards.

The analytical engine was never built. All that remains are the volumes of descriptions and drawings, and a section of the mill and printer built by Babbage's son, who also had to concede defeat. It was, unfortunately for Charles Babbage, a lifetime of frustration to have conceived the basic building blocks of the modern computer a century before the technology existed to build it.

Introduction

Up to this point we have concentrated on circuits containing only resistance, which opposes the flow of current and then converts or dissipates power in the form of heat. Capacitance and inductance are two circuit properties that act differently from resistance in that they will charge or store the supplied energy and then return almost all the stored energy back to the circuit, rather than lose it in wasted heat. Inductance will be discussed in a following chapter.

Capacitance is the ability of a circuit or device to store electrical charge. A device or component specifically designed to have this capacity or capacitance is called a *capacitor.* A capacitor stores an electrical charge similar to a bucket holding water. Using the analogy throughout this chapter that a capacitor is similar to a bucket will help you gain a clear understanding of the capacitor's operation. For example, the capacitor holds charge in the same way that a bucket holds water. A larger capacitor will hold more charge and will take longer to charge, just as a larger bucket will hold more water and take longer to fill. A larger circuit resistance means a smaller circuit current, and therefore a longer capacitor charge time. Similarly, a smaller hose will have a greater water resistance producing a smaller water flow, and therefore the bucket will take a longer time to fill. Capacitors store electrons, and basically, the amount of electrons stored is a measure of the capacitor's capacitance.

Capacitor

Device that stores electric energy in the form of an electric field that exists within a dielectric (insulator) between two conducting plates each of which is connected to a lead. This device was originally called a condenser.

9-1   CAPACITOR CONSTRUCTION

Several years ago, capacitors were referred to as condensers, but that term is very rarely used today. Figure 9-1 illustrates the main parts and schematic symbol of the **capacitor.** Two leads are connected to two parallel metal conductive plates, which are separated by an insulating material known as a *dielectric.* It is called a dielectric because it exists between two (*di*) plates and when the capacitor is charged and electric field exists within it (*di-electric*). The conductive plates are normally made of metal foil, while the dielectric can be paper, air, glass, ceramic, mica, or some other form of insulator.

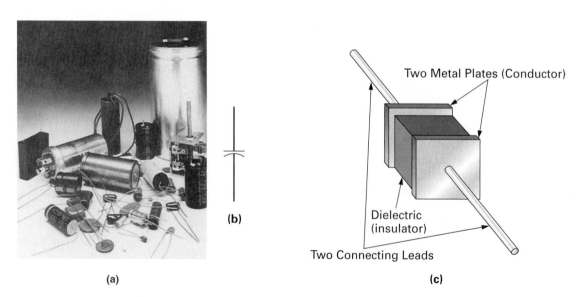

(a)

(b)

(c)

FIGURE 9-1   **Capacitor. (a) Physical Appearance. (Courtesy of Sencore) (b) Schematic Symbol. (c) Basic Construction.**

Use the following questions to test your understanding of Section 9-1.

1. What are the principal parts of a capacitor?
2. What is the name given to the insulating material between the two conductive plates?

9-2 CHARGING AND DISCHARGING A CAPACITOR

Like a secondary battery, a capacitor can be made to charge or discharge, and when discharging it will return all of the energy it consumed during charge.

9-2-1 Charging a Capacitor

Capacitance is the ability of a capacitor to store an electrical charge. Figure 9-2 illustrates how the capacitor stores an electric charge. This capacitor is shown as two plates with air acting as the dielectric.

In Figure 9-2(a), the switch is open, so no circuit current results. An equal number of electrons exist on both plates, so the voltmeter (VM) indicates zero, which means no potential difference exists across the capacitor.

In Figure 9-2(b), the switch is now closed and electrons travel to the positive side of the battery, away from the right-hand plate of the capacitor. This creates a positive right-hand capacitor plate, which results in an attraction of free electrons from the negative side of the battery to the left-hand plate of the capacitor. In fact, for every electron that leaves the right-hand capacitor plate and is attracted into the positive battery terminal, another electron leaves the negative side of the battery and travels to the left-hand plate of the capacitor. Current appears to be flowing from the negative side of the battery, around to the positive, through the capacitor. This is not really the case, because no electrons can flow through the insulator or gap between the plates, and although there appears to be one current flowing throughout the circuit, there are in fact two separate currents—one from the battery to the capacitor and the other from the capacitor to the battery.

A voltmeter across the capacitor will indicate an increase in the potential difference between the plates, and the capacitor is said to be charging toward, in this example, 5 V. This potential difference builds up across the two plates until the voltage across the capacitor is equal to the voltage of the battery. In this example, when the capacitor reaches a charge of 5 V, the capacitor will be equivalent to a 5 V battery, as seen in Figure 9-2(c). Once charged, there will be no potential difference between the battery and capacitor, and so circuit current flow will be zero when the capacitor is fully charged.

9-2-2 Discharging a Capacitor

If the capacitor is now disconnected from the circuit by opening switch 1, as shown in Figure 9-3(a), it will remain in its charged condition. If switch 2 is now closed, a path exists across the charged capacitor, as shown in Figure 9-3(b), and the excess of electrons on the left plate will flow through the conducting wire to the positive plate on the right side. The capacitor is now said to be *discharging*. When an equal number of electrons exists on both sides, the capacitor is said to be *discharged* since both plates have an equal charge and the potential difference across the capacitor will be zero, as shown in Figure 9-3(c).

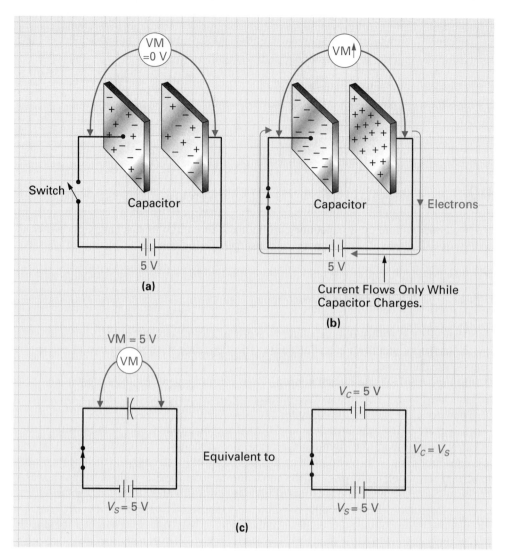

FIGURE 9-2 **Charging a Capacitor. (a) Uncharged Capacitor. (b) Charging Capacitor. (c) Charged Capacitor.**

Referring to figures 9-2 and 9-3 you can see that the capacitor's charge and discharge currents are in opposite directions. In both cases the current flow is always from one plate to the other and never exists through the dielectric insulator.

Use the following questions to test your understanding of Section 9-2.

1. True or false: When both plates of a capacitor have an equal charge, the capacitor is said to be charged.
2. If 2 μA of current flows into one plate of a capacitor and 2 μA flows out of the other plate, how much current is flowing through the dielectric of this working capacitor?

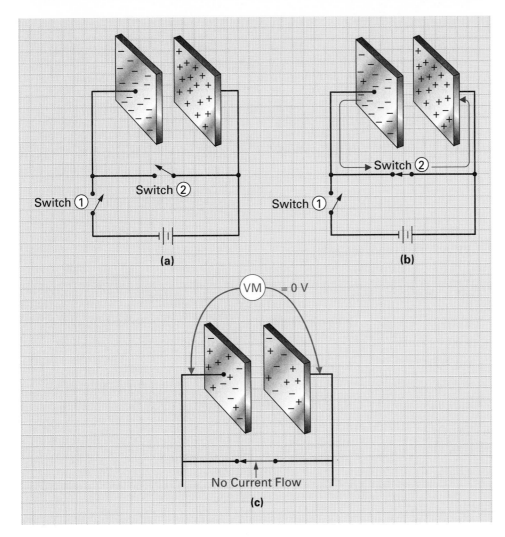

FIGURE 9-3 **Discharging a Capacitor. (a) Charged Capacitor. (b) Discharging Capacitor. (c) Discharged Capacitor.**

9-3   THE UNIT OF CAPACITANCE

Capacitance is the ability of a capacitor to store an electrical charge, and the unit of capacitance is the **farad** (F), named in honor of Michael Faraday's work in 1831 in the field of capacitance. A capacitor with the capacity of 1 farad (1 F) can store 1 coulomb of electrical charge (6.24×10^{18} electrons) if 1 volt is applied across the capacitor's plates, as seen in Figure 9-4.

A 1 F capacitor is a very large value and not frequently found in electronic equipment. Most values of capacitance found in electronic equipment are in the units between the microfarad ($\mu F = 10^{-6}$) and picofarad ($pF = 10^{-12}$). A microfarad is 1 millionth of a farad (10^{-6}). So if a 1 F capacitor can store 6.24×10^{18} electrons with 1 V applied, a 1 μF capacitor, which has 1 millionth the capacity of a 1 F capacitor, can store only 1 millionth of a coulomb, or $(6.24 \times 10^{18}) \times (1 \times 10^{-6}) = 6.24 \times 10^{12}$ electrons when 1 V is applied, as shown in Figure 9-5.

Farad
Unit of capacitance.

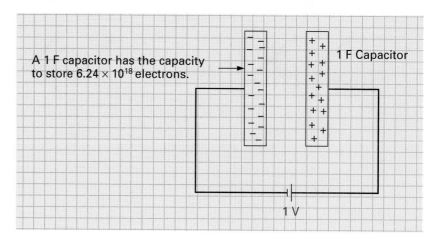

FIGURE 9-4   One Farad of Capacitance.

☐   **EXAMPLE:**

Convert the following to either microfarads or picofarads (whichever is more appropriate):

    a. 0.00002 F

    b. 0.00000076 F

    c. 0.00047×10^{-7} F

■   *Solution:*

    a. 20 μF

    b. 0.76 μF

    c. 47 pF

Since there is a direct relationship between capacitance, charge, and voltage, there must be a way of expressing this relationship in a formula.

$$\text{capacitance, } C \text{ (farads)} = \frac{\text{charge, } Q \text{ (coulombs)}}{\text{voltage, } V \text{ (volts)}}$$

where C = capacitance, in farads
Q = charge, in coulombs
V = voltage, in volts

FIGURE 9-5   One-Millionth of a Farad.

By transposition of the formula, we arrive at the following combinations for the same formula:

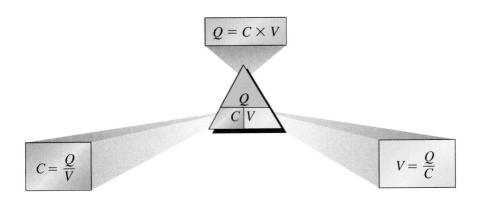

☐ **EXAMPLE:**

If a capacitor has the capacity to hold 36 C ($36 \times 6.24 \times 10^{18} = 2.25 \times 10^{20}$ electrons) when 12 V is applied across its plates, what is the capacitance of the capacitor?

■ *Solution:*

$$C = \frac{Q}{V}$$
$$= \frac{36 \text{ C}}{12 \text{ V}}$$
$$= 3 \text{ F}$$

☐ **EXAMPLE:**

How many electrons could a 3 µF capacitor store when 5 V is applied across it?

■ *Solution:*

$$Q = C \times V$$
$$= 3 \text{ µF} \times 5 \text{ V}$$
$$= 15 \text{ µC}$$

(15 microcoulombs is 15 millionths of a coulomb.) Since 1 C = 6.24×10^{18} electrons, 15 µC = $(15 \times 10^{-6}) \times 6.24 \times 10^{18} = 9.36 \times 10^{13}$ electrons.

☐ **EXAMPLE:**

If a capacitor of 2 F has stored 42 C of charge (2.63×10^{20} electrons), what is the voltage across the capacitor?

■ *Solution:*

$$V = \frac{Q}{C}$$
$$= \frac{42 \text{ C}}{2 \text{ F}}$$
$$= 21 \text{ V}$$

Use the following questions to test your understanding of Section 9-3.

1. What is the unit of capacitance?
2. State the formula for capacitance in relation to charge and voltage.
3. Convert 30,000 μF to farads.
4. If a capacitor holds 17.5 C of charge when 9 V is applied, what is the capacitance of the capacitor?

9-4   FACTORS DETERMINING CAPACITANCE

The capacitance of a capacitor is determined by three factors:

1. The plate area of the capacitor
2. The distance between the plates
3. The type of dielectric used

The formula that combines these three factors is

$$C = \frac{(8.85 \times 10^{-12}) \times K \times A}{d}$$

where C = capacitance, in farads (F)

8.85×10^{-12} is a constant

K = dielectric constant (see table 9-1)

A = plate area, in square meters (m^2)

d = distance between the plates, in meters (m)

This formula states that the capacitance of a capacitor is directly proportional to the dielectric constant (K) and the plates' area (A) and is inversely proportional to the dielectric thickness or distance between the plates (d).

TABLE 9-1   Dielectric Constants

MATERIAL	DIELECTRIC CONSTANT (K)[a]
Vacuum	1.0
Air	1.0006
Teflon	2.0
Wax	2.25
Paper	2.5
Amber	2.65
Rubber	3.0
Oil	4.0
Mica	5.0
Ceramic (low)	6.0
Bakelite	7.0
Glass	7.5
Water	78.0
Ceramic (high)	8000.

[a]The different material compositions can cause different values of K.

EXAMPLE:

What is the capacitance of a ceramic capacitor with a 0.3 m^2 plate area and a dielectric thickness of 0.0003 m?

Solution:

$$C = \frac{(8.85 \times 10^{-12}) \times K \times A}{d}$$

$$= \frac{(8.85 \times 10^{-12}) \times 6 \times 0.3 \text{ m}^2}{0.0003 \text{ m}}$$

$$= 5.31 \times 10^{-8}$$

$$= 0.0531 \text{ } \mu\text{F}$$

CALCULATOR SEQUENCE

Step	Keypad Entry	Display Response
1.	[8] [.] [8] [5] [E] [1] [2] [+/−]	8.85E-12
2.	[×]	
3.	[6]	6
4.	[×]	5.31E-11
5.	[0] [.] [3]	
6.	[÷]	1.59E-11
7.	[0] [.] [0] [0] [0] [3]	
8.	[=]	5.31E-8

SELF-TEST EVALUATION POINT FOR SECTION 9-4

Use the following questions to test your understanding of Section 9-4.

1. List the three variable factors that determine the capacitance of a capacitor.
2. State the formula for capacitance.
3. If a capacitor's plate area is doubled, the capacitance will _____.
4. If a capacitor's dielectric thickness is halved, the capacitance will _____.

9-5 DIELECTRIC BREAKDOWN AND LEAKAGE

Capacitors store a charge just as a container or tank stores water. The amount of charge stored by a capacitor is proportional to the capacitor's capacitance and the voltage applied across the capacitor ($Q = C \times V$). The charge stored by a fixed-value capacitor (C is fixed) therefore can be increased by increasing the voltage across the plates, as shown in Figure 9-6(a). If the voltage across the capacitor is increased further, the charge held by the capacitor will increase until the dielectric between the two plates of the capacitor breaks down and a spark jumps or arcs between the plates.

Using the water analogy shown in Figure 9-6(b), if the pressure of the water being pumped in is increased, the amount of water stored in the tank will also increase until a time is reached when the tank's seams at the bottom of the tank cannot contain the large amount of pressure and break down under strain. The amount of water stored in the tank is proportional to the pressure applied, just as the amount of charge stored in a capacitor is proportional to the amount of voltage applied.

The **breakdown voltage** of a capacitor is determined by the strength of the dielectric used. Table 9-2 illustrates some of the different strengths of many of the common dielectrics. As an example, let's consider a capacitor that uses 1 mm of air as a dielectric between its two plates. This particular capacitor can withstand any voltage up to 787 V. If the voltage is increased further, the dielectric will break down and current will flow between the plates, destroying the capacitor (air capacitors, however, can recover from ionization).

The ideal or perfect insulator should have a resistance equal to infinite ohms. Insulators or the dielectric used to isolate the two plates of a capacitor are not perfect and therefore

Breakdown Voltage

The voltage at which breakdown occurs in a dielectric or insulation.

(a)

(b)

FIGURE 9-6   Breakdown. (a) Voltage Increase, Charge Increases until Dielectric Breakdown. (b) Pressure Increase, Water Increases until Seams Break Down.

have some very high values of resistance. This means that some value of resistance exists between the two plates, as shown in Figure 9-7; although this value of resistance is very large, it will still allow a small amount of current to flow between the two plates (in most applications a few nanoamperes or picoamperes). This small current, referred to as **leakage current,** causes any charge in a capacitor to slowly, over a long period of time, discharge between the two plates. A capacitor should have a large leakage resistance to ensure the smallest possible leakage current.

Leakage Current

Small, undesirable flow of current through an insulator or dielectric.

SELF-TEST EVALUATION POINT FOR SECTION 9-5

Use the following questions to test your understanding of Section 9-5.

1. Which dielectric material has the best breakdown voltage figure?

2. True or false: Current flows through the dielectric at and below the breakdown voltage of the material.

TABLE 9-2   Dielectric Strengths

MATERIAL	DIELECTRIC STRENGTH (V/MM)
Air	787
Oil	12,764
Ceramic	39,370
Paper	49,213
Teflon	59,055
Mica	59,055
Glass	78,740

FIGURE 9-7   Capacitor Leakage.

FIGURE 9-8   **Capacitors in Parallel.**

(a)

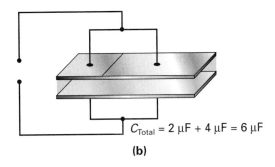

$C_{Total} = 2\ \mu F + 4\ \mu F = 6\ \mu F$

(b)

9-6   CAPACITORS IN COMBINATION

Like resistors, capacitors can be connected in either series or parallel. As you will see in this section, the rules for determining total capacitance for parallel- and series-connected capacitors are opposite to series- and parallel-connected resistors.

9-6-1   *Capacitors in Parallel*

In Figure 9-8(a), you can see a 2 μF and 4 μF capacitor connected in parallel with one another. As the top plate of capacitor *A* is connected to the top plate of capacitor *B* with a wire, and a similar situation occurs with the bottom plates, you can see that this is the same as if the top and bottom plates were touching one another, as shown in Figure 9-8(b). When drawn so that the respective plates are touching, the dielectric constant and plate separation is the same as shown in Figure 9-8(a), but now we can easily see that the plate area is actually increased. Consequently, if capacitors are connected in parallel, the effective plate area is increased; and since capacitance is proportional to plate area [$C\uparrow = (8.85 \times 10^{-12})$ $\times K \times A\uparrow/d$], the capacitance will also increase. Total capacitance is actually calculated by adding the plate areas, so total capacitance is equal to the sum of all the individual capacitances in parallel.

$$C_T = C_1 + C_2 + C_3 + C_4 + \cdots$$

☐   EXAMPLE:

Determine the total capacitance of the circuit in Figure 9-9(a). What will be the voltage drop across each capacitor?

■   *Solution:*

$$C_T = C_1 + C_2 + C_3$$
$$= 1\ \mu F + 0.5\ \mu F + 0.75\ \mu F$$
$$= 2.25\ \mu F$$

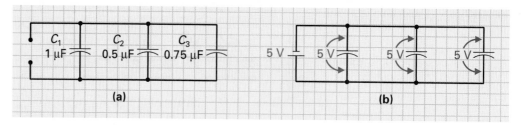

(a) (b)

FIGURE 9-9   Example of Parallel-Connected Capacitors.

As with any parallel-connected circuit, the source voltage appears across all the components. If, for example, 5 V is connected to the circuit of Figure 9-9(b), all the capacitors will charge to the same voltage of 5 V because the same voltage always exists across each section of a parallel circuit.

9-6-2   *Capacitors in Series*

In Figure 9-10(a), we have taken the two capacitors of 2 μF and 4 μF and connected them in series. Since the bottom plate of the *A* capacitor is connected to the top plate of the *B* capacitor, they can be redrawn so that they are touching, as shown in Figure 9-10(b).

The top plate of the *A* capacitor is connected to a wire into the circuit, and the bottom plate of *B* is connected to a wire into the circuit. This connection creates two center

FIGURE 9-10   Capacitors in Series.

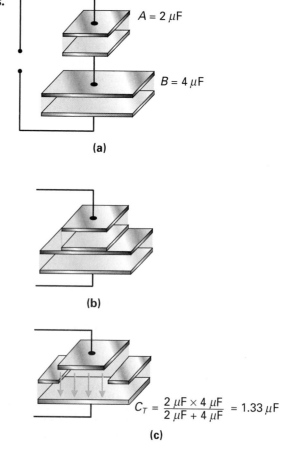

$A = 2 \, \mu F$

$B = 4 \, \mu F$

(a)

(b)

$$C_T = \frac{2 \, \mu F \times 4 \, \mu F}{2 \, \mu F + 4 \, \mu F} = 1.33 \, \mu F$$

(c)

plates that are isolated from the circuit and can therefore be disregarded, as shown in Figure 9-10(c). The first thing you will notice in this illustration is that the dielectric thickness ($d\uparrow$) has increased, causing a greater separation between the plates. The effective plate area of this capacitor has decreased, as it is just the area of the top plate only. Even though the bottom plate extends out further, the electric field can only exist between the two plates, so the surplus metal of the bottom plate has no metal plate opposite for the electric field to exist.

Consequently, when capacitors are connected in series the effective plate area is decreased ($A\downarrow$) and the dielectric thickness increased ($d\uparrow$), and both of these effects result in an overall capacitance decrease ($C\downarrow\downarrow = (8.85 \times 10^{-12}) \times K \times A\downarrow/d\uparrow$).

The plate area is actually decreased to the smallest individual capacitance connected in series, which in this example is the plate area of A. If the plate area were the only factor, then capacitance would always equal the smallest capacitor value. However, the dielectric thickness is always equal to the sum of all the capacitor dielectrics, and this factor always causes the total capacitance (C_T) to be less than the smallest individual capacitance when capacitors are connected in series.

The total capacitance of two or more capacitors in series therefore is calculated by using the following formulas: For two capacitors in series,

$$C_T = \frac{C_1 \times C_2}{C_1 + C_2}$$

(product-over-sum formula)

For more than two capacitors in series,

$$C_T = \frac{1}{(1/C_1) + (1/C_2) + (1/C_3) + \cdots}$$

(reciprocal formula)

EXAMPLE:

Determine the total capacitance of the circuit in Figure 9-11.

Solution:

$$C_T = \frac{1}{(1/C_1) + (1/C_2) + (1/C_3)}$$

$$= \frac{1}{(1/4 \ \mu F) + (1/2 \ \mu F) + (1/1 \ \mu F)}$$

$$= \frac{1}{1.75 \times 10^6} = 5.7143 \times 10^{-7}$$

$$= 0.5714 \ \mu F \quad \text{or} \quad 0.6 \ \mu F$$

FIGURE 9-11 **Example of Series-Connected Capacitors.**

The total capacitance for capacitors in series is calculated in the same way as total resistance when resistors are in parallel.

As with series-connected resistors, the sum of all of the voltage drops across the series-connected capacitors will equal the voltage applied (Kirchhoff's voltage law). With capacitors connected in series, the charged capacitors act as a voltage divider, and therefore the voltage-divider formula can be applied to capacitors in series.

$$V_{cx} = \frac{C_T}{C_x} \times V_T$$

where V_{cx} = voltage across desired capacitor
C_T = total capacitance
C_x = desired capacitor's value
V_T = total supplied voltage

EXAMPLE:

Using the voltage-divider formula, calculate the voltage dropped across each of the capacitors in Figure 9-11 if $V_T = 24$ V.

Solution:

$$V_{C1} = \frac{C_T}{C_1} \times V_T = \frac{0.5714 \ \mu F}{4 \ \mu F} \times 24 \ V = 3.4 \ V$$

$$V_{C2} = \frac{C_T}{C_2} \times V_T = \frac{0.5714 \ \mu F}{2 \ \mu F} \times 24 \ V = 6.9 \ V$$

$$V_{C3} = \frac{C_T}{C_3} \times V_T = \frac{0.5714 \ \mu F}{1 \ \mu F} \times 24 \ V = 13.7 \ V$$

$$V_T = V_{C1} + V_{C2} + V_{C3} = 3.4 + 6.9 + 13.7 = 24 \ V$$

(Kirchhoff's voltage law)

If the capacitor values are the same, as seen in Figure 9-12(a), the voltage is divided equally across each capacitor, as each capacitor has an equal amount of charge and therefore has half of the applied voltage (in this example, 3 V across each capacitor).

When the capacitor values are different, the smaller value of capacitor will actually charge to a higher voltage than the larger capacitor. In the example in Figure 9-12(b), the smaller capacitor is actually half the size of the other capacitor, and it has charged to twice

FIGURE 9-12   Voltage Drops across Series-Connected Capacitors.

the voltage. Since Kirchhoff's voltage law has to apply to this and every series circuit, you can easily calculate that the voltage across C_1 will equal 4 V and is twice that of C_2, which is 2 V. To understand this fully, we must first understand that although the capacitance is different, both capacitors have an equal value of coulomb charge held within them, which in this example is 8 μC.

$$Q_1 = C_1 \times V_1$$
$$= 2 \ \mu F \times 4 \ V = 8 \ \mu C$$
$$Q_2 = C_2 \times V_2$$
$$= 4 \ \mu F \times 2 \ V = 8 \ \mu C$$

This equal charge occurs because the same amount of current flow exists throughout a series circuit, so both capacitors are being supplied with the same number or quantity of electrons. The charge held by C_1 is large with respect to its small capacitance, whereas the same charge held by C_2 is small with respect to its larger capacitance.

If the charge remains the same (Q is constant) and the capacitance is small, the voltage drop across the capacitor will be large, because the charge is large with respect to the capacitance:

$$V{\uparrow} = \frac{Q}{C{\downarrow}}$$

On the other hand, for a constant charge, a large capacitance will have a small charge voltage because the charge is small with respect to the capacitance:

$$V{\downarrow} = \frac{Q}{C{\uparrow}}$$

We can apply the water analogy once more and imagine two series-connected buckets, one of which is twice the size of the other. Both are being supplied by the same series pipe, which has an equal flow of water throughout, and are consequently each holding an equal amount of water, for example, 1 gallon. The 1 gallon of water in the small bucket is large with respect to the size of the bucket, and a large amount of pressure exists within that bucket. The 1 gallon of water in the large bucket is small with respect to the size of the bucket, so a small amount of pressure exists within this bucket. The pressure within a bucket is similar to the voltage across a capacitor, and therefore a small bucket or capacitor will have a greater pressure or voltage associated with it, while a large bucket or capacitor will develop a small pressure or voltage.

To summarize capacitors in series, all the series-connected components will have the same charging current throughout the circuit, and because of this, two or more capacitors in series will always have equal amounts of coulomb charge. If the charge (Q) is equal, the voltage across the capacitor is determined by the value of the capacitor. A small capacitance will charge to a larger voltage ($V{\uparrow} = Q/C{\downarrow}$), whereas a large value of capacitance will charge to a smaller voltage ($V{\downarrow} = Q/C{\uparrow}$).

SELF-TEST EVALUATION POINT FOR SECTION 9-6

Use the following questions to test your understanding of Section 9-6.

1. If 2 μF, 3 μF, and 5 μF capacitors are connected in series, what will be the total circuit capacitance?
2. If 7 pF, 2 pF, and 14 pF capacitors are connected in parallel, what will be the total circuit capacitance?
3. State the voltage-divider formula as it applies to capacitance.
4. True or false: With resistors, the large value of resistor will drop a larger voltage, whereas with capacitors the smaller value of capacitor will actually charge to a higher voltage.

Capacitors come in a variety of shapes and sizes and can be either fixed or variable in their values of capacitance. Within these groups, capacitors are generally classified by the dielectric used between the plates.

A **fixed-value capacitor** is a capacitor whose capacitance value remains constant and cannot be altered. Fixed capacitors normally come in a disk or a tubular package and consist of metal foil plates separated by one of the following types of insulators (dielectric), which is the means by which we classify them, as shown in Figure 9-13.

Fixed-Value Capacitor
A capacitor whose value is fixed and cannot be varied.

9-7-1   *Variable-Value Capacitors*

Variable-value capacitors are the second basic type, and within this group, four types are commercially available. Like the fixed-value type of capacitors, they are classified by dielectric. The variable-value type uses a hand-rotated shaft to vary the effective plate area, and so capacitance. The adjustable type uses screw-in or screw-out mechanical adjustment to vary the distance between the plates, and so capacitance.

Variable-Value Capacitor
A capacitor whose value can be varied.

Air (Variable)

Figure 9-14(a) illustrates the construction of a typical air dielectric variable capacitor. With this type of capacitor, the effective plate area is adjusted to vary capacitance by causing a set of rotating plates (rotor) to mesh with a set of stationary plates (stator). When the rotor plates are fully out, the capacitance is minimum, and when the plates are fully in, the capacitance is maximum, because the maximum amount of rotor plate area is now opposite the stator plate, creating the maximum value of capacitance.

The plates are usually made of aluminum to prevent corrosion and the dielectric between the plates is air. The capacitor has to be carefully manufactured to ensure that the rotor plates do not touch the stator plates when the shaft is rotated and the plates interweave. Figure 9-14(b) illustrates a typical variable air capacitor.

In radio equipment, it is sometimes necessary to have two or more variable-value capacitors that have been constructed in such a way that a common shaft (rotor) runs through all the capacitors and varies their capacitance simultaneously, as shown in Figure 9-14(c). If you mechanically couple (gang) two or more variable capacitors so that they can all be operated from a single control, the component is known as a **ganged capacitor,** and the symbol for this arrangement is also shown in Figure 9-14(c).

Ganged Capacitor
Mechanical coupling of two or more capacitors, switches, potentiometers, or any other components so that the activation of one control will operate all sections.

Mica, Ceramic, and Plastic (Adjustable)

Figure 9-15 illustrates some of the typical packages for mica, ceramic, or plastic film types of adjustable capacitors, which are also referred to as *trimmers*. The adjustable capacitor generally has one stationary plate and one spring metal moving plate. The screw forces the spring metal plate closer or farther away from the stationary plate, varying the distance between the plates and so changing capacitance. The two plates are insulated from one another by either mica, ceramic, or plastic film, and the advantages of each are the same as for fixed-value capacitors.

These types of capacitors should only be adjusted with a plastic or nonmetallic alignment tool, because a metal screwdriver may affect the capacitance of the capacitor when nearby, making it very difficult to adjust for a specific value of capacitance.

Name	Construction	Approximate Range of Values and Tolerances	Characteristics	
Mica		1 pF–0.1 μF ±1% to ±5%	Lower voltage rating than other capacitors of the same size	Small Capacitor Values
Ceramic	Ceramic Dielectric / Lead Wire	*Low Dielectric K:* 1 pF–0.01 μF ±0.5% to ±10% *High Dielectric K:* 1 pF–0.1 μF ±10% to ±80%	Most popular small value capacitor due to lower cost than mica, and its ruggedness	
Paper	Lead to Inner Foil Sheet / Inner Foil / Lead to Outer Foil Sheet / Outer Foil	1 pF–1 μF ± 10%	Has a large plate area and therefore large capacitance for a small size	Large Capacitor Values
Plastic	Outside Foil / Inside Foil	1 pF–10 μF ± 5% to ±10%	Has almost completely replaced paper capacitors; has large capacitance values for small size and high voltage ratings	
Electrolytic (Aluminum and Tantalum)	(Tantalum has tantalum rather than aluminum foil plates) Aluminum Foil / Aluminum Oxide / Aluminum Foil / Gauze (Saturated with Electrolyte)	1 μF–1F ± 10% to ±50%	Most popular large value capacitor: large capacitance into small area, wide range of values. Disadvantages are: cannot be used in AC circuits as they are polarized; poor tolerances; low leakage resistance and so high leakage current. *Tantalum* advantages over aluminum include smaller size, longer life than aluminum, which has an approximate lifespan of 12 years. Disadvantages: 4 to 5 times the price.	

FIGURE 9-13   **Summary of Fixed-Value Capacitors.**

FIGURE 9-14   Variable Air-Type Capacitors. (a) Construction. (b) Physical Appearance. (Courtesy of Sencore.) (c) Ganged.

FIGURE 9-15   Adjustable Mica, Ceramic, or Plastic Capacitors. (a) Construction. (b) Physical Appearance. (Courtesy of Sencore.)

SELF-TEST EVALUATION POINT FOR SECTION 9-7

Use the following questions to test your understanding of Section 9-7.

1. List the five types of fixed-value capacitors.
2. Which fixed-value capacitor is the most popular in applications where:
   a. Large values are required?   b. Small values are required?
3. List the four basic types of variable capacitors.

9-8 CODING OF CAPACITANCE VALUES

The capacitor value, tolerance, and voltage rating need to be shown in some way on the capacitor's exterior. Presently, two methods are used: (1) alphanumeric labels and (2) color coding.

9-8-1 *Alphanumeric Labels*

Manufacturers today most commonly use letters of the alphabet and numbers (alphanumerics) printed on either the disk or tubular body to indicate the capacitor's specifications, as illustrated in Figure 9-16.

The tubular type is the easier of the two since the information is basically uncoded. The value of capacitance and unit, typically the microfarad (μF or MF), tolerance figure (preceded by ± or followed by %), and voltage rating (followed by a V for voltage) are printed on all sizes of tubular cases. The remaining letters or numbers are merely manufacturers' codes for case size, series, and the like.

With disk capacitors (dipped or molded), certain rules have to be applied when decoding the notations. Many capacitors of this type do not define the unit of capacitance; in this situation, try to locate a decimal point. If a decimal point exists, for example 0.01 or 0.001, the value is in microfarads (10^{-6}). If no decimal point exists, for example 50 or 220, the value is in picofarads (10^{-12}) and you must analyze the number in a little more detail.

☐ **EXAMPLE:**

What is the value of a capacitor if it is labeled 50, 50 V, ±5?

■ *Solution:*

Since no decimal point is present, the unit is in picofarads:

$$50 \text{ pF}, \quad 50 \text{ V}, \quad \pm5\%$$

If no decimal point is present and three digits exist and the last digit is a zero, the value is as stands and in picofarads. If the third digit is a number other than 0 (1 to 9), it is a multiplier and describes the number of zeros to be added to the picofarad value.

☐ **EXAMPLE:**

If two capacitors are labeled with the following coded values, how should they be interpreted?

  a. 220

  b. 104

Since no decimal point is present, they are both in picofarads.

a. If the last of the three digits is a zero, the value is as it stands.

$$\underbrace{2\ 2\ 0}_{\text{three-digit value}} = 220\ \text{pF}$$

b. If the third digit is a number from 1 to 9, it is a multiplier.

$$\underbrace{1\ 0}_{}\ \underbrace{4}_{}$$

two-digit value ——┘ └—— multiplier

So

$$10 \times 10^4 = 100{,}000\ \text{pF}$$

or

$$100{,}000 \times 10^{-12} = 0.1 \times 10^{-6} = 0.1\ \mu\text{F}$$

The tolerance of the capacitor is sometimes clearly indicated, for example, ± 5 or 10%; in other cases, a letter designation is used, such as

$$F = \pm 1\%$$
$$G = \pm 2\%$$
$$J = \pm 5\%$$
$$K = \pm 10\%$$
$$M = \pm 20\%$$
$$Z = -20\%, +80\%$$

Unfortunately, there does not seem to be a standard among capacitor manufacturers, which can cause confusion when trying to determine the value of capacitance. Therefore, if you are not completely sure, you should always measure the value or consult technical data or information sheets from the manufacturer.

SELF-TEST EVALUATION POINT FOR SECTION 9-8

Use the following questions to test your understanding of Section 9-8.

What are the following values of capacitance?
1. 470 ± 2
2. 0.47 ± 5

9-9   CAPACITIVE TIME CONSTANT

When a capacitor is connected across a dc voltage source, it will charge to a value equal to the voltage applied. If the charged capacitor is then connected across a load, the capacitor will then discharge through the load. The time it takes a capacitor to charge or discharge can be calculated if the circuit's resistance and capacitance are known. Let us now see how we can calculate a capacitor's charge time and discharge time.

9-9-1 DC Charging

When a capacitor is connected across a dc voltage source, such as a battery or power supply, current will flow and the capacitor will charge up to a value equal to the dc source voltage, as shown in Figure 9-17. When the charge switch is first closed, as seen in Figure 9-17(a), there is no voltage across the capacitor at that instant and therefore a potential difference exists between the battery and capacitor. This causes current to flow and begin charging the capacitor.

Once the capacitor begins to charge, the voltage across the capacitor does not instantaneously rise to 100 V. It takes a certain amount of time before the capacitor voltage is equal to the battery voltage. When the capacitor is fully charged no potential difference exists between the voltage source and the capacitor. Consequently, no more current flows in the circuit as the capacitor has reached its full charge, as seen in Figure 9-17(b). The amount of time it takes for a capacitor to charge to the supplied voltage (in this example, 100 V) is dependent on the circuit's resistance and capacitance value. If the circuit's resistance is increased, the opposition to current flow will be increased, and it will take the capacitor a longer period of time to obtain the same amount of charge because the circuit current available to charge the capacitor is less.

If the value of capacitance is increased, it again takes a longer time to charge to 100 V because a greater amount of charge is required to build up the voltage across the capacitor to 100 V.

The circuit's resistance (R) and capacitance (C) are the two factors that determine the charge time (τ). Mathematically, this can be stated as

$$\tau = R \times C$$

where τ = **time constant** (s)
R = resistance (Ω)
C = capacitance (F)

Time Constant

Time needed for either a voltage or current to rise to 63.2% of the maximum or fall to 36.8% of the initial value. The time constant of an RC circuit is equal to the product of R and C.

In this example, we are using a resistance of 1 MΩ and a capacitance of 2 μF, which means that the time constant is equal to

$$\tau = R \times C$$
$$= 2\ \mu\text{F} \times 1\ \text{M}\Omega$$
$$= (2 \times 10^{-6}) \times (1 \times 10^{6})$$
$$= 2\ \text{s}$$

Two seconds is the time, so what is the constant? The constant value that should be remembered throughout this discussion is "**63.2**."

Figure 9-18 illustrates the rise in voltage across the capacitor from 0 to a maximum of 100 V in five time constants (5 $\times$ 2 s = 10 s). So where does 63.2 come into all this?

FIGURE 9-17   Capacitor Charging. (a) Switch Is Closed and Capacitor Begins to Charge. (b) Capacitor Charged.

FIGURE 9-18   Charging Capacitor.

First time constant: In 1*RC* seconds (1 × *R* × *C* = 2 s), the capacitor will charge to 63.2% of the applied voltage (63.2% × 100 V = 63.2 V).

Second time constant: In 2*RC* seconds (2 × *R* × *C* = 4 s), the capacitor will charge to 86.5% of the applied voltage (86.5% × 100 V = 86.5 V).

Third time constant: In 3*RC* seconds (6 s), the capacitor will charge to 95% of the applied voltage (95% × 100 V = 95 V).

Fourth time constant: In 4*RC* seconds (8 s), the capacitor will charge to 98.2% of the applied voltage (98.2% × 100 V = 98.2 V).

Fifth time constant: In 5*RC* seconds (10 s), the capacitor is considered to be fully charged since the capacitor will have reached 99.3% of the applied voltage (99.3% × 100 V = 99.3V, approximately 100 V).

The voltage waveform produced by the capacitor acquiring a charge is known as an *exponential* waveform, and the voltage across the capacitor is said to rise exponentially. An exponential rise is also referred to as a *natural increase*. There are many factors that exponentially rise and fall. For example, we grow exponentially, in that there is quite a dramatic change in our height in the early years and then this increase levels off and reaches a maximum.

Before the switch is closed and even at the instant the switch is closed, the capacitor is not charged, which means that there is no capacitor voltage to oppose the supply voltage and, therefore, a maximum current of *V/R*, 100 V/1 MΩ = 100 μA flows. This current begins to charge the capacitor, and a potential difference begins to build up across the plates of the capacitor, and this voltage opposes the supply voltage, causing a decrease in charging current. As the capacitor begins to charge, less of a potential difference exists between the supply voltage and capacitor voltage and so the current begins to decrease.

To calculate the current at any time, we can use the formula

$$i = \frac{V_S - V_C}{R}$$

where i = instantaneous current
V_S = source voltage
V_C = capacitor voltage
R = resistance

For example, the current flowing in the circuit after one time constant will equal the source voltage, 100 V, minus the capacitor's voltage, which in one time constant will be 63.2% of the source voltage or 63.2 V, divided by the resistance.

$$i = \frac{V_S - V_C}{R}$$
$$= \frac{100\ V - 63.2\ V}{1\ M\Omega}$$
$$= 36.8\ \mu A$$

As the charging continues, the potential difference across the plates exponentially rises to equal the supply voltage, as seen in Figure 9-19(a), while the current exponentially falls to zero, as shown in Figure 9-19(b). The constant of 63.2 can be applied to the exponential fall of current from 100 μA to 0 μA in $5RC$ seconds.

When the switch was closed to start the charging of the capacitor, there was no charge on the capacitor; therefore, a maximum potential difference existed between the battery and capacitor, causing a maximum current flow of 100 μA ($I = V/R$).

First time constant: In $1RC$ seconds, the current will have exponentially decreased 63.2% (63.2% of 100 μA = 63.2 μA) to a value of 36.8 μA (100 μA − 63.2 μA). In the example of 2 μF and 1 MΩ, this occurs in 2 s.

Second time constant: In $2RC$ seconds ($2 \times R \times C = 4$ s), the current will decrease to 13.5 μA or 13.5%.

FIGURE 9-19   **Exponential Rise in Voltage and Fall in Current in a Charging Capacitive Circuit.**

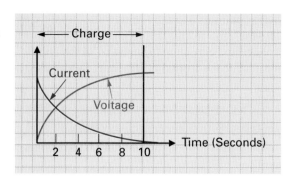

Third time constant: In 3*RC* seconds (6 s), the capacitor's charge current will decrease to 5 μA or 5%.

Fourth time constant: In 4*RC* seconds (8 s), the current will have decreased to 1.8 μA or 1.8%.

Fifth time constant: In 5*RC* seconds (10 s), the charge current is now 0.7 μA or 0.7%. At this time, the charge current is assumed to be zero and the capacitor is now charged to a voltage equal to the applied voltage.

Studying the exponential rise of the voltage and the exponential decay of current in a capacitive circuit, you will notice an interesting relationship. In a pure resistive circuit, the current flow through a resistor will be in step with the voltage across that same resistor, in that an increased current will cause a corresponding increase in voltage drop across the resistor. Voltage and current are consequently said to be *in step* or *in phase* with one another. With the capacitive circuit, the current flow in the circuit and voltage across the capacitor are

FIGURE 9-21   **Discharging Capacitor. (a) Voltage Waveform. (b) Current Waveform.**

CHAPTER 9 / CAPACITANCE AND CAPACITORS

not in step or in phase with one another. When the switch is closed to charge the capacitor, the current is maximum (100 μA), while the voltage across the capacitor is zero. After five time constants (10 s), the capacitor's voltage is now maximum (100 V) and the circuit current is zero, as seen in Figure 9-20. The circuit current flow is out of phase with the capacitor voltage, and this difference is referred to as a *phase shift*. In any circuit containing capacitance, current will lead voltage.

9-9-2   *DC Discharging*

Figure 9-21 illustrates the circuit, voltage, and current waveforms that occur when a charged capacitor is discharged from 100 V to 0 V. The 2 μF capacitor, which was charged to 100 V in 10 s (5RC), is discharged from 100 to 0 V in the same amount of time.

Looking at the voltage curve, you can see that the voltage across the capacitor decreases exponentially, dropping 63.2% to 36.8 V in 1RC seconds, another 63.2% to 13.5 V in 2RC seconds, another 63.2% to 5 V in 3RC seconds, and so on, until zero.

The current flow within the circuit is dependent on the voltage in the circuit, which is across the 2 μF capacitor. As the voltage decreases, the current will also decrease by the same amount ($I\downarrow = V\downarrow/R$).

$$\text{discharge switch closed: } I = \frac{V}{R} = \frac{100 \text{ V}}{1 \text{ M}\Omega} = 100 \text{ μA} \quad \text{maximum}$$

$$1RC \text{ (2) seconds: } I = \frac{V}{R} = \frac{36.8 \text{ V}}{1 \text{ M}\Omega} = 36.8 \text{ μA}$$

$$2RC \text{ (4) seconds: } I = \frac{V}{R} = \frac{13.5 \text{ V}}{1 \text{ M}\Omega} = 13.5 \text{ μA}$$

$$3RC \text{ (6) seconds: } I = \frac{V}{R} = \frac{5 \text{ V}}{1 \text{ M}\Omega} = 5.0 \text{ μA}$$

$$4RC \text{ (8) seconds: } I = \frac{V}{R} = \frac{1.8 \text{ V}}{1 \text{ M}\Omega} = 1.8 \text{ μA}$$

$$5RC \text{ (10) seconds: } I = \frac{V}{R} = \frac{0.7 \text{ V}}{1 \text{ M}\Omega} = 0.7 \text{ μA} \quad \text{zero}$$

SELF-TEST EVALUATION POINT FOR SECTION 9-9

Use the following questions to test your understanding of Section 9-9.

1. In one time constant, a capacitor will have charged to what percentage of the applied voltage?
2. In one time constant, a capacitor will have discharged to what percentage of its full charge?

9-10   AC CHARGE AND DISCHARGE

Let us begin by returning to our charged capacitor that was connected across a 100 V dc source, as shown in Figure 9-22(a). When a capacitor is connected across a dc source, it charges in five time constants and then the voltage across the capacitor will oppose the source voltage, causing current to stop. Since no current flows in this circuit, the capacitor is effectively acting as an open circuit when a constant dc voltage is applied.

If the 100 V source is reversed, as shown in Figure 9-22(b), the 100 V charge on the capacitor is now aiding the path of the battery instead of pushing or reacting against it. The

(a) Charge   Capacitor Is Charged
Circuit Current Is Zero   **(b)** Discharge   **(c)** Charge

(d) Discharge   **(e)** Charge   **(f)**

FIGURE 9-22   AC Charge and Discharge.

discharge current will flow in the opposite direction to the charge current, until the capacitor voltage is equal to 0 V, at which time it will begin to charge in the reverse direction, as shown in Figure 9-22(c).

If the battery source is once more reversed, as shown in Figure 9-22(d), the charge current will discharge the capacitor to 0 V and then begin once more to charge it in the reverse direction, as shown in Figure 9-22(e).

The result of this switching or alternating procedure is that there is current flow in the circuit at all times (except for the instant when the battery source is removed and the polarity reversed). The switching of a dc voltage source in this way has almost the same effect as if we were applying an alternating voltage, as shown in Figure 9-22(f).

Figure 9-23 summarizes the capacitor's reaction to a dc source and an ac source. When a dc voltage is applied across a capacitive circuit, as shown in Figure 9-23(a), the capacitor

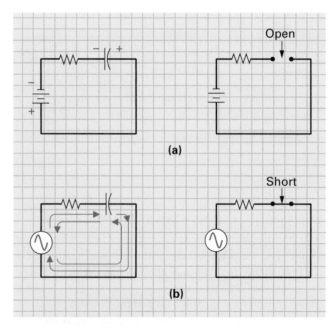

FIGURE 9-23   Capacitors' Reaction to DC and AC. (a) Capacitor Will Block DC. (b) Capacitor Will Pass AC.

CHAPTER 9 / CAPACITANCE AND CAPACITORS

will charge and then oppose the dc source, preventing current flow. Since no current can flow when the capacitor is charged, the capacitor acts like an open circuit, and it is this action that accounts for why capacitors are referred to as an "open to dc."

When an ac voltage is applied across a capacitive circuit, as shown in Figure 9-23(b), the continual reversal of the applied voltage will cause a continual charging and discharging of the capacitor, and therefore circuit current will always be present. Any type of ac source or in fact any fluctuating or changing dc source will cause a circuit current, and if current is present, the opposition is low. It is this action that accounts for why capacitors are referred to as a "short to ac."

This ability of a capacitor to block dc and seem to pass ac will be exploited and explained later in applications of capacitors.

SELF-TEST EVALUATION POINT FOR SECTION 9-10

Use the following questions to test your understanding of Section 9-10.

1. What differences occur when ac is applied to a capacitor rather than dc?
2. True or false: A capacitor's reaction to ac and dc accounts for why it is known as an ac short and dc block.

9-11   PHASE RELATIONSHIP BETWEEN CAPACITOR CURRENT AND VOLTAGE

Thinking of the applied alternating voltage as a dc source that is being continually reversed sometimes makes it easier to understand how a capacitor reacts to ac. Whether the applied voltage is dc or ac, the rule holds true in that a 90° phase shift or difference exists between circuit current and capacitor voltage. The exact relationship between ac current and voltage in a capacitive circuit is illustrated in Figure 9-24.

This phase shift that exists between the circuit current and the capacitor voltage is normally expressed in degrees. At 0°, the capacitor is fully discharged (0 V), and the source is supplying a maximum circuit charge current. From 0° to 90°, the capacitor will charge toward a maximum positive value, and this increase in capacitor voltage will oppose the source voltage, whose circuit charging current will slowly decrease to 0 A. At 90°, the capacitor is fully charged, and the circuit current is at 0 A, so the capacitor will return to the

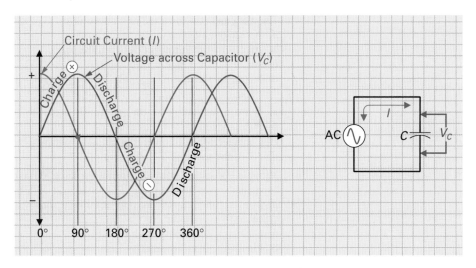

FIGURE 9-24   AC Current Leads Voltage in a Capacitive Circuit.

circuit the energy it consumed during the positive (+) charge cycle. This discharge circuit current is in the opposite direction to the positive charge current. At 180°, the capacitor is fully discharged (0 V), and the source is supplying a maximum circuit charge current. From 180° to 270°, the capacitor will charge toward a maximum negative value, and this negative increase in capacitor voltage will oppose the source voltage, whose circuit charging current will slowly decrease to 0 A. At 270°, the capacitor is fully charged and the circuit current is at 0 A, so the capacitor will discharge and return the energy it consumed during the negative (−) charge cycle. This discharge circuit current is in the opposite direction to the negative charge current.

Throughout this cycle, notice that the voltage across the capacitor follows the circuit current. This current, therefore, leads the voltage by 90°, and this 90° leading phase shift (current leading voltage) will occur only in a capacitive circuit.

SELF-TEST EVALUATION POINT FOR SECTION 9-11

Use the following questions to test your understanding of Section 9-11.

1. True or false: In a resistive circuit, the current leads the voltage by exactly 90°.
2. True or false: In a purely capacitive circuit, a 90° phase difference occurs between current and voltage.

REVIEW QUESTIONS

Multiple-Choice Questions

1. When a capacitor charges:
   a. The voltage across the plates rises exponentially
   b. The circuit current falls exponentially
   c. The capacitor charges to the source voltage in $5RC$ seconds
   d. All of the above

2. What is the capacitance of a capacitor if it can store 24 C of charge when 6 V is applied across the plates?
   a. 2 μF   b. 3 μF   c. 4.7 μF   d. None of the above

3. The capacitance of a capacitor is directly proportional to:
   a. The plate area
   b. The distance between the plates
   c. The constant of the dielectric used
   d. Both (a) and (c)
   e. Both (a) and (b)

4. The capacitance of a capacitor is inversely proportional to:
   a. The plate area
   b. The distance between the plates
   c. The dielectric used
   d. Both (a) and (c)
   e. Both (a) and (b)

5. The breakdown voltage of a capacitor is determined by:
   a. The type of dielectric used
   b. The size of the capacitor plates
   c. The wire gauge of the connecting leads
   d. Both (a) and (b)

6. Total series capacitance of two capacitors is calculated by:
   a. Using the product-over-sum formula
   b. Using the voltage-divider formula
   c. Using the series resistance formula on the capacitors
   d. Adding all the individual values

7. Total parallel capacitance is calculated by:
   a. Using the product-over-sum formula
   b. Using the voltage-divider formula
   c. Using the parallel resistance formula on capacitors
   d. Adding all the individual values

8. The mica and ceramic fixed capacitors have:
   a. An arrangement of stacked plates
   b. An electrolyte substance between the plates
   c. An adjustable range
   d. All of the above

9. An electrolytic capacitor:
   a. Is the most popular large-value capacitor
   b. Is polarized
   c. Can have either aluminum or tantalum plates
   d. All of the above

10. Variable capacitors normally achieve a large variation in capacitance by varying _____, while adjustable trimmer capacitors only achieve a small capacitance range by varying _____ .
    a. Dielectric constant, plate area
    b. Plate area, plate separation
    c. Plate separation, dielectric constant
    d. Plate separation, plate area

11. In one time constant, a capacitor will charge to _____ of the source voltage.
 a. 86.5%   **b.** 63.2%   **c.** 99.3%   **d.** 98.2%

12. When ac is applied across a capacitor, a _____ phase shift exists between circuit current and capacitor voltage.
 a. 45°   **b.** 60°   **c.** 90°   **d.** 63.2°

13. A capacitor consists of:
 a. Two insulated plates separated by a conductor
 b. Two conductive plates separated by a conductor
 c. Two conductive plates separated by an insulator
 d. Two conductive plates separated by a conductive spacer

14. A 47 µF capacitor charged to 6.3 V will have a stored charge of:
 a. 296.1 µC   **d.** All of the above
 b. 2.96×10^{-4} C   **e.** Both (a) and (b)
 c. 0.296 mC

Practice Problems

15. If a 10 µF capacitor is charged to 10 V, how many coulombs of charge has it stored?

16. Calculate the electric field strength within the dielectric of a capacitor that is charged to 6 V and the dielectric thickness is 32 µm (32 millionths of a meter).

17. If a 0.006 µF capacitor has stored 125×10^{-6} C of charge, what potential difference would appear across the plates?

18. Calculate the capacitance of the capacitor that has the following parameter values: $A = 0.008$ m²; $d = 0.00095$ m; the dielectric used is paper.

19. Calculate the total capacitance if the following are connected in:
 a. Parallel: 1.7 µF, 2.6 µF, 0.03 µF, 1200 pF
 b. Series: 1.6 µF, 1.4 µF, 4 µF

20. If three capacitors of 0.025 µF, 0.04 µF, and 0.037 µF are connected in series across a 12 V source, as shown in Figure 9-25, what would be the voltage drop across each?

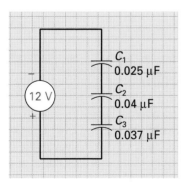

FIGURE 9-25

21. Give the value of the following alphanumeric capacitor value codes:
 a. 104   **b.** 125   **c.** 0.01   **d.** 220

22. Give the value of the following color-coded capacitor values:
 a. Yellow, violet, black   **b.** Orange, yellow, violet

23. What would be the time constant of the following RC circuits?
 a. $R = 6$ kΩ, $C = 14$ µF
 b. $R = 12$ MΩ, $C = 1400$ pF
 c. $R = 170$ Ω, $C = 24$ µF
 d. $R = 140$ kΩ, $C = 0.007$ µF

24. If 10 V were applied across all the RC circuits in Question 23, what would be the voltage across each capacitor after one time constant, and how much time would it take each capacitor to fully charge?

25. In one application, a capacitor is needed to store 25 µC of charge and will always have 125 V applied across its terminals. Calculate the capacitance value needed.

Capacitive Circuits, Testing, and Applications

10

Let's Toss for It!

In 1938, Bill Hewlett and Dave Packard, close friends and engineering graduates at Stanford University, set up shop in the one-car garage behind the Packards' rented home in Palo Alto, California. In the garage, the two worked on what was to be the first product of their lifetime business together, an electronic oscillating instrument that represented a breakthrough in technology and was specifically designed to test sound equipment.

Late in the year, the oscillator (designated the 200A "because the number sounded big") was presented at a West Coast meeting of the Institute of Radio Engineers (now the Institute of Electrical and Electronics Engineers—IEEE) and the orders began to roll in. Along with the first orders for the 200A was a letter from Walt Disney Studios asking the two to build a similar oscillator covering a different frequency range. The Model 200B was born shortly thereafter, and Disney purchased eight to help develop the unique sound system for the classic film *Fantasia*. By 1940, the young company had outgrown the garage and moved into a small rented building nearby.

Over the years, the company continued a steady growth, expanding its product line to more than 10,000, including computer systems and peripheral products, test and measuring instruments, hand-held calculators, medical electronic equipment, and systems for chemical analysis. Employees have increased from 2 to almost 82,000, and the company is one of the 100 largest industrial corporations in America, with a net revenue of $8.1 billion. Whose name should go first was decided by the toss of a coin on January 1, 1939, and the outcome was Hewlett-Packard or HP.

Introduction

In Chapter 9 you discovered that when a capacitor was connected to a dc circuit, it charged up until it was at a value equal to the applied dc voltage, and then from that point on current was zero. When a capacitor was connected to an ac circuit, however, the continual reversal of the applied voltage resulted in a continuous charge and discharge of the capacitor, so circuit current was always present. This ability of the capacitor makes it an extremely useful device in ac circuits.

In this chapter we will first study capacitive reactance, which is the opposition a capacitor offers to current flow. Then we will study the ac circuit characteristics of a series *RC* circuit and parallel *RC* circuit. To complete our study of capacitors, in the final two sections in this chapter we will discuss the testing of capacitors and then some of the more common applications of capacitors.

10-1 CAPACITIVE REACTANCE

Resistance (R), by definition, is the opposition to current flow with the dissipation of energy and is measured in ohms. Capacitors oppose current flow like a resistor, but a resistor dissipates energy, whereas a capacitor stores energy (when it charges) and then gives back its energy into the circuit (when it discharges). Because of this difference, a new term had to be used to describe the opposition offered by a capacitor. **Capacitive reactance (X_C),** by definition, is the opposition to current without the dissipation of energy and is also measured in ohms.

If capacitive reactance is basically opposition, it is inversely proportional to the amount of current flow. If a large current is within a circuit, the opposition or reactance must be low ($I\uparrow$, $X_C\downarrow$). Conversely, a small circuit current will be the result of a large opposition or reactance ($I\downarrow$, $X_C\uparrow$).

When a dc source is connected across a capacitor, current will flow only for a short period of time ($5RC$ seconds) to charge the capacitor. After this time, there is no further current flow. Consequently, the capacitive reactance or opposition offered by a capacitor to dc is infinite (maximum).

Alternating current is continuously reversing in polarity, resulting in the capacitor continuously charging and discharging. This means that charge and discharge currents are always flowing around the circuit, and if we have a certain value of current, we must also have a certain value of reactance or opposition.

Initially, when the capacitor's plates are uncharged, they will not oppose or react against the charging current and therefore maximum current will flow ($I\uparrow$) and the reactance will be very low ($X_C\downarrow$). As the capacitor charges, it will oppose or react against the charge current, which will decrease ($I\downarrow$), so the reactance will increase ($X_C\uparrow$). The discharge current is also highest at the start of discharge ($I\uparrow$, $X_C\downarrow$) as the voltage of the charged capacitor is also high; but as the capacitor discharges, its voltage decreases and the discharge current will also decrease ($I\downarrow$, $X_C\uparrow$).

To summarize, at the start of a capacitor charge or discharge, the current is maximum, so the reactance is low. This value of current then begins to fall to zero, so the reactance increases.

If the applied alternating current is at a high frequency, as shown in Figure 10-1(a), it is switching polarity more rapidly than a lower frequency and there is very little time between the start of charge and discharge. As the charge and discharge currents are largest at the beginning of the charge and discharge of the capacitor, the reactance has very little time to build up and oppose the current, which is why the current is a high value and the capacitive reactance is small at higher frequencies. With lower frequencies, as shown in Figure 10-1(b), the applied alternating current is switching at a slower rate, and therefore the reactance, which is low at the beginning, has more time to build up and oppose the current.

Capacitive reactance is therefore inversely proportional to frequency:

$$\text{capacitive reactance } (X_C) \propto \frac{1}{f \text{ (frequency)}}$$

Frequency, however, is not the only factor that determines capacitive reactance. Capacitive reactance is also inversely proportional to the value of capacitance. If a larger capacitor value is used a longer time is required to charge the capacitor ($t\uparrow = C\uparrow R$), which means that current will be present for a longer period of time, so the overall current will be large ($I\uparrow$); consequently, the reactance must be small ($X_C\downarrow$). On the other hand, a small capacitance value will charge in a small amount of time ($t\downarrow = C\downarrow R$) and the current is present for only a short period of time. The overall current will therefore be small ($I\downarrow$), indicating a large reactance ($X_C\uparrow$).

$$\text{capacitive reactance } (X_C) \propto \frac{1}{C \text{ (capacitance)}}$$

FIGURE 10-1 **Capacitive Reactance Is Inversely Proportional to Frequency.**

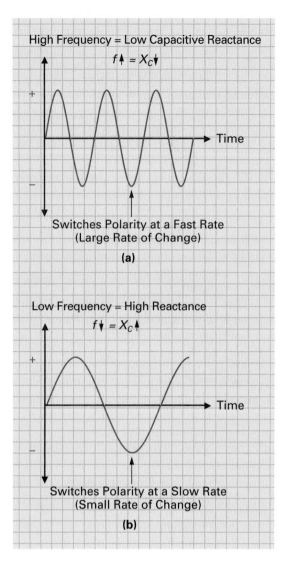

High Frequency = Low Capacitive Reactance

$f\!\uparrow = X_C\!\downarrow$

Time

Switches Polarity at a Fast Rate
(Large Rate of Change)

(a)

Low Frequency = High Reactance

$f\!\downarrow = X_C\!\uparrow$

Time

Switches Polarity at a Slow Rate
(Small Rate of Change)

(b)

Capacitive reactance (X_C) is therefore inversely proportional to both frequency and capacitance and can be calculated by using the formula

$$X_C = \frac{1}{2\pi f C}$$

where X_C = capacitive reactance, in ohms
2π = constant
f = frequency, in hertz
C = capacitance, in farads

☐ **EXAMPLE:**

Calculate the reactance of a 2 µF capacitor when a 10 kHz sine wave is applied.

Solution:

Step	Keypad Entry	Display Response
1.	②	2.0
2.	✕	
3.	π	3.1415927
4.	✕	6.283185
5.	① ⓪ EE ③	10E3
6.	✕	6.2831.8
7.	② EE ⑥ +/−	2.-06
8.	=	0.1256637
9.	1/x	7.9577

$$X_C = \frac{1}{2\pi fC}$$

$$= \frac{1}{2 \times \pi \times 10 \text{ kHz} \times 2 \text{ μF}} = 8 \ \Omega$$

SELF-TEST EVALUATION POINT FOR SECTION 10-1

Use the following questions to test your understanding of Section 10-1.

1. Define *capacitive reactance.*

2. State the formula for capacitive reactance.

3. Why is capacitive reactance inversely proportional to frequency and capacitance?

4. If $C = 4$ μF and $f = 4$ kHz, calculate X_C.

10-2   SERIES *RC* CIRCUIT

In a purely resistive circuit, as shown in Figure 10-2(a), the current flowing within the circuit and the voltage across the resistor are in phase with one another. In a purely capacitive circuit, as shown in Figure 10-2(b), the current flowing in the circuit leads the voltage across the capacitor by 90°.

> *Purely resistive:* 0° phase shift (*I* is in phase with *V*)
> *Purely capacitive:* 90° phase shift (*I* leads *V* by 90°)

If we connect a resistor and capacitor in series, as shown in Figure 10-3(a), we have probably the most commonly used electronic circuit, which has many applications. The voltage across the resistor (V_R) is always in phase with the circuit current (I), as can be seen in Figure 10-3(b), because maximum points and zero crossover points occur at the same time. The voltage across the capacitor (V_C) lags the circuit current by 90°.

Since the capacitor and resistor are in series, the same current is supplied to both components; Kirchhoff's voltage law can be applied, which states that the sum of the voltage drops around a series circuit is equal to the voltage applied (V_S). The voltage drop across the resistor (V_R) and the voltage drop across the capacitor (V_C) are out of phase with one another, which means that their peaks occur at different times. The signal for the applied voltage (V_S) is therefore obtained by adding the values of V_C and V_R at each instant in time, plotting the results, and then connecting the points with a line; this is represented in Figure 10-3(b) by the shaded waveform.

10-2-1   *Voltage*

Figure 10-4(a), (b), and (c) repeat our previous *RC* series circuit with waveforms and a vector diagram. In Figure 10-4(b), the current peak flowing in the series *RC* circuit occurs at 0°

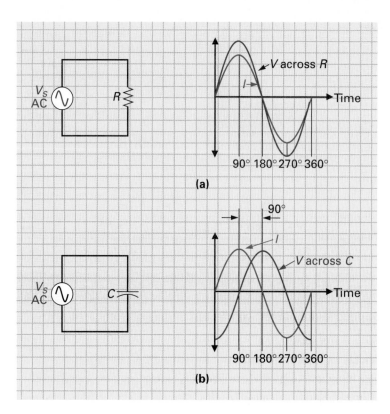

(a)

(b)

FIGURE 10-2 Phase Relationships between *V* and *I*. (a) Resistive Circuit: Current and Voltage Are in Phase. (b) Capacitive Circuit: Current Leads Voltage by 90°.

and will be used as a reference. The voltage across the resistor (V_R) is in phase or coincident with the current (I).

The voltage across the capacitor (V_C) is, as shown in Figure 10-4(b), 90° out of phase (lagging) with the circuit's current. Since the ohmic values of the resistor (R) and the capacitor (X_C) are equal, the voltage drop across both components is the same. The V_R and V_C waveforms are subsequently equal in size.

The source voltage (V_S) is, by Kirchhoff's voltage law, equal to the sum of the series voltage drops (V_C and V_R). However, since these voltages are not in phase with one another, we cannot simply add the two together. The source voltage (V_S) will be the sum of both V_C and V_R at a particular time. Studying the waveforms in Figure 10-4(b), you will notice that peak source voltage will occur at 45°. The angle theta (θ) formed between circuit current (I)

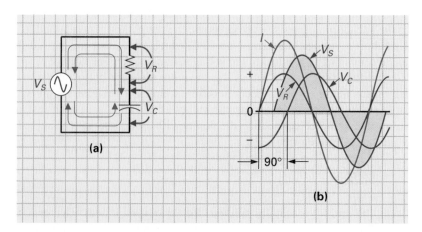

(a)

(b)

FIGURE 10-3 *RC* Series Circuit. (a) Circuit. (b) Waveforms.

CHAPTER 10 / CAPACITIVE CIRCUITS, TESTING, AND APPLICATIONS

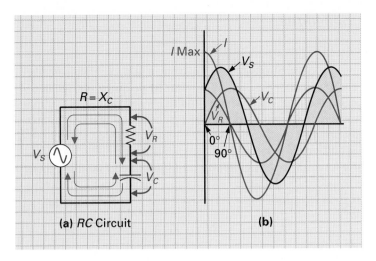

FIGURE 10-4 *RC* **Series Circuit Analysis.**

and source voltage (V_S) will always be less than 90°, and in this example is equal to $-45°$ because the voltage drops across R and C are equal due to R and X_C being of the same ohmic value.

The Pythagorean theorem for right-angle triangles states that if you take the square of a (V_R) and add it to the square of b (V_C), the square root of the result will equal c (the source voltage, V_S).

$$V_S = \sqrt{V_R{}^2 + V_C{}^2}$$

By transposing the formula according to the rules of algebra we can calculate any unknown if two variables are known.

$$V_C = \sqrt{V_S{}^2 - V_R{}^2}$$
$$V_C = \sqrt{V_S{}^2 - V_R{}^2}$$

☐ **EXAMPLE:**

Calculate the source voltage applied across an *RC* series circuit if $V_R = 12$ V and $V_C = 8$ V.

■ *Solution:*

$$
\begin{aligned}
V_S &= \sqrt{V_R{}^2 + V_C{}^2} \\
&= \sqrt{12\ \text{V}^2 + 8\ \text{V}^2} \\
&= \sqrt{144 + 64} \\
&= 14.42\ \text{V}
\end{aligned}
$$

CALCULATOR SEQUENCE

Step	Keypad Entry	Display Response
1.	① ②	12.0
2.	x^2	144.0
3.	+	
4.	⑧	8.0
5.	x^2	64.0
6.	=	208.0
7.	$\sqrt{x}$	14.42220

10-2-2   *Impedance*

Impedance (*Z*)

Measured in ohms, it is the total opposition a circuit offers to current flow (reactive and resistive).

Since resistance is the opposition to current with the dissipation of heat, and reactance is the opposition to current without the dissipation of heat, a new term is needed to describe the total resistive and reactive opposition to current. **Impedance** (designated *Z*) is also measured in ohms and is the total circuit opposition to current flow. It is a combination of resistance (*R*) and reactance (X_C); however, in our capacitive and resistive circuit, a phase shift or difference exists, and just as V_C and V_R cannot be merely added together to obtain V_S, *R* and X_C cannot be simply summed to obtain *Z*.

If the current within a series circuit is constant (the same throughout the circuit), the resistance of a resistor (*R*) or reactance of a capacitor (X_C) will be directly proportional to the voltage across the resistor (V_R) or the capacitor (V_C).

$$V_R\updownarrow = I \times R\updownarrow, \qquad V_C\updownarrow = I \times X_C\updownarrow$$

The Pythagorean theorem can be used to calculate the total opposition or impedance (*Z*) to current flow, taking into account both *R* and X_C.

$$Z = \sqrt{R^2 + X_C{}^2}$$

$$R = \sqrt{Z^2 - X_C{}^2}$$
$$X_C = \sqrt{Z^2 - R^2}$$

▢ EXAMPLE:

Calculate the total impedance of a series *RC* circuit if $R = 27\ \Omega$, $C = 0.005\ \mu F$, and the source frequency = 1 kHz.

▢ *Solution:*

The total opposition (*Z*) or impedance is equal to

$$Z = \sqrt{R^2 + X_C{}^2}$$

R is known, but X_C will need to be calculated.

$$X_C = \frac{1}{2\pi f C}$$

$$= \frac{1}{2 \times \pi \times 1\ \text{kHz} \times 0.005\ \mu F}$$
$$= 31.8\ k\Omega$$

Since $R = 27\ \Omega$ and $X_C = 31.8\ k\Omega$, then

$$Z = \sqrt{R^2 + X_C{}^2}$$
$$= \sqrt{27\ \Omega^2 + 31.8\ k\Omega^2}$$
$$= \sqrt{729 + 1 \times 10^9}$$
$$= 31.8\ k\Omega$$

As you can see in this example, the small resistance of 27 Ω has very little effect on the circuit's total opposition or impedance, due to the relatively large capacitive reactance of 31,800 Ω.

Calculate the total impedance of a series *RC* circuit if $R = 45$ kΩ and $X_C = 45$ Ω.

■ *Solution:*

$$Z = \sqrt{R^2 + X_C^2}$$
$$= \sqrt{45 \text{ k}\Omega^2 + 45^2}$$
$$= 45 \text{ k}\Omega$$

In this example, the relatively small value of X_C had very little effect on the circuit's opposition or impedance, due to the large circuit resistance.

Calculate the total impedance of a series *RC* circuit if $X_C = 100$ Ω and $R = 100$ Ω.

■ *Solution:*

$$Z = \sqrt{R^2 + X_C^2}$$
$$= \sqrt{100^2 + 100^2}$$
$$= 141.4 \text{ } \Omega$$

In this example *R* was equal to X_C.

We can define the total opposition or impedance in terms of Ohm's law, in the same way as we defined resistance.

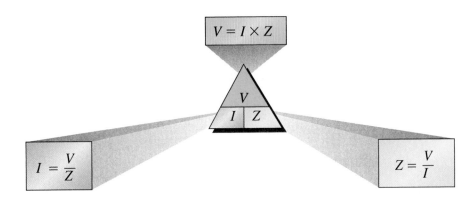

By transposition, we can arrive at the usual combinations of Ohm's law.

10-2-3 *Phase Angle or Shift (θ)*

In a purely resistive circuit, the total opposition (*Z*) is equal to the resistance of the resistor, so the phase shift (θ) is equal to 0°.

In a purely capacitive circuit, the total opposition (*Z*) is equal to the capacitive reactance (X_C) of the capacitor, so the phase shift (θ) is equal to −90°. When a circuit contains both resistance and capacitive reactance, the total opposition or impedance has a phase shift that is between 0 and 90°.

The phase angle, θ, is equal to

$$\theta = \text{invtan} \frac{X_C}{R}$$

This formula will determine by what angle Z leads R. Since X_C/R is equal to V_C/V_R, the phase angle can also be calculated if V_R and V_C are known.

$$\theta = \text{invtan} \frac{V_C}{V_R}$$

This formula will determine by what angle V_R leads V_S.

▢ EXAMPLE:

Calculate the phase shift or angle in two different series RC circuits if:

    a. $V_R = 12$ V, $V_C = 8$ V
    b. $R = 27$ Ω, $X_C = 31.8$ kΩ

▢ Solution:

    a. $\theta = \text{invtan} \dfrac{V_C}{V_R}$

       $= \text{invtan} \dfrac{8 \text{ V}}{12 \text{ V}}$

       $= 33.7°$ (V_R leads V_S by $33.7°$)

    b. $\theta = \text{invtan} \dfrac{X_C}{R}$

       $= \text{invtan} \dfrac{31.8 \text{ kΩ}}{27 \text{ Ω}}$

       $= 89.95°$ (R leads Z by $89.95°$)

10-2-4 Power

In this section we examine power in a series ac circuit. Let us begin with a simple resistive circuit and review the power formulas used previously.

Purely Resistive Circuit

In Figure 10-5 you can see the current, voltage, and power waveforms generated by applying an ac voltage across a purely resistive circuit. The applied voltage causes current to flow around the circuit, and the electrical energy is converted into heat energy. This heat or power is dissipated and lost and can be calculated by using the power formula

$$P = V \times I$$
$$P = I^2 \times R$$
$$P = \frac{V^2}{R}$$

Voltage and current are in phase with one another in a resistive circuit, and instantaneous power is calculated by multiplying voltage by current at every instant through $360°$ ($P = V \times I$). The sinusoidal power waveform is totally positive, because a positive voltage multiplied by a positive current gives a positive value of power, and a negative voltage multiplied by a negative current will also produce a positive value of power. For these reasons, a resistor is said to generate a positive power waveform, which you may have noticed is twice the frequency of the voltage and current waveforms; two power cycles occur in the same time as one voltage and current cycle.

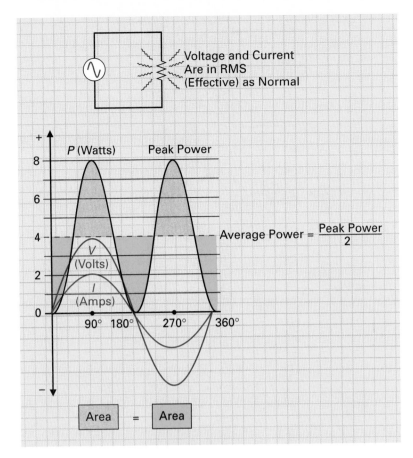

FIGURE 10-5 **Power in a Purely Resistive Circuit.**

The power waveform has been split in half, and this line that exists between the maximum point (8 W) and zero point (0 W) is the average value of power (4 W) that is being dissipated by the resistor.

Purely Capacitive Circuit

In Figure 10-6, you can see the current, voltage, and power waveforms generated by applying an ac voltage source across a purely capacitive circuit. As expected, the current leads the voltage by 90°, and the power wave is calculated by multiplying voltage by current, as before, at every instant through 360°. The resulting power curve is both positive and negative. During the positive alternation of the power curve, the capacitor is taking power as the capacitor charges. When the power alternation is negative, the capacitor is giving back the power it took as it discharges back into the circuit.

The average power dissipated is once again the value that exists between the maximum positive and maximum negative points and causes the area above this line to equal the area below. This average power level calculates out to be zero, which means that no power is dissipated in a purely capacitive circuit.

Resistive and Capacitive Circuit

In Figure 10-7 you can see the current, voltage, and power waveforms generated by applying an ac voltage source across a series-connected *RC* circuit. The current leads the voltage by some phase angle less than 90°, and the power waveform is once again determined by the product of voltage and current. The negative alternation of the power cycle indicates that the capacitor is discharging and giving back the power that it consumed during the charge.

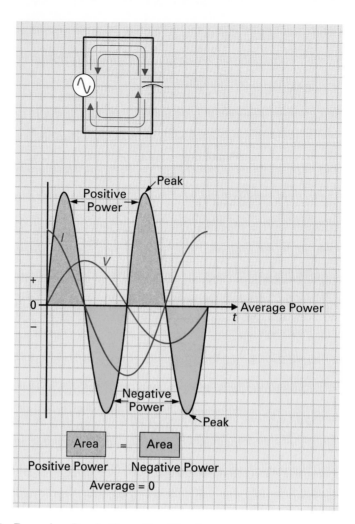

FIGURE 10-6   Power in a Purely Capacitive Circuit.

The positive alternation of the power cycle is much larger than the negative alternation because it is the combination of both the capacitor taking power during charge and the resistor consuming and dissipating power in the form of heat. The average power being dissipated will be some positive value, due to the heat being generated by the resistor.

Power Factor

In a purely resistive circuit, all the energy supplied to the resistor from the source is dissipated in the form of heat. This form of power is referred to as **resistive power** (P_R) or **true power,** and is calculated with the formula

$$P_R = I^2 \times R$$

In a purely capacitive circuit, all the energy supplied to the capacitor is stored from the source and then returned to the source, without energy loss. This form of power is referred to as **reactive power** (P_X) or **imaginary power.**

$$P_X = I^2 \times X_C$$

When a circuit contains both capacitance and resistance, some of the supply is stored and returned by the capacitor and some of the energy is dissipated and lost by the resistor.

Resistive Power or True Power

The average power consumed by a circuit during one complete cycle of alternating current.

Reactive Power or Imaginary Power

Also called wattless power, it is the power value obtained by multiplying the effective value of current by the effective value of voltage and the sine of the angular phase difference between current and voltage.

FIGURE 10-7　Power in a Resistive and
Capacitive Circuit.

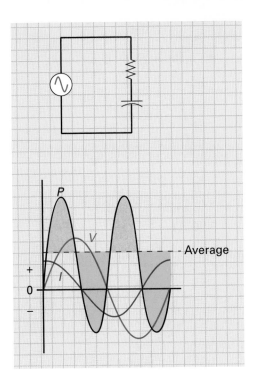

Applying the Pythagorean theorem once again, we can calculate apparent power by

$$P_A = \sqrt{P_R^2 + P_X^2}$$

where P_A = apparent power, in volt-amperes (VA)
　　　P_R = true power, in watts (W)
　　　P_X = reactive power, in volt-amperes reactive (VAR)

The **power factor** is a ratio of true power to apparent power and is a measure of the loss in a circuit. It can be calculated by using the formula

$$PF = \frac{\text{true power } (P_R)}{\text{apparent power } (P_A)}$$

Power Factor

Ratio of actual power to apparent power. A pure resistor has a power factor of 1 or 100% while a capacitor has a power factor of 0 or 0%.

A power factor of 1 indicates a maximum power loss (circuit is resistive), while a power factor of 0 indicates no power loss (circuit is capacitive). With circuits that contain both resistance and reactance, the power factor will be somewhere between zero (0 = reactive) and one (1 = resistive).

Since true power is determined by resistance and apparent power is dependent on impedance, the power factor can also be calculated by using the formula

$$PF = \frac{R}{Z}$$

The power factor can also be determined by the cosine of angle θ.

$$PF = \cos \theta$$

□ EXAMPLE:

Calculate the following for a series RC circuit if $R = 2.2$ kΩ, $X_C = 3.3$ kΩ, and $V_S = 5$ V.

a. Z	d. P_R	g. PF
b. I	e. P_X	
c. θ	f. P_A	

■ *Solution:*

a. $Z = \sqrt{R^2 + X_C^2}$

$= \sqrt{2.2 \text{ k}\Omega^2 + 3.3 \text{ k}\Omega^2}$

$= 3.96 \text{ k}\Omega$

b. $I = \dfrac{V_S}{Z} = \dfrac{5 \text{ V}}{3.96 \text{ k}\Omega} = 1.26 \text{ mA}$

c. $\theta = \text{invtan} \dfrac{X_C}{R} = \text{invtan} \dfrac{3.3 \text{ k}\Omega}{2.2 \text{ k}\Omega}$

$= \text{invtan } 1.5 = 56.3°$

d. True power $= I^2 + R$

$= 1.26 \text{ mA}^2 \times 2.2 \text{ k}\Omega$

$= 3.49 \text{ mW}$

e. Reactive power $= I^2 \times X_C$

$= 1.26 \text{ mA}^2 \times 3.3 \text{ k}\Omega$

$= 5.24 \times 10^{-3}$ or 5.24 mVAR

f. Apparent power $= \sqrt{P_R^2 + P_X^2}$

$= \sqrt{3.49 \text{ mW}^2 + 5.24 \text{ mW}^2}$

$= 6.29 \times 10^{-3}$ or 6.29 mVA

g. Power factor $= \dfrac{R}{Z} = \dfrac{2.2 \text{ k}\Omega}{3.96 \text{ k}\Omega} = 0.55$

or

$= \dfrac{P_R}{P_A} = \dfrac{3.49 \text{ mW}}{6.29 \text{ mW}} = 0.55$

or

$= \cos \theta = \cos 56.3° = 0.55$

CALCULATOR SEQUENCE

Step	Keypad Entry	Display Response
1.	[3] [.] [3] [E] [3]	3.3E3
2.	[÷]	
3.	[2] [.] [2] [E] [3]	2.2E3
4.	[=]	1.5
5.	[inv] [tan]	56.309932

Use the following questions to test your understanding of Section 10-2.

1. What is the phase relationship between current and voltage in a series RC circuit?
2. Define and state the formula for impedance.

10-3 PARALLEL *RC* CIRCUIT

Now that we have analyzed the characteristics of a series *RC* circuit, let us connect a resistor and capacitor in parallel.

10-3-1 *Voltage*

As with any parallel circuit, the voltage across all components in parallel is equal to the source voltage; therefore,

$$V_R = V_C = V_S$$

10-3-2 *Current*

In Figure 10-8(a), you will see a parallel circuit containing a resistor and a capacitor. The current through the resistor and capacitor is simply calculated by applying Ohm's law:

$$\text{(resistor current) } I_R = \frac{V_S}{R}$$

$$\text{(capacitor current) } I_C = \frac{V_S}{X_C}$$

Total current (I_T), however, is not as simply calculated. As expected, resistor current (I_R) is in phase with the applied voltage (V_S), as shown in the waveforms in Figure 10-8(b). Capacitor current will always lead the applied voltage by 90°.

Total current is therefore the vector sum of both the resistor and capacitor currents. Using the Pythagorean theorem, total current can be calculated by

$$I_T = \sqrt{I_R^2 + I_C^2}$$

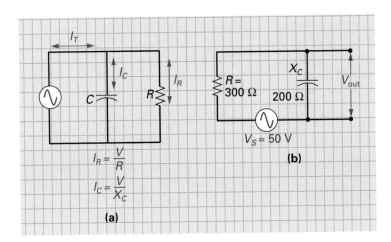

FIGURE 10-8 Parallel *RC* Circuit.

10-3-3   *Phase Angle*

The angle by which the total current (I_T) leads the source voltage (V_S) can be determined with either of the following formulas:

$$\theta = \text{invtan}\frac{I_C \text{ (opposite)}}{I_R \text{ (adjacent)}} \qquad \theta = \text{invtan}\frac{R}{X_C}$$

10-3-4   *Impedance*

Since the circuit is both capacitive and resistive, the total opposition or impedance of the parallel *RC* circuit can be calculated by

$$Z = \frac{V_S}{I_T}$$

The impedance of a parallel *RC* circuit is equal to the total voltage divided by the total current. Using basic algebra, this basic formula can be rearranged to express impedance in terms of reactance and resistance.

$$Z = \frac{R \times X_C}{\sqrt{R^2 + X_C^2}}$$

10-3-5   *Power*

With respect to power, there is no difference between a series circuit and a parallel circuit. The true power or resistive power (P_R) dissipated by an *RC* circuit is calculated with the formula

$$P_R = I_R^2 \times R$$

The imaginary power or reactive power (P_X) of the circuit can be calculated with the formula

$$P_X = I_C^2 \times X_C$$

The apparent power is equal to the vector sum of the true power and the reactive power.

$$P_A = \sqrt{P_R^2 + P_X^2}$$

As with series *RC* circuits the power factor is calculated as

$$PF = \frac{P_R \text{ (resistive power)}}{P_A \text{ (apparent power)}}$$

A power factor of 1 indicates a purely resistive circuit, while a power factor of 0 indicates a purely reactive circuit.

■ EXAMPLE:

Calculate the following for a parallel RC circuit in which $R = 24\,\Omega$, $X_C = 14\,\Omega$, and $V_S = 10\,V$.

a. I_R d. Z
b. I_C e. θ
c. I_T

■ Solution:

a. $I_R = \dfrac{V_S}{R} = \dfrac{10\,V}{24\,\Omega} = 416.66\,mA$

b. $I_C = \dfrac{V_S}{X_C} = \dfrac{10\,V}{14\,\Omega} = 714.28\,mA$

c. $I_T = \sqrt{I_R^2 + I_C^2}$
   $= \sqrt{416.66\,mA^2 + 714.28\,mA^2}$
   $= \sqrt{0.173 + 0.510}$
   $= 826.5\,mA$

d. $Z = \dfrac{V_S}{I_T} = \dfrac{10\,V}{826.5\,mA} = 12\,\Omega$

or

$= \dfrac{R \times X_C}{\sqrt{R^2 + X_C^2}} = \dfrac{24 \times 14}{\sqrt{24^2 + 14^2}}$
$= 12\,\Omega$

e. $\theta = \arctan\dfrac{I_C}{I_R} = \arctan\dfrac{714.28\,mA}{416.66\,mA}$
   $= \arctan 1.714 = 59.7°$

or

$\theta = \arctan\dfrac{R}{X_C} = \arctan\dfrac{24\,\Omega}{14\,\Omega}$
$= \arctan 1.714 = 59.7°$

CALCULATOR SEQUENCE

Step	Keypad Entry	Display Response
1.	[7] [1] [4] [.] [2] [8] [E] [3] [+/−]	714.28E-3
2.	[÷]	
3.	[4] [1] [6] [.] [6] [6] [E] [3] [+/−]	416.66E-3
4.	[=]	1.7142994
5.	[inv] [tan]	59.743762

SELF-TEST EVALUATION POINT FOR SECTION 10-3

Use the following questions to test your understanding of Section 10-3.

1. What is the phase relationship between current and voltage in a parallel RC circuit?
2. Could a parallel RC circuit be called a voltage divider?

10-4   TESTING CAPACITORS

Now that you have a good understanding as to how a capacitor should function, let us investigate how to diagnose a capacitor malfunction.

10-4-1   *The Ohmmeter*

A faulty capacitor may have one of three basic problems:

1. A short, which is easy to detect and is caused by a contact from plate to plate.
2. An open, which is again quite easy to detect and is normally caused by one of the leads becoming disconnected from its respective plate.
3. A leaky dielectric or capacitor breakdown, which is quite difficult to detect, as it may only short at a certain voltage. This problem is usually caused by the deterioration of the dielectric, which starts displaying a much lower dielectric resistance than it was designed for. The capacitor with this type of problem is referred to as a *leaky capacitor.*

Capacitors of 0.5 μF and larger can be checked by using an analog ohmmeter or a digital multimeter with a bar graph display by using the procedure shown in Figure 10-9.

Step 1:   Ensure that the capacitor is discharged by shorting the leads together.

Step 2:   Set the ohmmeter to the highest ohms range scale.

Step 3:   Connect the meter to the capacitor, observing the correct polarity if an electrolytic is being tested, and observe the bar graph display. The capacitor will initially be discharged, and therefore maximum current will flow from the meter battery to the capacitor. Maximum current means low resistance, which is why the meter's display indicates 0 Ω.

Step 4:   As the capacitor charges, it will cause current flow from the meter's battery to decrease, and consequently, the bar graph display will increase.

A good capacitor will cause the meter to react as just explained. A larger capacitance will cause the meter to increase slowly to infinity (∞), as it will take a longer time to charge, while a smaller value of capacitance will charge at a much faster rate, causing the meter to increase rapidly toward ∞. For this reason, the ohmmeter cannot be reliably used to check capacitors with values of less than 0.5 μF, because the capacitor charges up too quickly and the meter does not have enough time to respond.

Step 1: Discharge Capacitor

Use DMM's
Bar Graph
Display

Step 2: Set Digital Ohmmeter to
High Ohms Range

Step 3: Meter Deflects Rapidly
to 0 Ohms Initially

Bar Graph
Display
Increases

Step 4: Meter Should Then Return to
Infinity as the Capacitor Charges

FIGURE 10-9   Testing a Capacitor of More Than 0.5 μF.

A shorted capacitor will cause the meter to show zero ohms and remain in that position. An open capacitor will cause a maximum reading (infinite resistance) because there is no path for current to flow.

A leaky capacitor will cause the bar graph to reduce its bars to the left, and then return almost all the way back to ∞ if only a small current is still flowing and the capacitor has a small dielectric leak. If the meter bars only come back to halfway or a large distance away from infinity, a large amount of current is still flowing and the capacitor has a large dielectric leak (defect).

When using the ohmmeter to test capacitors, there are some other points that you should be aware of:

1. Electrolytics are noted for having a small yet noticeable amount of inherent leakage, and so do not expect the meter's bar display to move all the way to the right (∞ ohms). Most electrolytic capacitors that are still functioning normally will show a resistance of 200 kΩ or more.

2. Some ohmmeters utilize internal batteries of up to 15 V, so be careful not to exceed the voltage rating of the capacitor.

10-4-2 *The Capacitance Meter or Analyzer*

The ohmmeter check tests the capacitor under a low-voltage condition. This may be adequate for some capacitor malfunctions; however, a problem that often occurs with capacitors is that they short or leak at a high voltage. The ohmmeter test is also adequate for capacitors of 0.5 μF or greater; however, a smaller capacitor cannot be tested because its charge time is too fast for the meter to respond. The ohmmeter cannot check for high-voltage failure, for small-value capacitance, or if the value of capacitance has changed through age or extreme thermal exposure.

A **capacitance meter** or analyzer, which is illustrated in Figure 10-10, can totally check all aspects of a capacitor in a range of values from approximately 1 pF to 20 F.

Capacitance Meter
Instrument used to measure the capacitance of a capacitor or circuit.

FIGURE 10-10   Capacitor Analyzer.

Use the following questions to test your understanding of Section 10-4.

1. The analog ohmmeter or digital multimeter with bar graph can only be used to test capacitors that are _____ or more in value.

2. A _____ _____ tests a capacitor for value change, leakage, dielectric absorption, and equivalent series resistance.

10-5 APPLICATIONS OF CAPACITORS

There are many applications of capacitors, some of which will be discussed now; others will be presented later and in your course of electronic studies.

10-5-1 Combining AC and DC

Figure 10-11(a) and (b) show how the capacitor can be used to combine ac and dc. The capacitor is large in value (electrolytic typically) and can be thought of as a very large bucket that will fill or charge to a dc voltage level. The ac voltage will charge and discharge the capacitor, which is similar to pouring in and pulling out (alternating) more and less water. The resulting waveform is a combination of ac and dc that varies above and below an average dc level. In this instance, the ac is said to be superimposed on a dc level.

10-5-2 The Capacitive Voltage Divider

In Chapter 9 we illustrated how capacitors could be connected in series across a dc voltage source to form a voltage divider. This circuit is repeated in Figure 10-12(a). Figure 10-12(b) shows how two capacitors can be used to divide up an ac voltage source of 30 V rms. Like a resistive voltage divider, voltage drop is proportional to current opposition, which with a capacitor is called *capacitive reactance* (X_C). The larger the capacitive reactance ($X_C\uparrow$), the larger the voltage drop ($V\uparrow$). Since capacitive reactance is inversely proportional to the capacitor value and the frequency of the ac source, a change in input frequency will cause a change in the capacitor's capacitive reactance, and this will change the voltage drop across

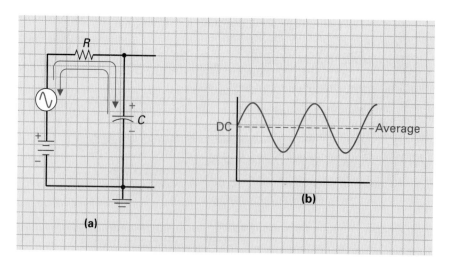

FIGURE 10-11 Superimposing AC on a DC Level.

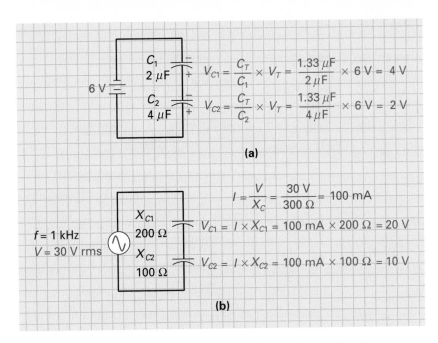

(a)

$$V_{C1} = \frac{C_T}{C_1} \times V_T = \frac{1.33\ \mu F}{2\ \mu F} \times 6\ V = 4\ V$$

$$V_{C2} = \frac{C_T}{C_2} \times V_T = \frac{1.33\ \mu F}{4\ \mu F} \times 6\ V = 2\ V$$

$$I = \frac{V}{X_C} = \frac{30\ V}{300\ \Omega} = 100\ mA$$

$$V_{C1} = I \times X_{C1} = 100\ mA \times 200\ \Omega = 20\ V$$

$$V_{C2} = I \times X_{C2} = 100\ mA \times 100\ \Omega = 10\ V$$

(b)

FIGURE 10-12 **Capacitive Voltage Divider. (a) DC Circuit. (b) AC Circuit.**

the capacitor. Although a 90° phase shift is present between voltage and current in a purely capacitive circuit (I leads V by 90°), there is no phase difference between the input voltage and the output voltages.

10-5-3   RC Filters

A **filter** is a circuit that allows certain frequencies to pass but blocks other frequencies. In other words, it filters out the unwanted frequencies but passes the wanted or selected ones.

There are two basic RC filters:

1. The **low-pass filter,** which can be seen in Figure 10-13(a); as its name implies, it passes the low frequencies but heavily **attenuates** the higher frequencies.

2. The **high-pass filter,** which can be seen in Figure 10-13(b); as its name implies, it allows the high frequencies to pass but heavily attenuates the lower frequencies.

With either the low- or high-pass filter, it is important to remember that capacitive reactance is inversely proportional to frequency ($X_C \propto 1/f$) as stated in the center of Figure 10-13.

With the low-pass filter shown in Figure 10-13(a), the output is connected across the capacitor. As the frequency of the input increases, the amplitude of the output decreases. At dc (0 Hz) and low frequencies, the capacitive reactance is very large ($X_C \uparrow = 1/f \downarrow$) with respect to the resistor. All the input will appear across the capacitor, because the capacitor and resistor form a voltage divider, as shown in Figure 10-14(a). As with any voltage divider, the larger opposition to current flow will drop the larger voltage. Since the output voltage is determined by the voltage drop across the capacitor, almost all the input will appear across the capacitor and therefore be present at the output.

If the frequency of the input increases, the reactance of the capacitor will decrease ($X_C \downarrow = 1/f \uparrow$), and a larger amount of the signal will be dropped across the resistor. As frequency increases, the capacitor becomes more of a short circuit (lower reactance), and the output, which is across the capacitor, decreases, as shown in Figure 10-14(b).

Below the circuit of the low-pass filter in Figure 10-13(a), you will see a graph known as the **frequency response curve** for the low-pass filter. This curve illustrates that as the frequency of the input increases the voltage at the output will decrease.

Filter

Network composed of resistor, capacitor, and inductors used to pass certain frequencies yet block others through heavy attenuation.

Low-Pass Filter

Network or circuit designed to pass any frequencies below a critical or cutoff frequency and reject or heavily attenuate all frequencies above.

Attenuate

To reduce in amplitude an action or signal.

High-Pass Filter

Network or circuit designed to pass any frequencies above a critical or cutoff frequency and reject or heavily attenuate all frequencies below.

Frequency Response Curve

A graph indicating how effectively a circuit or device responds to the frequency spectrum.

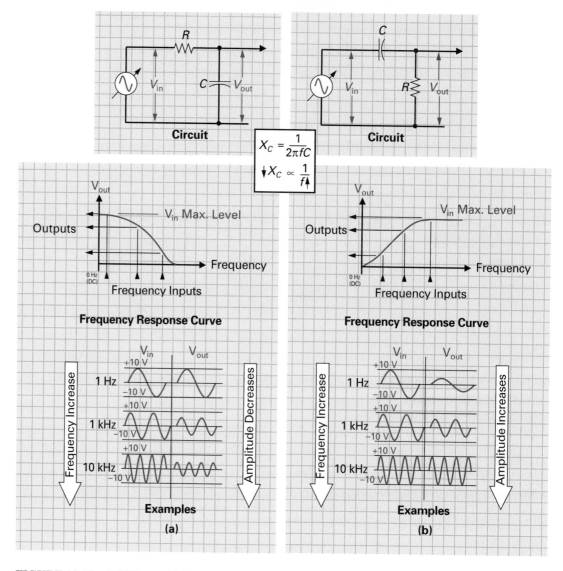

FIGURE 10-13 *RC* **Filters. (a) Low-Pass Filter. (b) High-Pass Filter.**

With the high-pass filter, seen in Figure 10-13(b), the capacitor and resistor have traded positions to show how the opposite effect for the low-pass filter occurs.

At low frequencies, the reactance will be high and almost all of the signal will be dropped across the capacitor. Very little signal appears across the resistor, and, consequently, the output, as shown in Figure 10-15(a). As the frequency of the input increases, the reactance of the capacitor decreases, allowing more of the input signal to appear across the resistor and therefore appear at the output, as shown in Figure 10-15(b).

Below the circuit of the high-pass filter in Figure 10-13(b), you will see the frequency response curve for the high-pass filter. This curve illustrates that as the frequency of the input increases the voltage at the output increases.

10-5-4   *The RC Integrator*

Integrator

Device that approximates and whose output is proportional to an integral of the input signal.

Up until now we have analyzed the behavior of an *RC* circuit only when a sine-wave input signal was applied. In both this and the following sections we demonstrate how an *RC* circuit will react to a square-wave input, and show two other important applications of capacitors. The term **integrator** is derived from a mathematical function in calculus. This combination

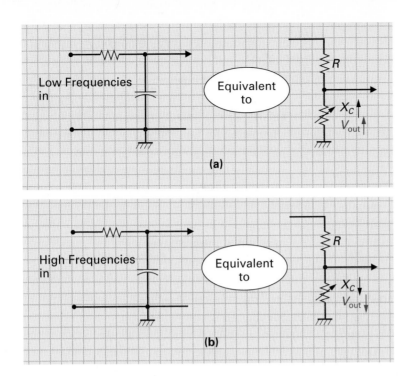

FIGURE 10-14 Low-Pass Filter. (a) Low-Pass Filter, High X_C. (b) Low-Pass Filter, Low X_C.

of R and C, in some situations, displays this mathematical function. Figure 10-16(a) illustrates an integrator circuit that can be recognized by the series connection of R and C, but mainly from the fact that the output is taken across the capacitor.

If a 10 V square wave is applied across the circuit, as shown in the waveforms in Figure 10-16(b), and the time constant of the RC combination calculates out to be 1 second, the

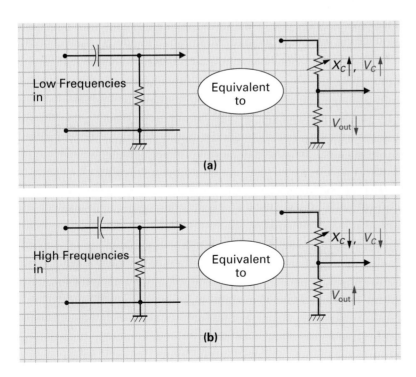

FIGURE 10-15 High-Pass Filter. (a) High X_C. (b) Low X_C.

FIGURE 10-16   *RC* Integrator. (a) Circuit. (b) Waveforms.

capacitor will charge when the square wave input is positive toward the applied voltage (10 V) and reaches it in five time constants (5 seconds). Since the positive alternation of the square wave lasts for 6 seconds, the capacitor will be fully charged 1 second before the positive alternation ends, as shown in Figure 10-17.

When the positive half-cycle of the square-wave input ends after 6 seconds, the input falls to 0 V and the circuit is equivalent to that shown in Figure 10-18(a). The 10 V charged capacitor now has a path to discharge and in five time constants (5 seconds) is fully discharged, as shown in Figure 10-18(b).

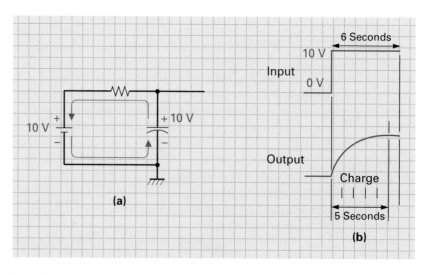

FIGURE 10-17   Integrator Response to Positive Step. (a) Equivalent Circuit. (b) Waveforms for Charge.

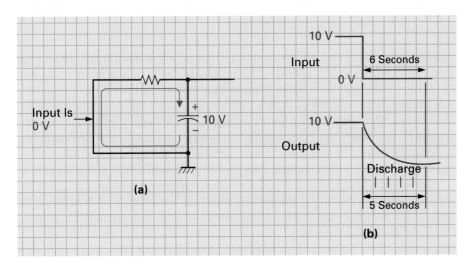

FIGURE 10-18 Integrator Response to Negative Step. (a) Equivalent Circuit. (b) Waveforms for Discharge.

If the same *RC* integrator circuit was connected to a square wave that has a 1-second positive half-cycle, the capacitor will not be able to charge fully toward 10 V. In fact, during the positive alternation of 1 second (one time constant), the capacitor will reach 63.2% of the applied voltage (6.32 V), and then during the 0 V half-cycle it will discharge to 63.2% of 6.32 V, to 2.33 V, as shown by the waveform in Figure 10-19. The voltage across the capacitor, and therefore the output voltage, will gradually build up and eventually level off to an average value of 5 V in about five time constants (5 seconds).

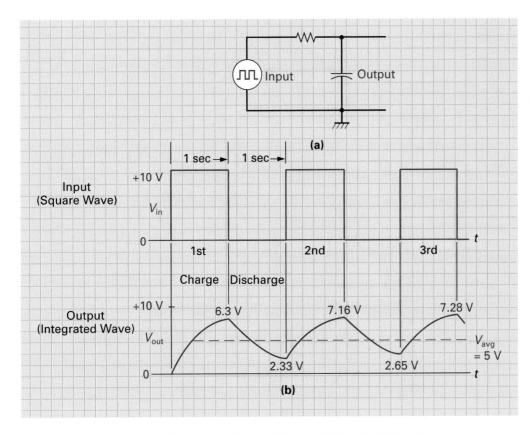

FIGURE 10-19 Integrator Response to Square Wave. (a) Circuit. (b) Waveforms.

10-5-5 *The RC Differentiator*

Figure 10-20(a) illustrates the **differentiator** circuit, which is the integrator's opposite. In this case the output is taken across the resistor instead of the capacitor, and the time constant is always short with respect to the input square wave period.

The differentiator output waveform, shown in Figure 10-20(b), is taken across the resistor and is the result of the capacitor's charge and discharge. When the square wave swings positive, the equivalent circuit is that shown in Figure 10-21(a). When the 10 V is initially applied (positive step of the square wave), all the voltage is across the resistor, and therefore at the output, as the capacitor cannot charge instantly. As the capacitor begins to charge, more of the voltage is developed across the capacitor and less across the resistor. The voltage across the capacitor exponentially increases and reaches 10 V in five time constants (5 × 1 ms = 5 ms), while the voltage across the resistor, and therefore the output, exponentially falls from its initial 10 V to 0 V in five time constants as shown in Figure 10-21(b), at which time all the voltage is across the capacitor and no voltage will be across the resistor.

When the positive half-cycle of the square wave ends and the input falls to zero, the circuit is equivalent to that shown in Figure 10-22(a). The negative plate of the capacitor is now applied directly to the output. Since the capacitor cannot instantly discharge, the output drops suddenly down to −10 V as shown in Figure 10-22(b). This is the voltage across the resistor, and therefore the output. The capacitor is now in series with the resistor and therefore has a path through the resistor to discharge, which it does in five time constants to 0 V.

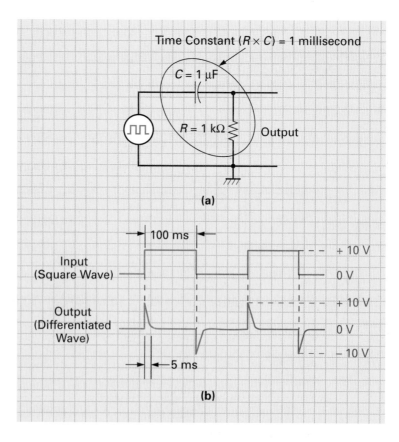

FIGURE 10-20 *RC* **Differentiator. (a) Circuit. (b) Waveforms.**

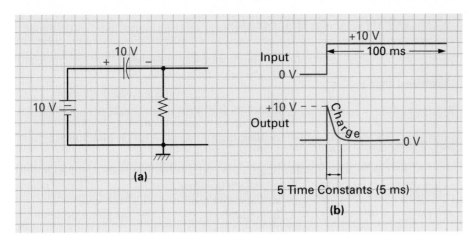

FIGURE 10-21 **Differentiator's Response to Positive Step. (a) Equivalent Circuit. (b) Waveforms.**

FIGURE 10-22 **Differentiator's Response to a Negative Step. (a) Equivalent Circuit. (b) Waveforms.**

SELF-TEST EVALUATION POINT FOR SECTION 10-5

Use the following questions to test your understanding of Section 10-5.

1. Give three circuit applications for a capacitor.
2. With an *RC* low-pass filter, the _____ is connected across the output. Whereas with an *RC* high-pass filter, the _____ is connected across the output.
3. An integrator has a _____ time constant compared to the period of the input square wave.
4. A _____ circuit produces positive and negative spikes at the output when a square wave is applied at the input.

Multiple-Choice Questions

1. Capacitive reactance is inversely proportional to:
   a. Capacitance and resistance
   b. Frequency and capacitance
   c. Capacitance and impedance
   d. Both (a) and (c)

2. The impedance of an RC series circuit is equal to:
   a. The sum of R and X_C
   b. The square root of the sum of R^2 and X_C^2
   c. The square of the sum of R and X_C
   d. The sum of the square root of R and X_C

3. In a purely resistive circuit:
   a. The current flowing in the circuit leads the voltage across the capacitor by 90°.
   b. The circuit current and resistor voltage are in phase with one another.
   c. The current leads the voltage by 45°.
   d. The current leads the voltage by a phase angle between 0 and 90°.

4. In a purely capacitive circuit:
   a. The current flowing in the circuit leads the voltage across the capacitor by 90°.
   b. The circuit current and resistor voltage are in phase with one another.
   c. The current leads the voltage by 45°.
   d. The current leads the voltage by a phase angle between 0 and 90°.

5. In a series circuit containing both capacitance and resistance:
   a. The current flowing in the circuit leads the voltage across the capacitor by 90°.
   b. The circuit current and resistor voltage are in phase with one another.
   c. The current leads the voltage by 45°.
   d. Both (a) and (b).

6. In a series RC circuit, the source voltage is equal to:
   a. The sum of V_R and V_C
   b. The difference between V_R and V_C
   c. The vectoral sum of V_R and V_C
   d. The sum of V_R and V_C squared

7. As the source frequency is increased, the capacitive reactance will:
   a. Increase
   c. Be unaffected
   b. Decrease
   d. Increase, depending on harmonic content

8. The phase angle of a series RC circuit indicates by what angle V_S _____ V_R.
   a. Lags
   c. Leads or lags
   b. Leads
   d. None of the above

9. In a series RC circuit, the vector combination of R and X_C is the circuit's _____.
   a. Phase angle
   c. Source voltage
   b. Apparent power
   d. Impedance

10. In a parallel RC circuit, the total current is equal to:
    a. The sum of I_R and I_C
    b. The difference between I_R and I_C
    c. The vectoral sum of I_R and I_C
    d. The sum of I_R and I_C squared

11. _____ is the opposition offered by a capacitor to current flow without the dissipation of energy.
    a. Capacitive reactance
    d. Phase angle
    b. Resistance
    e. The power factor
    c. Impedance

12. _____ is the total reactive and resistive circuit opposition to current flow.
    a. Capacitive reactance
    d. Phase angle
    b. Resistance
    e. The power factor
    c. Impedance

13. _____ is the ratio of true (resistive) power to apparent power and is therefore a measure of the loss in a circuit.
    a. Capacitive reactance
    d. Phase angle
    b. Resistance
    e. The power factor
    c. Impedance

14. In a series RC circuit, the leading voltage will be measured across the:
    a. Resistor
    c. Source
    b. Capacitor
    d. Any of the choices are true.

15. In a series RC circuit, the lagging voltage will be measured across the:
    a. Resistor
    c. Source
    b. Capacitor
    d. Any of the choices are true.

Practice Problems

16. Calculate the capacitive reactance of the capacitor circuits with the following parameters:
    a. $f = 1$ kHz, $C = 2$ μF
    b. $f = 100$ Hz, $C = 0.01$ μF
    c. $f = 17.3$ MHz, $C = 47$ μF

17. In a series RC circuit, the voltage across the capacitor is 12 V and the voltage across the resistor is 6 V. Calculate the source voltage.

18. Calculate the impedance for the following series RC circuits:
    a. 2.7 MΩ, 3.7 μF, 20 kHz
    b. 350 Ω, 0.005 μF, 3 MHz
    c. $R = 8.6$ kΩ, $X_C = 2.4$ Ω
    d. $R = 4700$ Ω, $X_C = 2$ kΩ

19. In a parallel RC circuit with parameters of $V_S = 12$ V, $R = 4$ MΩ, and $X_C = 1.3$ kΩ, calculate:
    a. I_R
    c. I_T
    e. θ
    b. I_C
    d. Z

20. Calculate the total reactance in:
    a. A series circuit where $X_{C1} = 200$ Ω, $X_{C2} = 300$ Ω, $X_{C3} = 400$ Ω
    b. A parallel circuit where $X_{C1} = 3.3$ kΩ, $X_{C2} = 2.7$ kΩ

21. Calculate the capacitance needed to produce 10 kΩ of reactance at 20 kHz.

22. At what frequency will a 4.7 μF capacitor have a reactance of 2000 Ω?

FIGURE 10-23

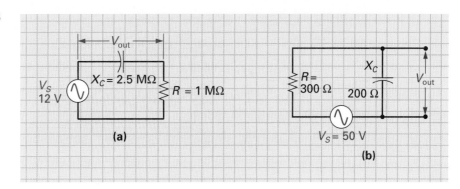

23. A series *RC* circuit contains a resistance of 40 Ω and a capacitive reactance of 33 Ω across a 24 V source.
    a. Sketch the schematic diagram.
    b. Calculate Z, I, V_R, V_C, I_R, I_C, and θ.

24. A parallel *RC* circuit contains a resistance of 10 kΩ and a capacitive reactance of 5 kΩ across a 100 V source.
    a. Sketch the schematic diagram.
    b. Calculate I_R, I_C, I_T, Z, V_R, V_C, and θ.

25. Calculate V_R and V_C for the circuits seen in Figure 10-23(a) and (b).

26. Calculate the impedance of the four circuits shown in Figure 10-24.

27. In Figure 10-25, the output voltage, since it is taken across the capacitor, will _____ the voltage across the resistor by _____ degrees.

28. If the positions of the capacitor and resistor in Figure 10-25 are reversed, the output voltage, since it is now taken across the resistor, will _____ the voltage across the capacitor by _____ degrees.

29. Calculate the resistive power, reactive power, apparent power, and power factor for the circuit seen in Figure 10-26; $V_{in} = 24$ V and $f = 35$ kHz.

30. Refer to Figure 10-26 and calculate the following:
    a. [Figure 10-26(a)] X_C, I, Z, I_R, θ, V_R, V_C
    b. [Figure 10-26(b)] V_R, V_C, I_R, I_C, I_T, Z, θ

FIGURE 10-24

FIGURE 10-25

(a)

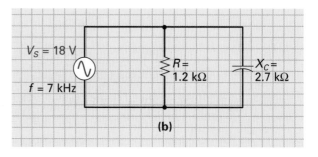

(b)

FIGURE 10-26

Electromagnetism and Electromagnetic Induction

11

The Great Experimenter

Michael Faraday was born to James and Margaret Faraday on September 22, 1791. At twenty-two, the gifted and engaging Faraday was at the right place at the right time, and with the right talents. He impressed the brilliant professor Humphry Davvy, who made him his research assistant at Count Rumford's Royal Institution. After only two years, Faraday was given a promotion and an apartment at the Royal Institution, and in 1821 he married. He lived at the Royal Institution for the rest of his active life, working in his laboratory, giving very dynamic lectures, and publishing over 450 scientific papers. Unlike many scientific papers, Faraday's papers would never use a calculus formula to explain a concept or idea. Instead, Faraday would explain all of his findings using logic and reasoning, so that a person trying to understand science did not need to be a scientist. It was this gift of being able to make even the most complex areas of science easily accessible to the student, coupled with his motivational teaching style, that made him so popular.

In 1855 he had written three volumes of papers on electromagnetism, the first dynamo, the first transformer, and the foundations of electrochemistry, a large amount on dielectrics and even some papers on plasma. The unit of capacitance is measured in *farads,* in honor of his work in these areas of science.

Faraday began two series of lectures at the Royal Institution, which have continued to this day. Michael and Sarah Faraday were childless, but they both loved children, and in 1860 Faraday began a series of Christmas lectures expressly for children, the most popular of which was called "The Chemical History of a Candle." The other lecture series was the "Friday Evening Discourses," of which he himself delivered over a hundred. These very dynamic, enlightening, and entertaining lectures covered areas of science or technology for the lay person, and were filled with demonstrations. On one evening in 1846, an extremely nervous and shy speaker ran off just moments before he was scheduled to give the Friday Evening Discourse. Faraday had to fill in, and to this day a tradition is still enforced whereby the lecturer for the Friday Evening Discourse is locked up for half an hour before the presentation with a single glass of whiskey.

Faraday was often referred to as "the great experimenter," and it was this consistent experimentation that led to many of his findings. He was fascinated by science and technology, and was always exploring new and sometimes dangerous horizons. In fact, in one of his reports he states, "I have escaped, not quite unhurt, from four explosions." When asked to comment on experimentation, his advice was to "Let your imagination go, guiding it by judgment and principle, but holding it in and directing it by experiment. Nothing is so good as an experiment which, while it sets an error right, gives you as a reward for your humility an absolute advance in knowledge."

Introduction

It was during a classroom lecture in 1820 that Danish physicist Hans Christian Oersted accidentally stumbled on an interesting reaction. As he laid a compass down on a bench he noticed that the compass needle pointed to an adjacent conductor that was carrying a current, instead of pointing to the earth's north pole. It was this discovery that first proved that magnetism and electricity were very closely related to one another. This phenomenon is now called *electromagnetism* since it is now known that any conductor carrying an electro or electrical current will produce a magnetic field.

In 1831, the English physicist Michael Faraday explored further Oersted's discovery of electromagnetism and found that the process could be reversed. Faraday observed that if a conductor was passed through a magnetic field, a voltage would be induced in the conductor and cause a current to flow. This phenomenon is referred to as *electromagnetic induction.*

In this chapter we examine the terms and characteristics of electromagnetism (electricity to magnetism) and electromagnetic induction (magnetism to electricity).

TIME LINE

In 1820, a Danish physicist, Hans C. Oersted (1777-1851), accidentally discovered an interesting phenomenon. Placing a compass near a current-carrying conductor, he noticed that the needle of the compass pointed to the conductor rather than north. He was quick to realize that electricity and magnetism were related, and in honor of his work, the unit oersted was adopted for the unit of magnetic field strength.

Electromagnetism

Relates to the magnetic field generated around a conductor when current is passed through it.

11-1 ELECTROMAGNETISM

The electron plays a very important role in magnetism. However, the real key or link between electricity and a magnetic field is motion. Anytime a charged particle moves, a magnetic field is generated. As a result, current flow, which is the movement of electrons, produces a magnetic field.

11-1-1 *Atomic Theory of Electromagnetism*

Every orbiting electron in motion around its nucleus generates a magnetic field. When electrons are forced to leave their parent atom by voltage and flow toward the positive polarity, they are all moving in the same direction and each electron's magnetic field will add to the next. The accumulation of all these electron fields will create a magnetic field around the conductor, as shown in Figure 11-1.

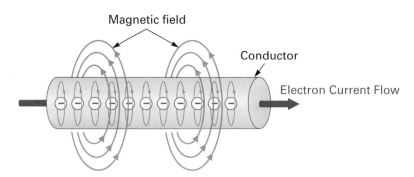

FIGURE 11-1   Electron Magnetic Fields.

SELF-TEST EVALUATION POINT FOR SECTION 11-1-1

Use the following questions to test your understanding of Section 11-1-1.

1. True or false: An electric field always encircles a current-carrying conductor.

2. What subatomic particle is said to have its own magnetic field?

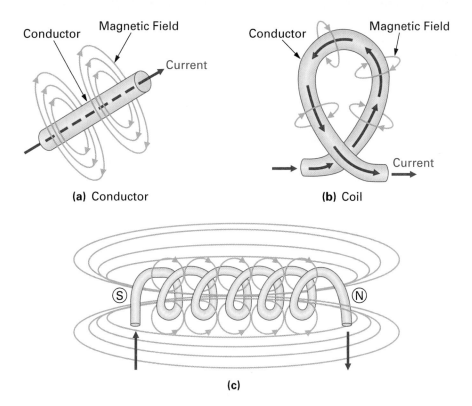

(a) Conductor   **(b)** Coil

(c)

FIGURE 11-2   **Conductor and Coil Electromagnetism.**

11-1-2   *DC, AC, and the Electromagnet*

A magnetic field results whenever a current flows through any piece of conductor or wire, as shown in Figure 11-2(a). If a conductor is wound to form a spiral, as illustrated in Figure 11-2(b), the conductor, which is now referred to as a **coil,** will, as a result of current flow, develop a magnetic field, which will sum or intensify within the coil. If many coils are wound in the same direction, an **electromagnet** is formed, as shown in Figure 11-2(c), which will produce a concentrated magnetic field whenever a current is passed through its coils.

The **left-hand rule** can be applied to electromagnets to determine which of the poles will be the north end of the electromagnet. This rule is illustrated in Figure 11-3, which shows that if you wrap the fingers of your left hand around the coil so that your fingers are pointing in the direction of current flow, your thumb will be pointing to the north end of the electromagnet.

Up until now, we have only been discussing current flow through a coil in one direction (dc). A dc voltage produces a fixed current in one direction and therefore generates a magnetic field in a coil of fixed polarity, as shown in Figure 11-4(a).

FIGURE 11-3   **Left-Hand Rule for Electromagnets.**

North ◄──

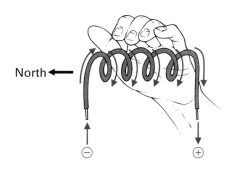

Coil

Number of turns of wire wound around a core to produce magnetic flux (an electromagnet) or to react to a changing magnetic flux (an inductor).

Electromagnet

A magnet consisting of a coil wound on a soft iron or steel core. When current is passed through the coil a magnetic field is generated and the core is strongly magnetized to concentrate the magnetic field.

Left-Hand Rule

If the fingers of the left hand are placed around the wire so that the thumb points in the direction of the electron flow, the fingers will be pointing in the direction of the magnetic field being produced by the conductor.

FIGURE 11-4   Electromagnets. (a) DC Electromagnet. (b) AC Electromagnet.

Alternating current (ac) is continually varying, and as the polarity of the magnetic field is dependent on the direction of current flow (left-hand rule), the magnetic field will also be alternating in polarity, as shown in Figure 11-4(b). Let us look at times 1 through 4 in Figure 11-4 in more detail.

Time 1: The alternating voltage has risen to a maximum positive level and causes current to flow in the direction seen in the circuit. This will cause a magnetic field with a south pole above and a north pole below.

Time 2: Between positions 1 and 2, the voltage, and therefore current, will decrease from a maximum positive value to zero. This will cause a corresponding collapse of the magnetic field from maximum (time 1) to zero (time 2).

Time 3: Voltage and current increase from zero to maximum negative between positions 2 and 3. The increase in current flow causes a similar increase or buildup of magnetic flux, producing a north pole above and south pole below.

Time 4: From time 3 to time 4, the current within the circuit diminishes to zero, and the magnetic field once again collapses. The cycle then repeats.

In regard to current flow, therefore, we can say that:

1. A direct current (dc) produces a constant magnetic field of a fixed polarity, for example, north–south.

2. An alternating current (ac) produces an alternating magnetic field, which continuously switches polarity, for example, north–south, south–north, north–south, and so on.

11-1-3 *Magnetic Terms*

The magnetic field strength of an electromagnet can be increased by increasing the number of turns in its coil, increasing the current through the electromagnet, or decreasing the length of the coil ($H = I \times N/l$). The field strength can be increased further if an iron core is placed within the electromagnet, as illustrated in Figure 11-5(a). This is because an iron core has less *reluctance* or opposition to the magnetic lines of force than air, so the *flux density (B)* is increased. Another way of saying this is that the *permeability* or conductance of magnetic lines of force within iron is greater than that of air, and if the permeability of iron is large, its reluctance must be small. The symbol for an iron-core electromagnet is seen in Figure 11-5(b).

The **B–H curve** in Figure 11-6 illustrates the relationship between the two most important magnetic properties: flux density (*B*) and *magnetizing force (H)*. Figure 11-6(a) illustrates the B–H curve, while Figure 11-6(b) illustrates the positive rising portion of the ac current that is being applied to the iron-core electromagnetic circuit in Figure 11-6(c).

The magnetizing force is actually equal to $H = I$ (current) $\times N$ (number of turns) /l (length of coil); but since the number of turns and length of coil are fixed for the coil being used, the magnetizing force (*H*) is proportional to the current (*I*) applied, which is shown

B–H Curve

Curve plotted on a graph to show successive states during magnetization of a ferromagnetic material.

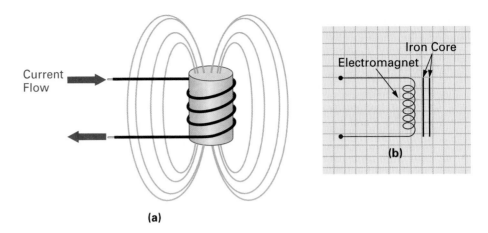

(a)

FIGURE 11-5   Iron-Core Electromagnet. (a) Operation. (b) Symbol.

in Figure 11-6(b). The positive rise of the current from zero to maximum positive is applied through the electromagnet and will produce a corresponding bloom or buildup in magnetic flux at a rate indicated by the *B–H* curve shape in Figure 11-6(a).

It is important to note that as the magnetizing force (current) is increased, there are three distinct stages in the change of flux density or magnetic flux out.

Stage 1: Up to this point, the increase in flux density is slow, as a large amount of force is required to begin alignment of the molecule magnets.

Stage 2: Increase in flux density is now rapid and almost linear as the molecule magnets are aligning easily.

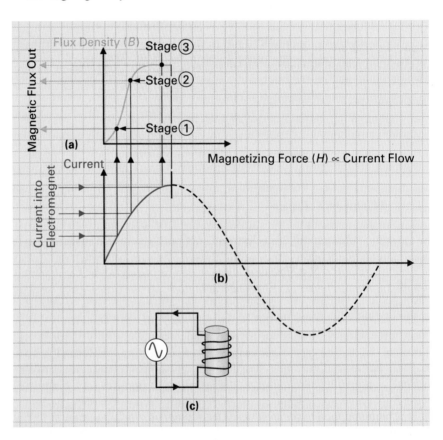

FIGURE 11-6   (a) Flux Density versus Magnetizing Force Curve (*B–H* Curve). (b) Current Applied to Circuit. (c) Circuit.

CHAPTER 11 / ELECTROMAGNETISM AND ELECTROMAGNETIC INDUCTION

Stage 3: In this state the molecule magnets cannot be magnetized any further because they are all fully aligned and no more flux density can be easily obtained. This is called the **saturation point,** and it is the state of magnetism beyond which an electromagnet is incapable of further magnetic strength. It is the point beyond which the *B–H* curve is a straight, horizontal line, indicating no change.

Saturation can easily be described by a simple analogy of a sponge. A dry sponge can only soak up a certain amount of water. As it continues to absorb water, a point will be reached where it will have soaked up the maximum amount of water possible. At this point, the sponge is said to be saturated with water, and no matter how much extra water you supply, it cannot hold any more.

The electromagnet is saturated at stage 3 and cannot produce any more magnetic flux, even though more magnetizing force is supplied as the sine wave continues on to its maximum positive level.

Looking at these three stages and the *B–H* curve that is produced, you can see that, in fact, the magnetization (setting up of the magnetic field, *B*) lags the magnetizing force (*H*) because of molecular friction. This lag or time difference between magnetizing force (*H*) and flux density (*B*) is known as **hysteresis.**

Figure 11-7(a) illustrates what is known as a *hysteresis loop,* which is formed when you plot magnetizing force (*H*) against flux density (*B*) through a complete cycle of alternating current, as seen in Figure 11-7(b). Initially, when the electric circuit switch is open,

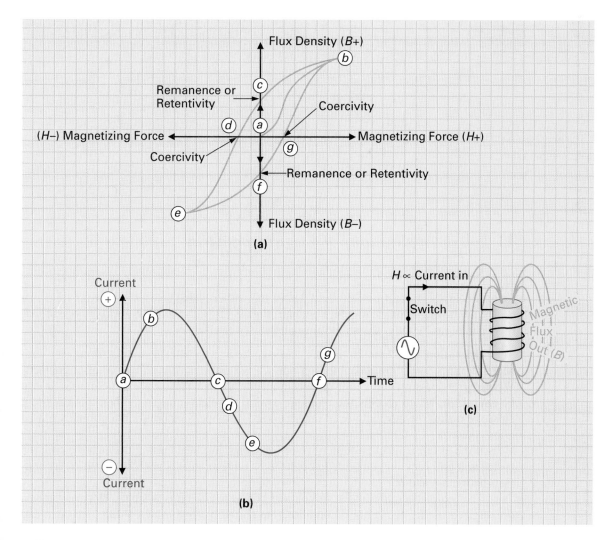

FIGURE 11-7   *B–H* Hysteresis Loop.

the iron core is unmagnetized. Therefore, both H and B are zero at point a. When the switch is closed, as seen in Figure 11-7(c), the current [Figure 11-7(b)] is increased and flux density [Figure 11-7(a)] increases until saturation point b is reached. This part of the waveform (from point a to b) is exactly the same as the B–H curve, discussed previously. The current continues on beyond saturation point b, but the flux density cannot increase beyond saturation.

At point c, the magnetizing force (current) is zero and B (flux density) falls to a value c that is the positive magnetic flux remaining after the removal of the magnetizing force (H). This particular value of flux density is termed **remanence or retentivity.**

The current or magnetizing force now reverses, and the amount of current in the reverse direction that causes flux density to be brought down from c to zero (point d) after the core has been saturated is termed the **coercive force.** The current, and therefore magnetizing force, continues on toward a maximum negative until saturation in the opposite magnetic polarity occurs at point e.

At point f, the magnetizing force (current) is zero and B falls to remanence, which is the negative magnetic flux remaining after the removal of the magnetizing force H. The value of current between f and g is the coercive force needed in the reverse direction to bring flux density down to zero (point g).

Remanence or Retentivity

Amount a material remains magnetized after the magnetizing force has been removed.

Coercive Force

Magnetizing force needed to reduce the residual magnetism within a material to zero.

Use the following questions to test your understanding of Section 11-1-3.

Define the following:

1. Flux density
2. Magnetizing force
3. True or false: The hysteresis loop is formed when you plot H against B through a complete cycle of ac.
4. What is magnetic saturation?

11-1-4 *Applications of Electromagnetism*

Relay

A **relay** is an electromechanical device that either makes (closes) or breaks (opens) a circuit by moving contacts together or apart. Figure 11-8(a) shows the normally open (NO) relay and Figure 11-8(b) shows the normally closed (NC) relay.

Operation. In both cases, the relay consists basically of an electromagnet connected to lines x and y, a movable iron arm known as the armature, and a set of contacts. When current passes from x to y, the electromagnet generates a magnetic field, or is said to be **energized,** which attracts the armature toward the electromagnet. When this occurs, it closes or makes the normally open relay contacts and opens or breaks the normally closed relay contacts.

If the electromagnet is deenergized by discontinuing the current through the coil, the spring will pull back the armature to open the NO relay contacts or close the NC relay contacts between A and B.

The "normal" condition for the contacts between A and B is when the electromagnet is deenergized. In the deenergized condition, the normally open relay contacts are open and the normally closed relay contacts are closed.

Relay

Electromechanical device that opens or closes contacts when a current is passed through a coil.

Energized

Being electrically connected to a voltage source so that the device is activated.

CHAPTER 11 / ELECTROMAGNETISM AND ELECTROMAGNETIC INDUCTION

FIGURE 11-8   Relays (a) Single-Pole, Single-Throw (SPST), Normally Open (NO) Relay (Contacts Are Open until Activated). (b) Single-Pole, Single-Throw (SPST), Normally Closed (NC) Relay (Contacts Are Closed until Activated).

The two relays discussed so far are actually single-pole, single-throw relays, as they have one movable contact (single pole) and one stationary contact that the pole can be thrown to, as shown in Figure 11-8. There are actually four basic configurations for relays and all are illustrated in Table 11-1(a). Variations on these basic four can come in all shapes and sizes, with one relay controlling sometimes several sets of contacts. Table 11-1(b) shows several different styles and packages of relays that are available.

Applications of a relay.   The relay is generally used in two basic applications:

1. To enable one master switch to operate several remote or difficultly placed contact switches, as illustrated in Figure 11-9. When the master switch is closed, the relay is energized, closing all its contacts and turning on all the lights. The advantage of this is twofold in that, first, the master switch can turn on three lights at one time, which saves time for the operator, and second, only one set of wires need be taken from the master switch to the lights, rather than three sets for all three lights.

2. The second basic application of the relay is to enable a switch in a low-voltage circuit to operate relay contacts in a high-voltage circuit, as shown in Figure 11-10. The operator activates the switch in the safer low-voltage circuit, which will energize the relay, closing its contacts and connecting the more dangerous high voltage to the motor.

TABLE 11-1 Relay Types.

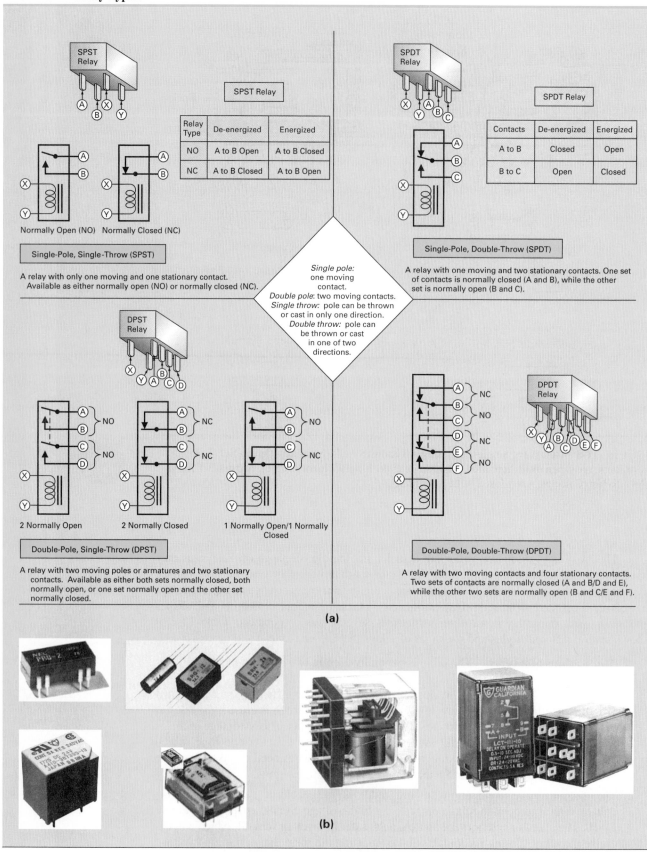

SPST Relay

Relay Type	De-energized	Energized
NO	A to B Open	A to B Closed
NC	A to B Closed	A to B Open

Normally Open (NO)   Normally Closed (NC)

Single-Pole, Single-Throw (SPST)

A relay with only one moving and one stationary contact. Available as either normally open (NO) or normally closed (NC).

Single pole: one moving contact. *Double pole:* two moving contacts. *Single throw:* pole can be thrown or cast in only one direction. *Double throw:* pole can be thrown or cast in one of two directions.

SPDT Relay

Contacts	De-energized	Energized
A to B	Closed	Open
B to C	Open	Closed

Single-Pole, Double-Throw (SPDT)

A relay with one moving and two stationary contacts. One set of contacts is normally closed (A and B), while the other set is normally open (B and C).

2 Normally Open   2 Normally Closed   1 Normally Open/1 Normally Closed

Double-Pole, Single-Throw (DPST)

A relay with two moving poles or armatures and two stationary contacts. Available as either both sets normally closed, both normally open, or one set normally open and the other set normally closed.

Double-Pole, Double-Throw (DPDT)

A relay with two moving contacts and four stationary contacts. Two sets of contacts are normally closed (A and B/D and E), while the other two sets are normally open (B and C/E and F).

(a)

(b)

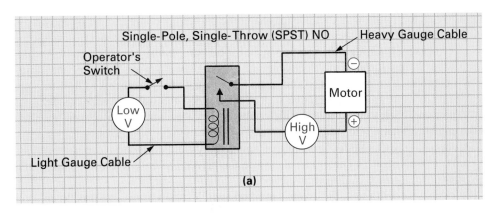

FIGURE 11-9   One Master Switch Operating Several Remotes. (a) Schematic. (b) Pictorial.

FIGURE 11-10   Low-Current Switch Enabling a High-Current Circuit. (a) Schematic. (b) Pictorial.

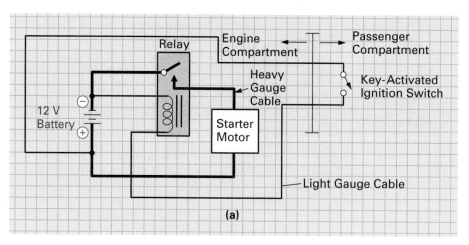

Solenoid

Coil and movable iron core that when energized by an alternating or direct current will pull the core into a central position.

(b)

FIGURE 11-11   Automobile Starter Circuit.

As an application, Figure 11-11 shows an automobile starter circuit. In this circuit a relay is being used to supply the large dc current needed to activate a starter motor in an automobile. When the ignition switch is engaged in the passenger compartment by the driver, current flows through a light-gauge wire from the negative side of the battery, through the relay's electromagnet, through the ignition switch, and back to the positive side of the battery. This current flow through the electromagnet of the relay energizes the relay and closes the relay's contacts. Closing the relay's contacts makes a path for the current to flow through the heavy-gauge cable from the negative side of the battery, through the relay contacts and starter motor, and back to the positive side of the battery. The starter motor's output shaft spins the engine, causing it to start.

This application is a perfect example of how a relay can be used to close contacts in a heavy-current (heavy-gauge cable) circuit, while the driver has only to close contacts in a small-current (light-gauge cable) circuit.

If the relay were omitted, the driver's ignition switch would have to be used to connect the 12 V and large current to the starter motor. This would mean that:

Reed Relay or Switch

Relay that consists of two thin magnetic strips within a glass envelope with a coil wrapped around the envelope so that when it is energized the relay's contacts or strips will snap together, making a connection between the two leads attached to each of the reed strips.

1. Heavy-gauge, expensive cable would need to be connected in a longer path between starter motor and passenger compartment.

2. The ignition switch would need to be larger to handle the heavier current.

3. The driver would be in closer proximity to a more dangerous, high-current circuit.

Reed relays and switches.   The **reed relay or switch** consists of two flat magnetic strips mounted inside a capsule, which is normally made of glass. The reed relay and reed switch are illustrated in Figure 11-12(a) and (b). The reed relay differs from the reed switch

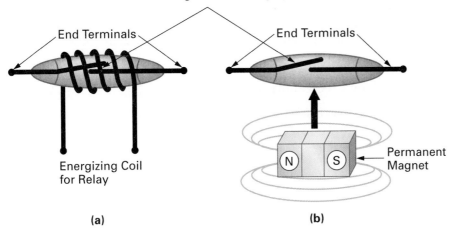

(a)   **(b)**

FIGURE 11-12   Reed Relay and Switch. (a) Reed Relay: Contacts Close When
Electromagnetic Coil Is Energized. (b) Reed Switch: Contacts Close When External Magnetic
Field Comes in Close Proximity, Due to Induced Magnetism in the Contacts.

in that a reed relay has its own energizing coil, while the reed switch needs an external magnetic force to operate.

    The reed relay and reed switch are both operated by a magnetic force that is provided
by a coil for the reed relay and by a separate permanent magnet for the reed switch. When
the magnetic force is present, opposite magnetic polarities are induced in the overlapping,
high-permeability reed blades, causing them to attract one another by induced magnetism
and snap together, thus closing the circuit. When the magnetic force is removed, the blades
spring apart due to spring tension and open the circuit.

    Referring to Figure 11-13 you will see a simple home security circuit. When the win-
dow is closed in the normal condition, the permanent magnet is directly adjacent to the reed
switch, and therefore its contacts are closed, allowing current to flow from the negative side
of the small 3 V battery, through the closed reed switch contacts, the relay's electromagnet,
and back to the positive side of the 3 V battery. The relay is a normally closed (NC) SPST.
As current is flowing through the coil of the relay, the contacts are open, preventing the large
positive and negative voltage from the 12 V battery from reaching the siren.

    If the window is forced and opened by an intruder, the permanent magnet will no
longer be in close proximity to the reed switch and the reed switch's contacts will open.
When the reed switch's contacts open, the relay's coil will no longer have current flowing

FIGURE 11-13   Reed Switch as Part of a Home Security System.

through it, and therefore the relay will deenergize and its contacts will return to their normal closed condition. A large current is now permitted to flow from the negative side of the large 12 V battery, through the relay's contacts, the siren, and back to the positive side of the battery. The siren will sound, as it now has 12 V applied to it, and alert the occupant of the home.

Solenoid-Type Electromagnet

Up to now, electromagnets have been used to close or open a set, or sets, of contacts to either make or break a current path. These electromagnets have used a stationary soft-iron core. Some electromagnets are constructed with movable iron cores, as shown in Figure 11-14, which can be used to open or block the passage of a gas or liquid through a valve. These are known as **solenoid**-type electromagnets.

When no current is flowing through the solenoid coil, no magnetic field is generated, so no magnetic force is exerted on the movable iron core, as shown in Figure 11-14(a), and therefore the compression spring maintains it in the up position, with the valve plug on the end of the core preventing the passage of either a liquid or gas through the valve (valve closed).

When a current flows through the electromagnet, the solenoid coil is energized, creating a magnetic field, as shown in Figure 11-14(b). Due to the influence of the coil's magnetic field, the movable soft-iron core will itself generate a magnetic field, as seen in the inset in Figure 11-14(b). This condition will create a north pole at the top of the solenoid coil and a south pole at the bottom of the movable core, and the resulting attraction will pull down the core (which is free to slide up and down), pulling with it the valve plug and opening the valve.

Solenoid-type electromagnets are actually constructed with the core partially in the coil and are used in washing machines to control water and in furnaces to control gas.

Solenoid

Coil and movable iron core that when energized by an alternating or direct current will pull the core into a central position.

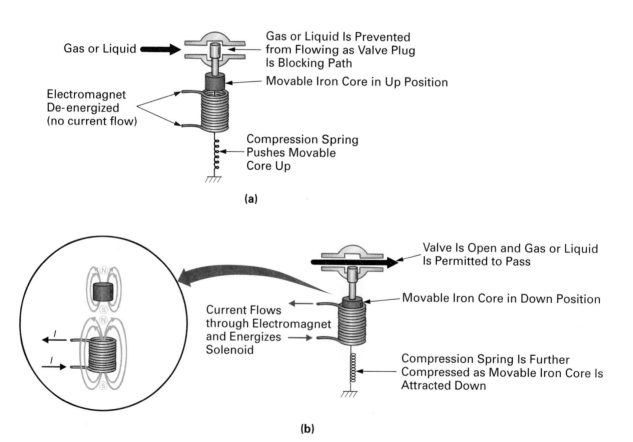

FIGURE 11-14   Solenoid-Type Electromagnet. (a) Deenergized. (b) Energized.

CHAPTER 11 / ELECTROMAGNETISM AND ELECTROMAGNETIC INDUCTION

Use the following questions to test your understanding of Section 11-1-4.

1. List two applications of the electromagnet.
2. What is the difference between an NO and an NC relay?
3. What is the difference between a reed relay and a reed switch?
4. Give an application for a conventional relay and one for the reed switch.

11-2   ELECTROMAGNETIC INDUCTION

Electromagnetic induction is the name given to the action that causes electrons to flow within a conductor when that conductor is moved through a magnetic field. Stated another way, electromagnetic induction is the voltage or emf induced or produced in a coil as the magnetic lines of force link with the turns of a coil. Since this phenomenon was first discovered by Michael Faraday, let us begin by studying **Faraday's Law.**

11-2-1   *Faraday's Law*

In 1831, Michael Faraday carried out an experiment in which he used a coil, a zero center ammeter (galvanometer), and a bar permanent magnet, as shown in Figure 11-15. Faraday's discoveries, which are illustrated in Figure 11-16(a) through (f), are that:

a. When the magnet is moved into a coil the magnetic lines of flux cut the turns of the coil. This action that occurs when the magnetic lines of flux link with a conductor is known as *flux linkage.* Whenever flux linkage occurs, an emf is induced in the coil known as an induced voltage, which causes current to flow within the circuit and the meter to deflect in one direction, for example, to the right. Faraday discovered, in fact, that if the magnet was moved into the coil, or if the coil was moved over the magnet, an emf or voltage was induced within the coil. What actually occurs is that the electrons within the coil are pushed to one end of the coil by the magnetic field, creating an abundance of electrons at one end of the coil (negative charge) and an absence of electrons at the other end of the coil (positive charge). This potential difference across the coil will produce a current flow if a complete path for current (closed circuit) exists.

b. When the magnet is stationary within the coil, the magnetic lines are no longer cutting the turns of the coil, and so there is no induced voltage and the meter returns to zero.

c. When the magnet is pulled back out of the coil, a voltage is induced that causes current to flow in the opposite direction to that of (a) and the meter deflects in the opposite direction, for example, to the left.

Electromagnetic Induction

The voltage produced in a coil due to relative motion between the coil and magnetic lines of force.

Faraday's Law

1. When a magnetic field cuts a conductor, or when a conductor cuts a magnetic field, an electric current will flow in the conductor if a closed path is provided over which the current can circulate.

2. Two other laws relate to electrolytic cells.

FIGURE 11-15   Faraday's Electromagnetic Induction Experiment Components.

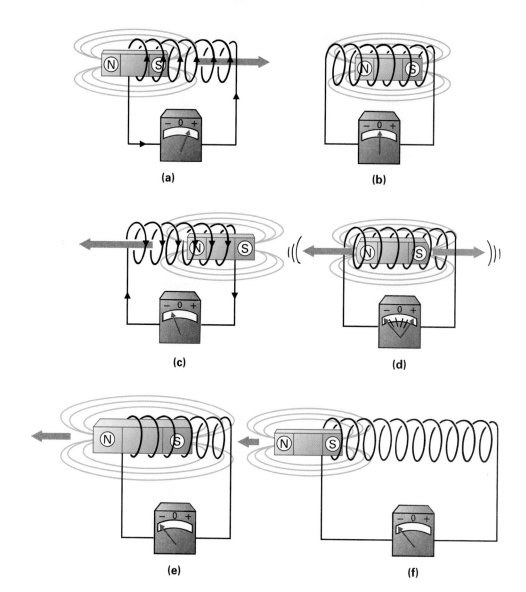

FIGURE 11-16   Faraday's Electromagnetic Induction Discoveries.

  d. If the magnet is moved into or out of the coil at a greater speed, the voltage induced also increases, and therefore so does current.

  e. If the size of the magnet and therefore the magnetic flux strength are increased, the induced voltage also increases.

  f. If the number of turns in the coil is increased, the induced voltage also increases.

  In summary, whenever there is relative motion or movement between the coil of a conductor and the magnetic lines of flux, a potential difference will be induced and this action is called *electromagnetic induction.* The magnitude of the induced voltage depends on the number of turns in the coil, rate of change of flux linkage, and flux density.

TIME LINE

Eduard W. Weber (1804-1891), a German physicist, made enduring contributions to the modern system of electric units, and magnetic flux is measured in webers in honor of his work.

11-2-2   *The Weber*

Consideration of Faraday's law enables us to take a closer look at the unit of flux (ϕ). The **weber** is equal to 10^8 magnetic lines of force, and from the electromagnetic induction point of view, if 1 weber of magnetic flux cuts a conductor for a period of 1 second, a voltage of 1 volt will be induced.

AC Generator

The ac **generator** or alternator is an example of a device that uses electromagnetic induction to generate electricity. If you stroll around your city or town during the day or night and try to spot every piece of equipment, appliance, or device that is running from the ac electricity supplied by generators from the electric power plant, it begins to make you realize how inactive, dark, and difficult our modern society might become without ac power.

When discussing Faraday's discoveries of electromagnetic induction in Figure 11-16, the coil or conductor remained stationary and the magnet was moved. It can be operated in the opposite manner so that the magnetic field remains stationary and the coil or conductor is moved. As long as the magnetic lines of force have relative motion with respect to the conductor, an emf will be induced in the coil. Figure 11-17(a) illustrates a piece of wire wound to form a coil and attached to a galvanometer; its needle rests in the center position, indicating zero current. If the conductor remains stationary within the magnetic lines of flux being generated by the permanent magnet, there is no emf or voltage induced into the wire and so no current flow through the circuit and meter.

If the conductor is moved past the permanent magnet so that it cuts the magnetic lines of flux, as seen in Figure 11-17(b), an emf is generated within the conductor, which is known as an induced voltage, and this will cause current to flow through the wire in one direction and be indicated on the meter by the deflection of the needle to the left.

If the conductor is moved in the opposite direction past the magnetic field, it will induce a voltage of the opposite polarity and cause the meter to deflect to the right, as shown in Figure 11-17(c).

The value or amount of induced voltage is indicated by how far the meter deflects and this voltage is dependent on three factors:

1. The speed at which the conductor passes through the magnetic field
2. The strength or flux density of the magnetic field
3. The number of turns in the coil

(a)   (b)   (c)

FIGURE 11-17   **(a) Stationary Conductor, No Induced Voltage. (b) Moving Conductor, Induced Voltage. (c) Moving Conductor, Induced Voltage.**

(a)

(b)

FIGURE 11-18   AC Generator. (a) 250 kW Diesel Generator. (b) Small (2 kW) Camping Generator.

If the speed at which the conductor passes through the magnetic field or the strength of the magnetic field or the number of turns of the coil is increased, then the induced voltage will also increase. This is merely a repetition of Faraday's law, and in this case we moved the conductor instead of the magnetic field, but the results were the same.

Basic generator.   The basic generator action makes use of Faraday's discoveries of electromagnetic induction. In Chapter 8 the physical appearance of a large 700,000 kW power plant generator was shown. Figure 11-18 illustrates some smaller generators that are mobile and can therefore be used in remote locations.

Figure 11-19 illustrates the basic generator's construction. The basic principle of operation is that the mechanical drive energy input will produce ac electrical energy out by means of electromagnetic induction.

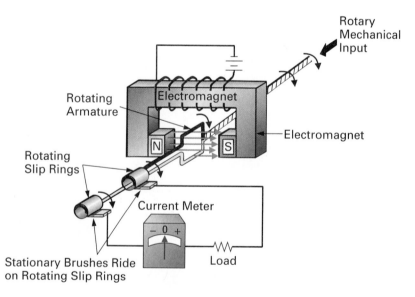

FIGURE 11-19   Basic Generator Construction.

A loop of conductor, known as an **armature,** is rotated continually through 360° by a mechanical drive. This armature resides within a magnetic field produced by a dc electromagnet. Voltage will be induced into the armature and will appear on slip rings, which are also being rotated. A set of stationary brushes rides on the rotating slip rings and picks off the generated voltage and applies this voltage across the load. This voltage will cause current to flow within the circuit and be indicated by the zero center ammeter.

Let's now take a closer look at the armature as it sweeps through 360°, or one complete revolution. Figure 11-20 illustrates four positions of the armature as it rotates through 360° in the clockwise direction.

Position 1: At this instant, the armature is in a position such that it does not cut any magnetic lines of force. The induced voltage in the armature conductor is equal to 0 V and there is no current flow through the circuit.

Position 2: As the conducting armature moves from position 1 to position 2, you can see that more and more magnetic lines of flux will be cut, and the induced emf in the armature (being coupled off by the brushes from the slip rings) will also increase to a maximum value. The current flow throughout the circuit will rise to a maximum as the voltage increases, and this can be seen by the zero center ammeter deflection to the right. From the ac waveform in Figure 11-20, you can see the sinusoidal increase from zero to a maximum positive as the armature is rotated from 0 to 90°.

Position 3: The armature continues its rotation from 90 to 180°, cutting through a maximum quantity and then fewer magnetic lines of force. The induced voltage decreases from the maximum positive at 90° to 0 V at 180°. At the 180° position, as with the 0° position, the armature is once again perpendicular to the magnetic field, and so no lines are cut and the induced voltage is equal to 0 V.

Position 4: From 180 to 270°, the armature is still moving in a clockwise direction, and as it travels toward 270°, it cuts more and more magnetic lines of force. The direction of the cutting action between 0 and 90° causes a positive induced voltage; and since the cutting position between 180 and 270° is the reverse, a negative induced voltage will result in the armature causing current flow in the opposite direction, as indicated by the deflection of the zero center ammeter to the left. The voltage induced when the armature is at 270° will be equal but opposite in polarity to the voltage generated when the armature was at the 90° position. The current will therefore also be equal in value but opposite in its direction of flow.

From position 4 (270°), the armature turns to the 360° or 0° position, which is equivalent to position 1, and the induced voltage decreases from maximum negative to zero. The cycle then repeats.

To summarize, in Figure 11-20, one complete revolution of the mechanical energy input causes one complete cycle of the ac electrical energy output, with a sine wave being generated as a result of circular motion.

Armature
Rotating or moving component of a magnetic circuit.

SELF-TEST EVALUATION POINT FOR SECTION 11-2

Use the following questions to test your understanding of Section 11-2.

1. Define *electromagnetic induction.*
2. Briefly describe Faraday's and Lenz's laws in relation to electromagnetic induction.
3. What waveform shape does the ac generator produce?

FIGURE 11-20 **360° Generator Operation.**

CHAPTER 11 / ELECTROMAGNETISM AND ELECTROMAGNETIC INDUCTION

REVIEW QUESTIONS

Multiple-Choice Questions

1. Electromagnetism was first discovered by:
   a. Hans Christian Oersted   c. James Watt
   b. Heinrich Hertz   d. James Clark Maxwell

2. Every current-carrying conductor generates a/an:
   a. Electric field   c. Both (a) and (b)
   b. Magnetic field   d. None of the above

3. If a greater number of loops of a conductor in a coil are added, a _____ magnetic field is generated.
   a. Weaker   b. Stronger

4. If a smaller current is passed through a coil, a _____ magnetic field is generated.
   a. Weaker   b. Stronger

5. Flux density is equal to the magnetic flux divided by area and is measured in _____.
   a. Webers   c. Maxwells
   b. Teslas   d. Oersteds

6. An electromagnet can be used in the application of:
   a. A relay   c. A circuit breaker
   b. An electric bell   d. All of the above

7. Relays can be used to:
   a. Allow one master switch to enable several others
   b. Allow several switches to enable one master
   c. Allow a switch in a low-current circuit to close contacts in a high-current circuit
   d. Both (a) and (b)
   e. Both (a) and (c)

8. The starter relay in an automobile is used to:
   a. Allow one master switch to enable several others
   b. Allow several switches to enable one master
   c. Allow a switch in a low-current circuit to close contacts in a high-current circuit
   d. Both (a) and (b)

9. The _____ uses an electromagnet around two flat magnetic strips mounted inside a glass capsule.
   a. Reed switch   c. Reed relay
   b. Magnetic circuit breaker   d. Starter relay

10. The reed switch could be used in a/an:
    a. Home security system
    b. Automobile starter
    c. Magnetic-type circuit breaker
    d. All of the above

11. An electromagnet is also known as a:
    a. Coil   d. Both (a) and (c)
    b. Solenoid   e. Both (a) and (b)
    c. Resistor

12. A normally open relay (NO) will have:
    a. Contacts closed until activated
    b. Contacts open until activated
    c. All contacts permanently open
    d. All contacts permanently closed

13. Direct current produces a/an _____ magnetic field of _____ polarity.
    a. Alternating, unchanging   c. Constant, unchanging
    b. Constant, alternating   d. Both (a) and (c)

14. Electromagnetism:
    a. Is the magnetism resulting from electrical current flow
    b. Is the electrical voltage resulting in a coil from the relative motion of a magnetic field
    c. Both (a) and (b)
    d. None of the above

15. Electromagnetic induction:
    a. Is the magnetism resulting from electrical current flow
    b. Is the electrical voltage resulting in a coil from the relative motion of a magnetic field
    c. Both (a) and (b)
    d. None of the above

16. The _____ plots magnetizing force against flux density through a complete cycle of alternating current.
    a. B–H curve   c. Power curve
    b. Coercive force   d. Hysteresis loop

17. The current induced in a coil due to the change in the magnetic flux is such as to oppose the cause producing it. This law was discovered by:
    a. Heinrich Lenz   c. Michael Faraday
    b. Guglielmo Marconi   d. Joseph Henry

18. An ac generator uses _____ to generate ac electricity.
    a. Electromagnetism   c. Magnetism
    b. Electromagnetic induction   d. None of the above

19. The generator converts _____ energy into _____ energy.
    a. Electrical, electrical   c. Chemical, electrical
    b. Mechanical, electrical   d. None of the above

20. The loop of conductor rotated through 360° in a generator is known as a/an:
    a. Electromagnet   c. Armature
    b. Field coil   d. Both (a) and (b)

Inductance and Inductors

12

The Wizard of Menlo Park

Thomas Alva Edison was born to Samuel and Nancy Edison on February 11, 1847. As a young boy he had a keen and inquisitive mind, yet he did not do well at school, so his mother, a former school teacher, withdrew him from school and tutored him at home. In later life he said that his mother taught him to read well and instilled a love for books that lasted the rest of his life. In fact, the inventor's personal library of more than ten thousand volumes is preserved at the Edison Laboratory National Monument in West Orange, New Jersey.

At the age of twenty-nine, after several successful inventions, Edison put into effect what is probably his greatest idea—the first industrial research laboratory.

Choosing Menlo Park in New Jersey, which was then a small rural village, Edison had a small building converted into a laboratory for his fifteen-member staff, and a house built for his wife and two small daughters. When asked to explain the point of this lab, Edison boldly stated that it would produce "a minor invention every ten days and a big thing every six months or so." At the time, most of the scientific community viewed his prediction as preposterous; however, in the next ten years Edison would be granted 420 patents, including those for the electric light bulb, the motion picture, the phonograph, the universal electric motor, the fluorescent lamp, and the medical fluoroscope.

Over one thousand patents were granted to Edison during his lifetime; his achievements at what he called his "invention factory" earned him the nickname "the wizard of Menlo Park." When asked about his genius he said, "Genius is two percent inspiration and ninety-eight percent perspiration."

Introduction

In Chapter 11 two important rules were discussed that relate to this chapter on inductance. The first was that a magnetic field will build up around any current-carrying conductor (electromagnetism), and the second was that a voltage will be induced into a conductor when it is subjected to a moving magnetic field. These two rules form the basis for a phenomenon that is covered in this chapter and is called *self-inductance* or more commonly *inductance*. Inductance, by definition, is the ability of a device to oppose a change in current flow, and the device designed specifically to achieve this function is called the *inductor*. There is, in fact, no physical difference between an inductor and an electromagnet, since they are both coils. The two devices are given different names because they are used in different applications even though their construction and principle of operation are the same. An electromagnet is used to generate a magnetic field in response to current, while an inductor is used to oppose any change in current.

In this chapter we examine all aspects of inductance and inductors, including inductive circuits, types, reactance, testing, and applications.

12-1   SELF-INDUCTION

In Figure 12-1 an external voltage source has been connected across a coil, forcing a current through the coil. This current will generate a magnetic field that will increase in field strength in a short time from zero to maximum, expanding from the center of the conductor (electromagnetism). The expanding magnetic lines of force have relative motion with respect to the stationary conductor, so an induced voltage results (electromagnetic induction). The blooming magnetic field generated by the conductor is actually causing a voltage to be induced in the conductor that is generating the magnetic field. This effect of a current-carrying coil of conductor inducing a voltage within itself is known as **self-inductance.** This phenomenon was first discovered by Heinrich Lenz, who observed that the induced voltage causes an induced (bucking) current to flow in the coil, which opposes the source current producing it.

Figure 12-2(a) shows an inductor connected across a dc source. When the switch is closed, a circuit current will exist through the inductor and the resistor. As the current rises toward its maximum value, the magnetic field expands, and throughout this time of relative motion between field and conductor, an induced voltage will be present. This induced voltage will produce an induced current to oppose the change in the circuit current.

When the current reaches its maximum, the magnetic field, which is dependent on current, will also reach a maximum value and then no longer expand but remain stationary. When the current remains constant, no change will occur in the magnetic field and therefore no relative motion will exist between the conductor and magnetic field, resulting in no

Self-Inductance

The property that causes a counter-electromotive force to be produced in a conductor when the magnetic field expands or collapses with a change in current.

FIGURE 12-1   Self-Induction of a Coil.

Expanding Magnetic Field

Increase in circuit current and, therefore, increase in magnetic field induces voltage in coil. This results in an induced current to oppose circuit current.

(a)

Stationary Magnetic Field

Stationary magnetic field and, therefore, no induced voltage or current to oppose circuit current.

(b)

Collapsing Magnetic Field

No circuit current and, therefore, magnetic field collapses and induces voltage in coil. This results in an induced current in same direction as circuit current.

(c)

FIGURE 12-2   **Self-Inductance. (a) Switch Closed. (b) Constant Circuit Current. (c) Switch Opened.**

induced voltage or current to oppose circuit current, as shown in Figure 12-2(b). The coil has accepted electrical energy and is storing it in the form of a magnetic energy field, just as the capacitor stored electrical energy in the form of an electric field.

If the switch is put in position *B*, as shown in Figure 12-2(c), the current from the battery will be zero, and the magnetic field will collapse as it no longer has circuit current to support it. As the magnetic lines of force collapse, they cut the conducting coils, causing relative motion between the conductor and magnetic field. A voltage is induced in the coil, which will produce an induced current to flow in the same direction as the circuit current was flowing before the switch was opened. The coil is now converting the magnetic field energy into electrical energy and returning the original energy that it stored.

After a short period of time, the magnetic field will have totally collapsed, the induced voltage will be zero, and the induced current within the circuit will therefore also no longer be present.

This induced voltage is called a **counter emf** or *back emf.* It opposes the applied emf (or battery voltage). The ability of a coil or conductor to induce or produce a counter emf within itself as a result of a change in current is called *self-inductance,* or more commonly **inductance** (symbolized *L*). The unit of inductance is the **henry** (H), named in honor of Joseph Henry, an American physicist, for his experimentation within this area of science. The inductance of an inductor is 1 henry when a current change of 1 ampere per second causes an induced voltage of 1 volt. Inductance is therefore a measure of how much counter emf (induced voltage) can be generated by an inductor for a given amount of current change through that same inductor.

This counter emf or induced voltage can be calculated by the formula

$$V_{\text{ind}} = L \times \frac{\Delta I}{\Delta t}$$

where L = inductance, in henrys (H)
ΔI = increment of change of current (I)
Δt = increment of change with respect to time (t)

Counter emf (Counter Electromotive Force)

Abbreviated "counter emf," or "back emf," it is the voltage generated in an inductor due to an alternating or pulsating current and is always of opposite polarity to that of the applied voltage.

Inductance

Property of a circuit or component to oppose any change in current as the magnetic field produced by the change in current causes an induced countercurrent to oppose the original change.

Henry

Unit of inductance.

A larger inductance ($L\uparrow$) will create a larger induced voltage ($V_{ind}\uparrow$), and if the rate of change of current with respect to time is increased ($\Delta I/\Delta t\uparrow$), the induced voltage or counter emf will also increase ($V_{ind}\uparrow$).

EXAMPLE:

What voltage is induced across an inductor of 4 H when the current is changing at a rate of:

   a. 1 A/s?

   b. 4 A/s?

Solution:

   a. $V_{ind} = L \times \dfrac{\Delta I}{\Delta t} = 4\text{ H} \times 1\text{ A/s} = 4\text{ V}$

   b. $V_{ind} = L \times \dfrac{\Delta I}{\Delta t} = 4\text{ H} \times 4\text{ A/s} = 16\text{ V}$

The faster the coil current changes, the larger the induced voltage.

SELF-TEST EVALUATION POINT FOR SECTION 12-1

Use the following questions to test your understanding of Section 12-1.

1. Define self-induction.

2. What is counter emf, and how can it be calculated?

3. Calculate the voltage induced in a 2 mH inductor if the current is increasing at a rate of 4 kA/s.

12-2 THE INDUCTOR

Inductor

Coil of conductor used to introduce inductance into a circuit.

An **inductor** is basically an electromagnet, as its construction and principle of operation are the same. We use the two different names because they have different applications. The purpose of the electromagnet or solenoid is to generate a magnetic field, while the purpose of an inductor or coil is to oppose any change of circuit current.

In Figure 12-3 a steady value of direct current is present within the circuit and the inductor is creating a steady or stationary magnetic field. If the current in the circuit is suddenly increased (by lowering the circuit resistance), the change in the expanding magnetic field will induce a counter emf within the inductor. This induced voltage will oppose the source voltage from the battery and attempt to hold current at its previous low level.

The counter emf cannot completely oppose the current increase, for if it did, the lack of current change would reduce the counter emf to zero. Current therefore incrementally increases up to a new maximum, which is determined by the applied voltage and the circuit resistance ($I = V/R$). Once the new higher level of current has been reached and remains constant, there will no longer be a change. This lack of relative motion between field and conductor will no longer generate a counter emf, so the current will remain at its new higher constant value.

This effect also happens in the opposite respect. If current decreases (by increasing circuit resistance), the magnetic lines of force will collapse because of the reduction of cur-

FIGURE 12-3   Inductor's Ability to Oppose Current Change.

rent and induce a voltage in the inductor, which will produce an induced current in the same direction as the circuit current. These two combine and tend to maintain the current at the higher previous constant level. Circuit current will fall, however, as the induced voltage and current are only present during the change (in this case the decrease from the higher current level to the lower); and once the new lower level of current has been reached and remains constant, the lack of change will no longer induce a voltage or current. So the current will then remain at its new lower constant value.

The inductor is therefore an electronic component that will oppose any changes in circuit current, and this ability or behavior is referred to as *inductance*. Since the current change is opposed by a counter emf, inductance may also be defined as the ability of a device to induce a counter emf within itself for a change in current.

SELF-TEST EVALUATION POINT FOR SECTION 12-2

Use the following questions to test your understanding of Section 12-2.

1. What is the difference between an electromagnet and an inductor?

2. True or false: The inductor will oppose any changes in circuit current.

12-3   FACTORS DETERMINING INDUCTANCE

The inductance of an inductor is determined by four factors:

1. Number of turns
2. Area of the coil
3. Length of the coil
4. Core material used within the coil

All the four factors described can be placed in a formula to calculate inductance.

$$L = \frac{N^2 \times A \times \mu}{l}$$

where L = inductance, in henrys (H)

N = number of turns

A = cross-sectional area, in square meters (m^2)

μ = permeability

l = length of core, in meters (m)

Refer to Figure 12-4(a) and (b) and calculate the inductance of each.

☐ *Solution:*

a. $L = \dfrac{5^2 \times 0.01 \times (6.28 \times 10^{-5})}{0.001} = 15.7 \text{ mH}$

b. $L = \dfrac{10^2 \times 0.1 \times (1.1 \times 10^{-4})}{0.1} = 11 \text{ mH}$

Length is 0.001 m

Area = 0.01 m²
N = 5 Turns
Nickel Core, so $\mu = 6.28 \times 10^{-5}$

(a)

Length is 0.1 m

Area = 0.1 m²
N = 10 Turns
Cast Iron Core, so $\mu = 1.1 \times 10^{-4}$

(b)

FIGURE 12-4   Inductor Examples.

SELF-TEST EVALUATION POINT FOR SECTION 12-3

Use the following questions to test your understanding of Section 12-3.

1. List the four factors that determine the inductance of an inductor.
2. State the formula for inductance.

12-4   INDUCTORS IN COMBINATION

Inductors oppose the change of current in a circuit and so are treated in a manner similar to resistors connected in combination. Two or more inductors in series merely extend the coil length and increase inductance. Inductors in parallel are treated in a manner similar to resistors, with the total inductance being less than that of the smallest inductor's value.

12-4-1 *Inductors in Series*

When inductors are connected in series with one another, the total inductance is calculated by summing all the individual inductances.

$$L_T = L_1 + L_2 + L_3 + \ldots$$

☐ **EXAMPLE:**

Calculate the total inductance of the circuit shown in Figure 12-5.

■ *Solution:*

$$
\begin{aligned}
L_T &= L_1 + L_2 + L_3 \\
&= 5 \text{ mH} + 7 \text{ mH} + 10 \text{ mH} \\
&= 22 \text{ mH}
\end{aligned}
$$

FIGURE 12-5   Inductors in Series.

12-4-2 *Inductors in Parallel*

When inductors are connected in parallel with one another, the reciprocal (two or more inductors) or product-over-sum (two inductors) formula can be used to find total inductance, which will always be less than the smallest inductor's value.

$$L_T = \dfrac{1}{(1/L_1) + (1/L_2) + (1/L_3) + \ldots}$$

$$L_T = \dfrac{L_1 \times L_2}{L_1 + L_2}$$

☐ **EXAMPLE:**

Determine L_T for the circuits in Figure 12-6(a) and (b).

■ *Solution:*

a. Reciprocal formula:

$$
\begin{aligned}
L_T &= \frac{1}{(1/L_1) + (1/L_2) + (1/L_3)} \\
&= \frac{1}{(1/10 \text{ mH}) + (1/5 \text{ mH}) + (1/20 \text{ mH})} \\
&= 2.9 \text{ mH}
\end{aligned}
$$

b. Product over sum:

$$L_T = \frac{L_1 \times L_2}{L_1 + L_2}$$
$$= \frac{10\ \mu\text{H} \times 2\ \mu\text{H}}{10\ \mu\text{H} + 2\ \mu\text{H}}$$
$$= \frac{20 \times 10^{-12}\text{H}}{12\ \mu\text{H}} = 1.67\ \mu\text{H}$$

FIGURE 12-6   Inductors in Parallel.

Use the following questions to test your understanding of Section 12-4.

1. True or false: To calculate total inductance, inductors can be treated in the same manner as capacitors.
2. State the formula for calculating total inductance in:
   a. A series circuit
   b. A parallel circuit
3. Calculate the total circuit inductance if 4 mH and 2 mH are connected:
   a. In series
   b. In parallel

12-5   TYPES OF INDUCTORS

As with resistors and capacitors, inductors are basically divided into the two categories of fixed-value and variable-value inductors, as shown by the symbols in Figure 12-7(a) and (b). Within these two categories, inductors are generally classified by the type of core material used. Figure 12-7(c) shows a variety of different types.

12-5-1   *Fixed-Value Inductors*

Fixed-Value Inductor

An inductor whose value cannot be changed.

With **fixed-value inductors,** the inductance value remains constant and cannot be altered. They are usually made of solid copper wire with an insulating enamel around the conductor. This enamel or varnish on the conductor is needed since the turns of a coil are generally wound on top of one another. The three major types of fixed-value inductors on the market today are:

    1. Air core   2. Iron core   3. Ferrite core

FIGURE 12-7 Inductor Types. (a) Fixed-Value Inductor Symbol. (b) Variable-Value Inductor Symbol. (c) Physical Appearance of Inductors. (Left photo courtesy of Sencore. Right photo courtesy of Coilcraft Inc., Cary, Illinois 60013.)

Air Core

Figure 12-8 illustrates some typical air-core fixed-value inductor types. The insulated copper wire can be wound on nonmagnetic materials such as plastic, ceramic, or Bakelite, which due to their nonmagnetic properties will have no effect on the inductance value. These materials act as a support or form for the inductor, whose wire size may be too small to support itself. If a heavier or more rigid wire size is used and it can support itself, there will be no need for a nonmagnetic form.

FIGURE 12-8 Air-Core Fixed-Value Inductors. (a) Air-Core Inductor. (Courtesy of Dale Electronics.) (b) PCB Inductor. (c) Symbol.

Inductors of this type, which make use of a nonmagnetic form or do not have a form, are known as **air-core inductors.** Air-core inductors have the advantage that their inductance value does not vary with current, as do iron- and ferrite-core inductors. Their typical low values (below 10 μH) normally find application in high-frequency equipment, such as AM and FM radio, TV, and other communications transmitter and receiver circuits.

Iron Core

Figure 12-9 illustrates a typical **iron-core inductor.** The name **choke** is used interchangeably with the word *inductor,* because choke basically describes an inductor's behavior. As we discovered previously, the inductor will oppose any change of current flow, whether it is in the form of pulsating dc or ac, because of its self-inductance, which generates a counter emf to limit or choke the flow of current.

An iron core has a higher permeability figure than air, which means that it will concentrate the magnetic lines of force and therefore increase inductance.

$$L\uparrow = \frac{N^2 \times A \times \mu\uparrow}{l}$$

The EI **ferrous,** or compound of iron, core [Figure 12-9(b)] has two sections, one of which resembles the letter E and the other I. The conductor is wrapped around the center section of the E to form the coil. These inductors or chokes can be made with the highest inductance values, up to the hundreds of henrys, and are used in dc and low-frequency ac circuits. Figure 12-9(c) shows the schematic symbol for an iron-core inductor.

Ferrite Core

Figure 12-10 illustrates two typical **ferrite-core inductors. Ferrite** is a chemical compound, which is basically powdered iron oxide and ceramic. Although its permeability is less than iron, it has a higher permeability than air and can therefore obtain a higher inductance value than air for the same number of turns.

The toroidal (doughnut shaped) inductor illustrated in Figure 12-10(a) has an advantage over the cylinder-shaped core in that the magnetic lines of force do not have to pass through air. With a cylindrical core, the magnetic lines of force merge out of one end of the cylinder, propagate or travel through air, and then enter in the other end of the cylinder. The

(a)

FIGURE 12-9 **Iron-Core Fixed-Value Inductors. (a) Physical Appearance. (Courtesy of Coilcraft.) (b) Construction. (c) Symbol.**

FIGURE 12-10 Ferrite-Core Fixed-Value Inductors. (a) Toroidal Inductor. (Courtesy of Coilcraft.) (b) Molded Inductor. (Courtesy of Dale Electronics.) (c) Schematic Symbol.

nonmagnetic air, which has a low permeability, will reduce the strength of the field and therefore the inductance value:

$$L\downarrow = \frac{N^2 \times A \times \mu\downarrow}{l}$$

If one could bend around either end of the cylinder to make a toroid, the magnetic lines of force emerging from one end of the inductor's core would pass from ferrite to ferrite, which has a high permeability.

Toroidal ferrite-core inductors have a greater inductance than the cylindrical-type inductors; however, they are more expensive to manufacture. They are used in both low- and high-frequency applications.

The molded inductor, shown in Figure 12-10(b), consists of a spiral metal film deposited on a cylindrical ferrite core. The entire assembly is encapsulated, and the available values range from 1.2 μH to 10 mH, with maximum currents of about 70 mA. Due to its small size, the value is often color coded onto the body instead of using alphanumeric or typographical labels, which can be used on all of the other larger inductors. Figure 12-10(c) shows the schematic symbol for a ferrite-core inductor.

12-5-2 *Variable-Value Inductors*

With **variable-value inductors,** the inductance value can be changed by adjusting the position of the movable core with respect to the stationary coil. There is basically only one type of variable-value inductor in wide use today: the ferrite core. Figure 12-11 illustrates a typical ferrite-core variable inductor. A screw adjustment moves a sliding ferrite core or slug farther in or out of the stationary coil. If the slug is all the way out, the permeability figure is

Variable-Value Inductor
An inductor whose value can be varied.

low, because an air core is being used ($L\downarrow \propto \mu\downarrow$). If the slug is adjusted and screwed into the coil, the air is replaced by the ferrite slug, which has a higher permeability and, therefore, inductance value ($L\uparrow \propto \mu\uparrow$).

Variable inductors should only be adjusted with a plastic or nonmetallic alignment tool since any metal object in the vicinity of the inductor will interfere with the magnetic field and change the inductance of the inductor. Variable-value inductors are used extensively in radio circuits.

FIGURE 12-11 **Ferrite-Core Variable-Value Inductors. (a) Construction. (b) Physical Appearance. (Courtesy of Sencore.) (c) Symbol.**

SELF-TEST EVALUATION POINT FOR SECTION 12-5

Use the following questions to test your understanding of Section 12-5.

1. List the three fixed-value inductor types.
2. Under which category would a fixed-value inductor with a nonmagnetic core come?
3. What is a ferrite compound?
4. What advantage does the toroid-shaped inductor have over a cylindrical inductor?
5. What type of variable-value inductor is the only one in wide use today?
6. What factor is varied to change the value of the variable inductor?

12-6   INDUCTIVE TIME CONSTANT

Inductors will not have any effect on a steady value of direct current (dc) from a dc voltage source. If, however, the dc is changing (pulsating), the inductor will oppose the change whether it is an increase or decrease in direct current, because a change in current causes the magnetic field to expand or contract, and in so doing it will cut the coil of the inductor and induce a voltage that will counter the applied emf.

12-6-1   *DC Current Rise*

Figure 12-12(a) illustrates an inductor (L) connected across a dc source (battery) through a switch and series-connected resistor. When the switch is closed, current will flow and the magnetic field will begin to expand around the inductor. This field cuts the coils of the inductor and induces a counter emf to oppose the rise in current. Current in an inductive circuit, therefore, cannot rise instantly to its maximum value, which is determined by Ohm's

(a)

(b) Current Rise

FIGURE 12-12 **DC Inductor Current Rise.**

law ($I = V/R$). Current will in fact take a time to rise to maximum, as graphed in Figure 12-12(b), due to the inductor's ability to oppose change.

It will actually take five time constants (5τ) for the current in an inductive circuit to reach maximum value. This time can be calculated by using the formula

$$\tau = \frac{L}{R} \quad \text{seconds}$$

The constant to remember is the same as before: 63.2%. In one time constant ($1 \times L/R$) the current in the *RL* circuit will have reached 63.2% of its maximum value. In two time constants ($2 \times L/R$), the current will have increased 63.2% of the remaining current, and so on, through five time constants.

For example, if the maximum possible circuit current is 100 mA and an inductor of 4 H is connected in series with a resistor of 2 Ω, the current will increase as shown in Figure 12-13.

Referring back to the $\tau = L/R$ formula, you will notice that how quickly an inductor will allow the current to rise to its maximum value is proportional to the inductance and inversely proportional to the resistance. A larger inductance increases the strength of the magnetic field, so the opposition or counter emf increases, and the longer it takes for current to rise to a maximum ($\tau\uparrow = L\uparrow/R$). If the circuit resistance is increased, the maximum current will be smaller, and a smaller maximum is reached more quickly than a higher ($\tau\downarrow = L/R\uparrow$).

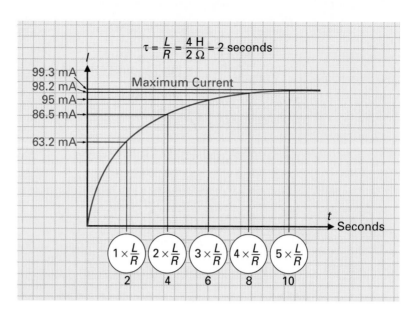

$$\tau = \frac{L}{R} = \frac{4\text{ H}}{2\,\Omega} = 2 \text{ seconds}$$

99.3 mA
98.2 mA
95 mA
86.5 mA

63.2 mA

Maximum Current

$1 \times \dfrac{L}{R}$ $2 \times \dfrac{L}{R}$ $3 \times \dfrac{L}{R}$ $4 \times \dfrac{L}{R}$ $5 \times \dfrac{L}{R}$

2   4   6   8   10

t
Seconds

FIGURE 12-13   **Exponential Current Rise.**

12-6-2   *DC Current Fall*

When the inductor's dc source of current is removed, as shown in Figure 12-14(a) by placing the switch in position *B,* the magnetic field will collapse and cut the coils of the inductor, inducing a voltage and causing a current to flow in the same direction as the original source current. This current will exponentially decay, or fall from the maximum to zero level, in five time constants ($5 \times L/R = 5 \times 4/2 = 10$ seconds), as shown in Figure 12-14(b).

▢   **EXAMPLE:**

Calculate the circuit current at each of the five time constants if a 12 V dc source is connected across a series *RL* circuit, and $R = 60\ \Omega$ and $L = 24$ mH. Plot the results on a graph showing current against time.

■   *Solution:*

$$\text{maximum current, } I_{\max} = \frac{V_S}{R} = \frac{12\text{ V}}{60\,\Omega} = 200 \text{ mA}$$

$$\text{time constant, } \tau = \frac{L}{R} = \frac{24\text{ mH}}{60\,\Omega} = 400\ \mu\text{s}$$

At one time constant (400 μs after source voltage is applied), the current will be

$$I = 63.2\% \text{ of } I_{\max}$$
$$= 0.632 \times 200 \text{ mA} = 126.4 \text{ mA}$$

At two time constants (800 μs after source voltage is applied):

$$I = 86.5\% \text{ of } I_{\max}$$
$$= 0.865 \times 200 \text{ mA} = 173 \text{ mA}$$

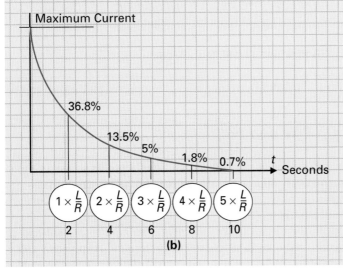

FIGURE 12-14　Exponential Current Fall.

At three time constants (1200 μs or 1.2 ms):

$$I = 95\% \text{ of } I_{max}$$
$$= 0.95 \times 200 \text{ mA} = 190 \text{ mA}$$

At four time constants (1.6 ms):

$$I = 98.2\% \text{ of } I_{max}$$
$$= 0.982 \times 200 \text{ mA} = 196.4 \text{ mA}$$

At five time constants (2 ms):

$$I = 99.3\% \text{ of } I_{max}$$
$$= 0.993 \times 200 \text{ mA} = 198.6 \text{ mA, approximately maximum (200 mA)}$$

See Figure 12-15.

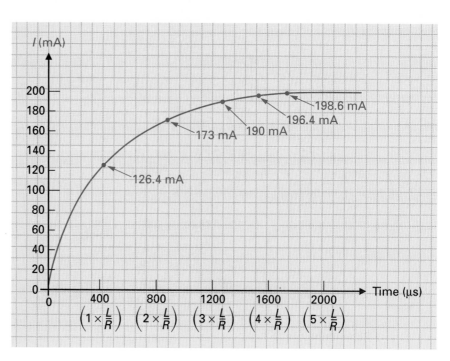

FIGURE 12-15   Exponential Current Rise Example.

12-6-3   *AC Rise and Fall*

If an alternating (ac) voltage is applied across an inductor, as shown in Figure 12-16(a), the inductor will continuously oppose the alternating current because it is always changing. Figure 12-16(b) shows the phase relationship between the voltage across an inductor or counter emf and the circuit current.

The current in the circuit causes the magnetic field to expand and collapse and cut the conducting coils, resulting in an induced counter emf. At points X and Y, the steepness of the current waveform indicates that the current will be changing at its maximum rate, and therefore the opposition or counter emf will also be maximum. When the current is at

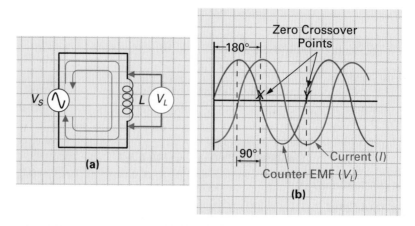

FIGURE 12-16   The Voltage across an Inductor Leads the Circuit Current by 90° in an Inductive Circuit.

CHAPTER 12 / INDUCTANCE AND INDUCTORS

its maximum positive or negative value, it has a very small or no rate of change (flat peaks). Therefore, the opposition or counter emf should be very small or zero, as can be seen by the waveforms. The counter emf is, therefore, said to be 90° out of phase with the circuit current.

To summarize, we can say that the voltage across the inductor (V_L) or counter emf leads the circuit current (I) by 90°.

SELF-TEST EVALUATION POINT FOR SECTION 12-6

Use the following questions to test your understanding of Section 12-6.

1. How does the inductive time constant relate to the capacitive time constant?
2. True or false: The greater the value of the inductor, the longer it would take for current to rise to a maximum.
3. True or false: A constant dc level is opposed continuously by an inductor.
4. What reaction does an inductor have to ac?

12-7   INDUCTIVE REACTANCE

Reactance is the opposition to current flow without the dissipation of energy, as opposed to resistance, which is the opposition to current flow with the dissipation of energy.

Inductive reactance (X_L) is the opposition to current flow offered by an inductor without the dissipation of energy. It is measured in ohms and can be calculated by using the formula:

Inductive Reactance

Measured in ohms, it is the opposition to alternating or pulsating current flow without the dissipation of energy.

$$X_L = 2\pi \times f \times L$$

where X_L = inductive reactance, in ohms (Ω)
2π = 2π radians, 360° or 1 cycle
f = frequency, in hertz (Hz)
L = inductance, in henrys (H)

Inductive reactance is proportional to frequency ($X_L \propto f$) because a higher frequency (fast-switching current) will cause a greater amount of current change, and a greater change will generate a larger counter emf, which is an opposition or reactance against current flow. When 0 Hz is applied to a coil (dc), there exists no change, so the inductive reactance of an inductor to dc is zero ($X_L = 2\pi \times 0 \times L = 0$).

Inductive reactance is also proportional to inductance because a larger inductance will generate a greater magnetic field and subsequent counter emf, which is the opposition to current flow.

Ohm's law can be applied to inductive circuits just as it can be applied to resistive and capacitive circuits. The current flow in an inductive circuit (I) is proportional to the voltage applied (V), and inversely proportional to the inductive reactance (X_L). Expressed mathematically,

$$I = \frac{V}{X_L}$$

EXAMPLE:

Calculate the current flowing in the circuit illustrated in Figure 12-17.

FIGURE 12-17

Solution:

The current can be calculated by Ohm's law and is a function of the voltage and opposition, which in this case is inductive reactance.

$$I = \frac{V}{X_L}$$

However, we must first calculate X_L:

$$
\begin{aligned}
X_L &= 2\pi \times f \times L \\
&= 6.28 \times 50 \text{ kHz} \times 15 \text{ mH} \\
&= 4710 \ \Omega \text{ or } 4.71 \text{ k}\Omega
\end{aligned}
$$

Current is therefore equal to

$$I = \frac{V}{X_L} = \frac{10 \text{ V}}{4.71 \text{ k}\Omega} = 2.12 \text{ mA}$$

EXAMPLE:

What opposition or inductive reactance will a motor winding or coil offer if $V = 12$ V and $I = 4.5$ A?

Solution:

$$X_L = \frac{V}{I} = \frac{12 \text{ V}}{4.5 \text{ A}} = 2.66 \ \Omega$$

SELF-TEST EVALUATION POINT FOR SECTION 12-7

Use the following questions to test your understanding of Section 12-7.

1. Define and state the formula for inductive reactance.
2. Why is inductive reactance proportional to frequency and the inductance value?
3. How does inductive reactance relate to Ohm's law?
4. True or false: Inductive reactance is measured in henrys.

12-8 SERIES *RL* CIRCUIT

In a purely resistive circuit, as seen in Figure 12-18, the current flowing within the circuit and the voltage across the resistor are in phase with one another. In a purely inductive circuit, as shown in Figure 12-19, the current will lag the applied voltage by 90°. If we connect a resistor and inductor in series, as shown in Figure 12-20(a), we will have the most common combination of *R* and *L* used in electronic equipment.

12-8-1 *Voltage*

The voltage across the resistor and inductor shown in Figure 12-20(a) can be calculated by using Ohm's law:

$$V_R = I \times R$$

$$V_L = I \times X_L$$

As with any series circuit, we have to apply Kirchhoff's voltage law when calculating the value of applied or source voltage (V_S), which, due to the phase difference between V_R and V_L, is the vector sum of all the voltage drops. Applying the Pythagorean theorem, we arrive at a formula for source voltage.

$$V_S = \sqrt{V_R{}^2 + V_L{}^2}$$

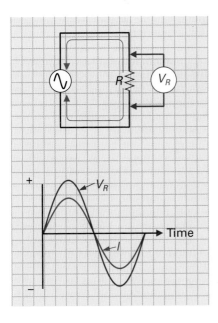

FIGURE 12-18   Purely Resistive Circuit: Current and Voltage Are in Phase.

FIGURE 12-19   Purely Inductive Circuit: Current Lags Applied Voltage by 90°.

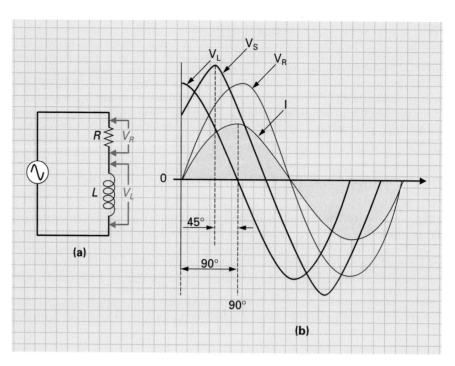

FIGURE 12-20   Series *RL* Circuit.

As with any formula with three quantities, if two are known, the other can be calculated by simply rearranging the formula to

$$V_S = \sqrt{V_R^2 + V_L^2}$$

$$V_R = \sqrt{V_S^2 - V_L^2}$$

$$V_L = \sqrt{V_S^2 - V_R^2}$$

■   **EXAMPLE:**

Calculate V_R, V_L, and V_S for the circuit shown in Figure 12-21.

■   *Solution:*

$$V_R = I \times R$$
$$= 100 \text{ mA} \times 55 \text{ } \Omega$$
$$= 5.5 \text{ V}$$
$$V_L = I \times X_L$$
$$= 100 \text{ mA} \times 26 \text{ } \Omega$$
$$= 2.6 \text{ V}$$
$$V_S = \sqrt{V_R^2 + V_L^2}$$
$$= \sqrt{5.5 \text{ V}^2 + 2.6 \text{ V}^2}$$
$$= 6 \text{ V}$$

FIGURE 12-21   Voltage in a Series *RL* Circuit.

12-8-2　Impedance (Z)

Impedance is the total opposition to current flow offered by a circuit with both resistance and reactance. It is measured in ohms and can be calculated by using Ohm's law:

$$Z = \frac{V}{I}$$

The impedance of a series *RL* circuit is equal to the square root of the sum of the squares of resistance and reactance, and by rearrangement, X_L and R can also be calculated if the other two values are known:

$$Z = \sqrt{R^2 + X_L^2}$$

$$R = \sqrt{Z^2 - X_L^2}$$

$$X_L = \sqrt{Z^2 - R^2}$$

■ **EXAMPLE:**

Referring back to Figure 12-21, calculate *Z*.

■ *Solution:*

$$\begin{aligned} Z &= \sqrt{R^2 + X_L^2} \\ &= \sqrt{55^2 + 26^2} \\ &= 60.8 \ \Omega \end{aligned}$$

12-8-3　*Phase Shift*

If a circuit is purely resistive, the phase shift (θ) between the source voltage and circuit current is zero. If a circuit is purely inductive, voltage leads current by 90°; therefore, the phase shift is +90°. If the resistance and inductive reactance are equal, the phase shift will equal +45°.

The phase shift in an inductive and resistive circuit is the degrees of lead between the source voltage (V_S) and current (I). Mathematically, it can be expressed as

$$\theta = \arctan \frac{X_L}{R}$$

As the current is the same in both the inductor and resistor in a series circuit, the voltage drops across the inductor and resistor are directly proportional to reactance and resistance:

$$V_R \updownarrow = I \ (\text{constant}) \times R \updownarrow, \quad V_L \updownarrow = I \ (\text{constant}) \times X_L \updownarrow$$

The phase shift can also be calculated by using the voltage drop across the inductor and resistor:

$$\theta = \arctan \frac{V_L}{V_R}$$

Referring back to Figure 12-21, calculate the phase shift between source voltage and circuit current.

■ *Solution:*

Since the ratio of X_L/R and V_L/V_R are known for this example circuit, either of the phase shift formulas can be used.

$$\theta = \arctan \frac{X_L}{R}$$
$$= \arctan \frac{26\ \Omega}{55\ \Omega}$$
$$= \arctan 0.4727$$
$$= 25.3°$$

or

$$\theta = \arctan \frac{V_L}{V_R}$$
$$= \arctan \frac{2.6\ \text{V}}{5.5\ \text{V}}$$
$$= \arctan 0.4727$$
$$= 25.3°$$

The source voltage in this example circuit leads the circuit current by 25.3°.

12-8-4 *Power*

Purely Resistive Circuit

Figure 12-22 illustrates the current, voltage, and power waveforms produced when applying an ac voltage across a purely resistive circuit. Voltage and current are in phase, and true power (P_R) in watts can be calculated by multiplying current by voltage ($P = V \times I$).

The sinusoidal power waveform is totally positive, as a positive voltage multiplied by a positive current produces a positive value of power, and a negative voltage multiplied by a negative current also produces a positive value of power. For this reason, the resistor is said to develop a positive power waveform that is twice the frequency of the voltage or current waveform.

The average value of power dissipated by a purely resistive circuit is the halfway value between maximum and zero, in this example, 4 watts.

Purely Inductive Circuit

The pure inductor, like the capacitor, is a reactive component, which means that it will consume power without the dissipation of energy. The capacitor holds its energy in an electric field, while the inductor consumes and holds its energy in a magnetic field and then releases it back into the circuit.

The power curve alternates equally above and below the zero line, as seen in Figure 12-23. During the first positive power half-cycle, the circuit current is on the increase, to maximum (point A), the magnetic field is building up, and the inductor is storing electrical energy. When the circuit current is on the decline between A and B, the magnetic field begins to collapse and self-induction occurs and returns electrical energy back into the circuit. The power alternation is both positive when the inductor is consuming power and negative when the inductor is returning the power back into the circuit. As the positive and negative power alternations are equal but opposite, the average power dissipated is zero.

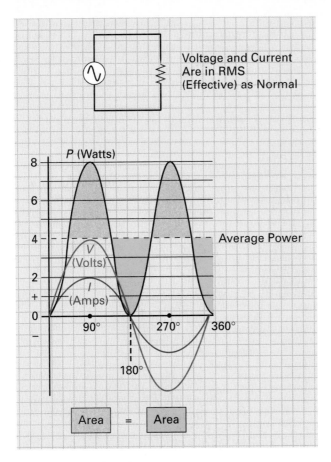

FIGURE 12-22 **Purely Resistive Circuit.**

Resistive and Inductive Circuit

An inductor is different from a capacitor in that it has a small amount of resistance no matter how pure the inductor. For this reason, inductors will never have an average power figure of zero, because even the best inductor will have some value of inductance and resistance within it, as seen in Figure 12-24. The reason an inductor has resistance is that it is simply a piece of wire, and any piece of wire has a certain value of resistance, as well as inductance. This coil resistance should be, and normally is, very small and can usually be ignored; however, in some applications even this small resistance can prevent the correct operation of a circuit, so a value or term had to be created to specify the differences in the quality of inductor available.

The **quality factor** (Q) of an inductor is the ratio of the energy stored in the coil by its inductance to the energy dissipated in the coil by the resistance; therefore, the higher the Q, the better the coil is at storing energy rather than dissipating it:

$$\text{quality } (Q) = \frac{\text{energy stored}}{\text{energy dissipated}}$$

The energy stored is dependent on the inductive reactance (X_L) of the coil, and the energy dissipated is dependent on the resistance (R) of the coil. The quality factor of a coil or inductor can therefore also be calculated by using the formula

$$Q = \frac{X_L}{R}$$

Quality Factor

Quality factor of an inductor or capacitor; it is the ratio of a component's reactance (energy stored) to its effective series resistance (energy dissipated).

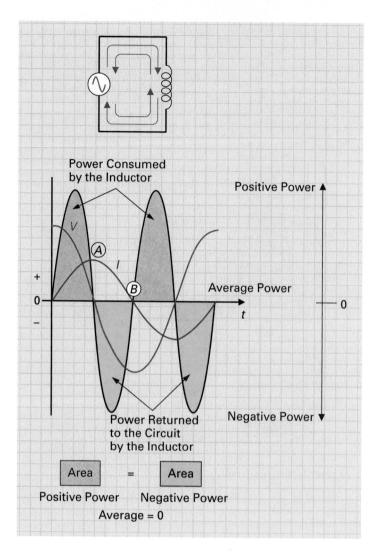

FIGURE 12-23   Purely Inductive Circuit.

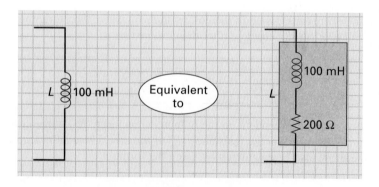

FIGURE 12-24   Resistance within an Inductor.

EXAMPLE:

Calculate the quality factor Q of a 22 mH coil connected across a 2 kHz, 10 V source if its internal coil resistance is 27 Ω.

Solution:

$$Q = \frac{X_L}{R}$$

The reactance of the coil is not known but can be calculated by the formula

$$X_L = 2\pi f L$$
$$= 2\pi \times 2 \text{ kHz} \times 22 \text{ mH} = 276.5 \ \Omega$$

Therefore,

$$Q = \frac{X_L}{R} = \frac{276.5 \ \Omega}{27 \ \Omega} = 10.24$$

Inductors, therefore, will never appear as pure inductance, but rather as an inductive and resistive (RL) circuit, and the resistance within the inductor will dissipate *true power.*

Figure 12-25 illustrates a circuit containing R and L and the power waveforms produced when $R = X_L$; the phase shift (θ) is equal to 45°.

The positive power alternation, which is above the zero line, is the combination of the power dissipated by the resistor and the power consumed by the inductor while circuit current was on the rise. The negative power alternation is the power that was given back to the circuit by the inductor while the inductor's magnetic field was collapsing and returning the energy that was consumed.

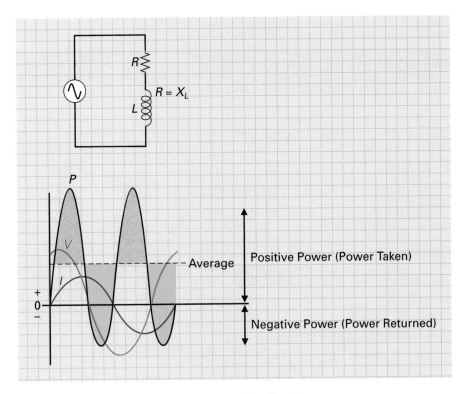

FIGURE 12-25   **Power in a Resistive and Inductive Circuit.**

Power Factor

When a circuit contains both inductance and resistance, some of the energy is consumed and then returned by the inductor (reactive or imaginary power), and some of the energy is dissipated and lost by the resistor (resistive or true power).

Apparent power is the power that appears to be supplied to the load and is the vector sum of both the reactive and true power; it can be calculated by using the formula

$$P_A = \sqrt{P_R^2 + P_X^2}$$

where P_A = apparent power, in volt-amperes (VA)
P_R = true power, in watts (W)
P_X = reactive power, in volt-amperes reactive (VAR)

The power factor is a ratio of the true power to the apparent power and is therefore a measure of the loss in a circuit.

$$PF = \frac{\text{true power } (P_R)}{\text{apparent power } (P_A)}$$

or

$$PF = \frac{R}{Z}$$

or

$$PF = \cos\theta$$

■ **EXAMPLE:**

Calculate the following for a series RL circuit if $R = 40$ kΩ, $L = 450$ mH, $f = 20$ kHz, and $V_S = 6$ V:

a. X_L

b. Z

c. I

d. θ

e. Apparent power

f. PF

■ *Solution:*

a. $X_L = 2\pi f L = 2\pi \times 20 \text{ kHz} \times 450 \text{ mH}$
   $= 56.5 \text{ k}\Omega$

b. $Z = \sqrt{R^2 + X_L^2}$
   $= \sqrt{40 \text{ k}\Omega^2 + 56.5 \text{ k}\Omega^2} = 69.23 \text{ k}\Omega$

c. $I = \dfrac{V_S}{Z} = \dfrac{6 \text{ V}}{69.23 \text{ k}\Omega} = 86.6 \text{ μA}$

d. $\theta = \arctan \dfrac{X_L}{R} = \arctan \dfrac{56.5 \text{ k}\Omega}{40 \text{ k}\Omega} = 54.7°$

e. Apparent power $= \sqrt{(\text{true power})^2 + (\text{reactive power})^2}$

$$P_R = I^2 \times R = 86.6 \ \mu A^2 \times 40 \ k\Omega = 300 \ \mu W$$
$$P_X = I^2 \times X_L = 86.6 \ \mu A^2 \times 56.5 \ k\Omega = 423.7 \ \mu VAR$$
$$P_A = \sqrt{P_R^2 + P_X^2}$$
$$= \sqrt{300 \ \mu W^2 + 423.7 \ \mu W^2} = 519.2 \ \mu VA$$

f. $PF = \dfrac{P_R}{P_A} = \dfrac{300 \ \mu W}{519.2 \ \mu W} = 0.57$

$\quad = \dfrac{R}{Z} = \dfrac{40 \ k\Omega}{69.23 \ k\Omega} = 0.57$

$\quad = \cos \theta = \cos 54.7° = 0.57$

SELF-TEST EVALUATION POINT FOR SECTION 12-8

Use the following questions to test your understanding of Section 12-8.

1. True or false: In a purely inductive circuit, the current will lead the applied voltage by 90°.
2. Calculate the applied source voltage V_S in an R_L circuit where $V_R = 4$ V and $V_L = 2$ V.
3. Define and state the formula for impedance when R and X_L are known.
4. If $R = X_L$, the phase shift will equal _____.
5. What is positive power?
6. Define Q and state the formula when X_L and R are known.
7. What is the difference between true and reactive power?
8. State the power factor formula.

12-9 PARALLEL *RL* CIRCUIT

Now that we have seen the behavior of resistors and inductors in series, let us analyze the parallel *RL* circuit.

12-9-1 *Current*

In Figure 12-26(a) you will see a parallel combination of a resistor and inductor. The voltage across both components is equal because of the parallel connection, and the current through each branch is calculated by applying Ohm's law.

$$(\text{resistor current}) \ I_R = \frac{V_S}{R}$$

$$(\text{inductor current}) \ I_L = \frac{V_S}{X_L}$$

Total current (I_T) is equal to the vector combination of the resistor current and inductor current, as shown in Figure 12-26(b).

$$I_T = \sqrt{I_R^2 + I_L^2}$$

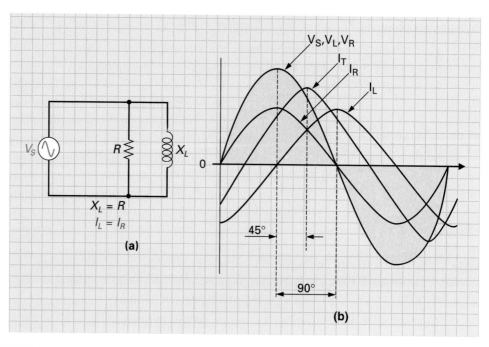

FIGURE 12-26   **Parallel *RL* Circuit.**

12-9-2   *Phase Angle*

The angle by which the total current (I_T) leads the source voltage (V_S) can be determined with either of the following formulas:

$$\theta = \arctan \frac{I_L}{I_R}$$

or

$$\theta = \arctan \frac{R}{X_L}$$

12-9-3   *Impedance*

The total opposition or impedance of a parallel *RL* circuit can be calculated by

$$Z = \frac{V_S}{I_T}$$

Using basic algebra, this formula can be rearranged to express impedance in terms of reactance and resistance.

$$Z = \frac{R \times X_L}{\sqrt{R^2 + X_L^2}}$$

12-9-4 Power

As with series *RL* circuits, resistive power and reactive power can be calculated by

$$P_R = I_R^2 \times R$$

$$P_X = I_L^2 \times X_L$$

The apparent power of the circuit is calculated by

$$P_A = \sqrt{P_R^2 + P_X^2}$$

and, finally, the power factor is equal to

$$PF = \cos\theta = \frac{P_R}{P_A}$$

☐ **EXAMPLE:**

Calculate the following for a parallel *RL* circuit if $R = 45\ \Omega$, $X_L = 1100\ \Omega$, and $V_S = 24$ V:

   a. I_R c. I_T e. θ

   b. I_L d. Z

▦ *Solution:*

a. $I_R = \dfrac{V_S}{R} = \dfrac{24\text{ V}}{45\ \Omega} = 533.3$ mA

b. $I_L = \dfrac{V_S}{X_L} = \dfrac{24\text{ V}}{1100\ \Omega} = 21.8$ mA

c. $I_T = \sqrt{I_R^2 + I_L^2}$
$= \sqrt{533.3\text{ mA}^2 + 21.8\text{ mA}^2} = 533.7$ mA

d. $Z = \dfrac{R \times X_L}{\sqrt{R^2 + X_L^2}} = \dfrac{45\ \Omega \times 1100\ \Omega}{\sqrt{45\ \Omega^2 + 1100\ \Omega^2}}$

$= \dfrac{49.5\text{ k}\Omega}{1100.9\ \Omega} = 44.96\ \Omega$

e. $\theta = \arctan\dfrac{R}{X_L} = \arctan\dfrac{45\ \Omega}{1100\ \Omega} = 2.34°$

   Therefore, I_T lags V_S by 2.34°.

SELF-TEST EVALUATION POINT FOR SECTION 12-9

Use the following questions to test your understanding of Section 12-9.

1. True or false: When *R* and *L* are connected in parallel with one another, the voltage across each will be different and dependent on the values of *R* and X_L.

2. State the formula for calculating total current (I_T) when I_R and I_L are known.

12-10 TESTING INDUCTORS

Basically, only three problems can occur with inductors:

1. An open
2. A complete short
3. A section short (value change)

12-10-1 *Open [Figure 12-27(a)]*

This problem can be isolated with an ohmmeter. Depending on the winding's resistance, the coil should be in the range of zero to a few hundred ohms. An open accounts for 75% of all defective inductors.

FIGURE 12-27 Defective Inductors.

FIGURE 12-28 Capacitor and Inductor Analyzer.

CHAPTER 12 / INDUCTANCE AND INDUCTORS

12-10-2 *Complete or Section Short [Figure 12-27(b)]*

A coil with one or more shorted turns or a complete short can be checked with an ohmmeter and thought to be perfectly good because of the normally low resistance of a coil, as it is just a piece of wire. But if it is placed in a circuit with a complete or section short present, it will not function effectively as an inductor, if at all. For these checks, an **inductor analyzer** needs to be used like the one seen in Figure 12-28, which can be used to check capacitance and inductance. Complete or section shorts account for 25% of all defective inductors.

Inductor Analyzer
A test instrument designed to test inductors.

SELF-TEST EVALUATION POINT FOR SECTION 12-10

Use the following questions to test your understanding of Section 12-10.

1. How could the following inductor malfunctions be recognized?
   a. An open
   b. A complete or section short
2. Which inductor malfunction accounts for almost 75% of all failures

---

12-11 APPLICATIONS OF INDUCTORS

12-11-1 RL *Filters*

The **RL filter** will achieve results similar to the *RC* filter in that it will pass some frequencies and block others, as seen in Figure 12-29. The inductive reactance of the coil and the resistance of the resistor form a voltage divider. Since inductive reactance is proportional to frequency ($X_L \propto f$), the inductor will drop less voltage at lower frequencies ($f\downarrow$, $X_L\downarrow$, $V_L\downarrow$) and more voltage at higher frequencies ($f\uparrow$, $X_L\uparrow$, $V_L\uparrow$).

With the low-pass filter shown in Figure 12-29(a) the output is developed across the resistor. If the frequency of the input is low, the inductive reactance will be low, so almost all the input will be developed across the resistor and applied to the output. If the frequency of the input increases, the inductor's reactance will increase, resulting in almost all the input being dropped across the inductor and none across the resistor and therefore the output.

With the high-pass filter, shown in Figure 12-29(b), the inductor and resistor have been placed in opposite positions. If the frequency of the input is low, the inductive reactance will be low, so almost all the input will be developed across the resistor and very little will appear across the inductor and therefore at the output. If the frequency of the input is high, the inductive reactance will be high, resulting in almost all of the input being developed across the inductor and therefore appearing at the output.

RL Filter
A selective circuit of resistors and inductors which offers little or no opposition to certain frequencies while blocking or alternating other frequencies.

12-11-2 RL *Integrator*

In the preceding section on *RL* filters we saw how an *RL* circuit reacted to a sine-wave input of different frequencies. In this section and the following we will see how an *RL* circuit will react to a square-wave input, and show two other important applications of inductors. In the **RL integrator,** the output is taken across the resistor, as seen in the circuit in Figure 12-30(a). The output shown in Figure 12-30(b) is the same as the previously described *RC* integrator's output.

The inductor's 10 V counter emf opposes the sudden input change from 0 to 10 V, and if 10 V is across the inductor, 0 V must be across the resistor (Kirchhoff's voltage law), and

RL Integrator
An *RL* circuit with an output proportionate to the integral of the input signal.

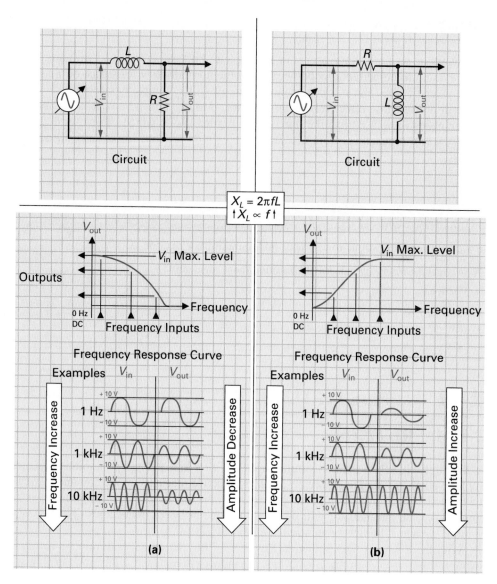

$$X_L = 2\pi fL$$
$$\uparrow X_L \propto f \uparrow$$

FIGURE 12-29 *RL* Filter. (a) Low-Pass Filter. (b) High-Pass Filter.

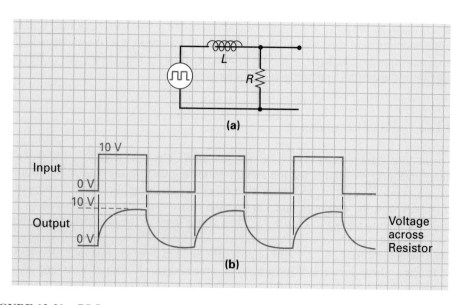

FIGURE 12-30 *RL* Integrator.

therefore appearing at the output. After five time constants ($5 \times L/R$), the inductor's current in the circuit will have built up to maximum (V_{in}/R), and the inductor will be an equivalent short circuit, because no change and consequently no back emf exist. All the input voltage will now be across the resistor and therefore at the output.

When the square-wave input drops to zero, the collapsing magnetic field will cause an induced voltage within the conductor, which will cause current to flow within the circuit for five time constants, whereupon it will reach 0.

As with the *RC* integrator, if the period of the square wave is decreased or if the time constant is increased, the output will reach an average value of half the input square wave's amplitude.

12-11-3 RL *Differentiator*

With the **RL differentiator,** the output is taken across the inductor, as shown in Figure 12-31, and is the same as the *RC* differentiator's output. When the square-wave input rises from 0 to 10 V, the inductor will generate a 10 V counter emf across it, and this 10 V will appear across the output. As the circuit current exponentially rises, the voltage across the inductor, and therefore at the output, will fall exponentially to 0 V after five time constants.

When the square wave input falls from 10 to 0 V, the collapsing magnetic field induces a counter emf. The sudden -10 V impulse at the output will decrease to 0 V as the circuit current decreases in five time constants.

RL Differentiator

An *RL* circuit whose output voltage is proportional to the rate of change of the input voltage.

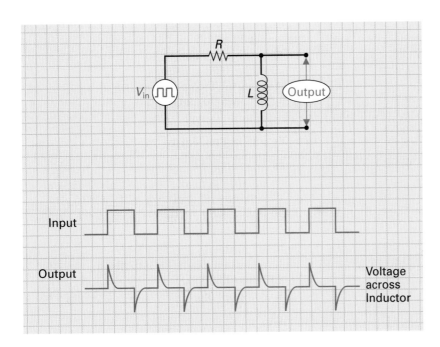

FIGURE 12-31   *RL* **Differentiator.**

SELF-TEST EVALUATION POINT FOR SECTION 12-11

Use the following questions to test your understanding of Section 12-11.

1. List three circuit applications of the inductor.
2. Will an *RL* filter act as a low-pass filter or high-pass filter, or can it be made to function as either?
3. What are the waveform differences between an integrator and differentiator?

Multiple-Choice Questions

1. Self-inductance is a process by which a coil will induce a voltage within _____.

   a. Another inductor
   b. Two or more close proximity inductors
   c. Itself
   d. Both (a) and (b)

2. The inductor stores electrical energy in the form of a (an) _____ field, just as a capacitor stores electrical energy in the form of a (an) _____ field.

   a. Electric, magnetic
   b. Magnetic, electric

3. The inductor is basically:

   a. An electromagnet
   b. A coil of wire
   c. A coil of conductor formed around a core material
   d. All of the above
   e. None of the above

4. The total inductance of a series circuit is:

   a. Less than the value of the smallest inductor
   b. Equal to the sum of all the inductance values
   c. Equal to the product over sum
   d. All of the above

5. The total inductance of a parallel circuit can be calculated by:

   a. Using the product-over-sum formula
   b. Using L divided by N for equal-value inductors
   c. Using the reciprocal resistance formula
   d. All of the above

6. Ferrite is a chemical compound, which is basically powdered:

   a. Iron oxide and ceramic
   b. Iron and steel
   c. Mylar and iron
   d. Gauze and electrolyte

7. The ferrite-core variable inductor varies inductance by changing:

   a. μ c. N
   b. l d. A

8. The time constant for a series inductive/resistive circuit is equal to:

   a. $L \times R$ c. V/R
   b. L/R d. $2\pi \times f \times L$

9. Inductive reactance (X_L) is proportional to:

   a. Time or period of the ac applied
   b. Frequency of the ac applied
   c. The stray capacitance that occurs due to the air acting as a dielectric between two turns of a coil
   d. The value of inductance
   e. Two of the above are true.

10. In a series RL circuit, the source voltage (V_S) is equal to:

    a. The square root of the sum of V_R^2 and V_L^2
    b. The vector sum of V_R and V_L
    c. $I \times Z$
    d. Two of the above are partially true.
    e. Answers (a), (b), and (c) are correct.

11. With an RL integrator, the output is taken across the:

    a. Inductor   c. Resistor
    b. Capacitor   d. Transformer's secondary

12. With an RL differentiator, the output is taken across the:

    a. Inductor   c. Resistor
    b. Capacitor   d. Transformer's secondary.

13. An inductor or choke between the input and output forms a:

    a. High-pass filter   b. Low-pass filter

14. An inductor or choke connected to ground or in shunt forms a:

    a. Low-pass filter   b. High-pass filter

15. When tested with an ohmmeter, an open coil would show:

    a. Zero resistance
    b. An infinite resistance
    c. A 100 Ω to 200 Ω resistance
    d. Both (b) and (c)

Practice Problems

16. Convert the following:

    a. 0.037 H to mH   c. 862 mH to H
    b. 1760 μH to mH   d. 0.256 mH to μH

17. Calculate the impedance (Z) of the following series RL combinations:

    a. 22 MΩ, 25 μH, $f = 1$ MHz
    b. 4 kΩ, 125 mH, $f = 100$ kHz
    c. 60 Ω, 0.05 H, $f = 1$ MHz

18. Calculate the total inductance of the following series circuits:

    a. 75 μH, 61 μH, 50 mH
    b. 8 mH, 4 mH, 22 mH

19. Calculate the total inductance of the following parallel circuits:

    a. 12 mH, 8 mH   b. 75 μH, 34 μH, 27 μH

20. Calculate the total inductance of the following series–parallel circuits:

    a. 12 mH in series with 4 mH, and both in parallel with 6 mH
    b. A two-branch parallel arrangement made up of 6 μH and 2 μH in series with one another, and 8 μH and 4 μH in series with one another

c. Two parallel arrangements in series with one another, made up of 1 µH and 2 µH in parallel and 4 µH and 15 µH in parallel

21. In a series *RL* circuit, if $V_L = 12$ V and $V_R = 6$ V, calculate:

    a. V_S **d.** Q
    b. I if $Z = 14$ kΩ   **e.** Power factor
    c. Phase angle

22. Calculate the impedance of the circuits seen in Figure 13-32.

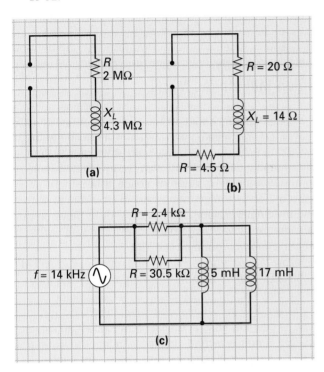

(a)

(b)

(c)

FIGURE 12-32   **Calculating Impedance.**

23. Referring to Figure 13-33, calculate the voltage across the inductor for all five time constants after the switch has been closed.

FIGURE 12-33   **Inductive Time Constant Example.**

24. Referring to Figure 13-34, calculate:

    a. L_T
    b. X_L
    c. Z
    d. I_{R1}, I_{L1}, and I_{L2}
    e. θ
    f. True power, reactive power, and apparent power
    g. Power factor

FIGURE 12-34   **Series *RL* Circuit.**

25. Referring to Figure 13-35, calculate:

    a. R_T **f.** I_{RT}, I_{LT}
    b. L_T **g.** I_T
    c. X_L **h.** θ
    d. Z **i.** Apparent power
    e. V_{RT}, V_{LT} **j.** PF

FIGURE 12-35   **Parallel *RL* Circuit.**

Transformers

13

Woolen Mill Makes Minis

At the age of 24 in 1950, Kenneth Olsen went to work at MIT's Digital Computer Laboratory as a research assistant. For the next seven years, Olsen worked on the SAGE (Semi Automatic Ground Environment) project, which was a new computer system designed to store the constantly changing data obtained from radars tracking long-range enemy bombers capable of penetrating American air space. IBM won the contract to build the SAGE network and Olsen traveled to the plant in New York to supervise production. After completing the SAGE project, Olsen was ready to move on to bigger and better things; however, he didn't want to go back to MIT and was not happy with the rigid environment at a large corporation like IBM.

In September 1957, Olsen and fellow MIT colleague Harlan Anderson leased an old brick woolen mill in Maynard, Massachusetts, and after some cleaning and painting they began work in their new company, which they named Digital Equipment Corporation or DEC. After three years, the company came out with their first computer, the PDP 1 (programmed data processor), and its ease of use, low cost, and small size made it an almost instant success. In 1965, the desktop PDP 8 was launched, and since in that era miniskirts were popular, it was probably inevitable that the small machine was nicknamed a minicomputer. In 1969, the PDP 11 was unveiled, and like its predecessors, this number-crunching data processor went on to be accepted in many applications, such as tracking the millions of telephone calls received on the 911 emergency line, directing welding machines in automobile plants, recording experimental results in laboratories, and processing all types of data in offices, banks, and department stores. Just 20 years after introducing low-cost computing in the 1960s, DEC was second only to IBM as a manufacturer of computers in the United States.

Introduction

When alternating current was introduced in Chapter 8, it was mentioned that an ac voltage could be stepped up to a larger voltage or stepped down to a smaller voltage by a device called a **transformer.** The transformer is an electrical device that makes use of electromagnetic induction to transfer alternating current from one circuit to another. The transformer consists of two inductors that are placed in very close proximity to one another. When an alternating current flows through the first coil or **primary winding** the inductor sets up a magnetic field. The expanding and contracting magnetic field produced by the primary cuts across the windings of the second inductor or **secondary winding** and induces a voltage in this coil.

By changing the ratio between the number of turns in the secondary winding to the number of turns in the primary winding, some characteristics of the ac signal can be changed or transformed as it passes from primary to secondary. For example, a low ac voltage could be stepped up to a higher ac voltage, or a high ac voltage could be stepped down to a lower ac voltage.

In this chapter we will examine the operation, characteristics, types, testing, and applications of transformers.

Transformer

Device consisting of two or more coils that are used to couple electric energy from one circuit to another, yet maintain electrical isolation between the two.

Primary Winding

First winding of a transformer that is connected to the source.

Secondary Winding

Output winding of a transformer that is connected across the load.

13-1 MUTUAL INDUCTANCE

As was discussed in the previous chapter, self-inductance is the process by which a coil induces a voltage within itself. The principle on which a transformer is based is an inductive effect known as **mutual inductance,** which is the process by which an inductor induces a voltage in another inductor.

Mutual Inductance

Ability of one inductor's magnetic lines of force to link with another inductor.

Figure 13-1 illustrates two inductors that are magnetically linked, yet electrically isolated from one another. As the alternating current continually rises, falls, and then rises in the opposite direction, a magnetic field will build up, collapse, and then build up in the opposite direction.

If a second inductor, or secondary coil (L_2), is in close proximity with the first inductor or primary coil (L_1), which is producing the alternating magnetic field, a voltage will be induced into the nearby inductor, which causes current to flow in the secondary circuit through the load resistor. This phenomenon is known as mutual inductance or transformer action.

As with self-inductance, mutual inductance is dependent on change. Direct current (dc) is a constant current and produces a constant or stationary magnetic field that does not

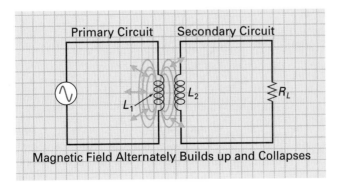

FIGURE 13-1 Mutual Inductance.

change, as reviewed in Figure 13-2(a). Alternating current, however, is continually varying, and as the polarity of the magnetic field is dependent on the direction of current flow (left-hand rule), the magnetic field will also be alternating in polarity, as seen in Figure 13-2(b); it is this continual building up and collapsing of the magnetic field that cuts the adjacent inductor's conducting coils and induces a voltage in the secondary circuit. Mutual induction is possible only with ac and cannot be achieved with dc due to the lack of change.

Self-induction is a measure of how much voltage an inductor can induce within itself. Mutual inductance is a measure of how much voltage is induced in the secondary coil due to the change in current in the primary coil.

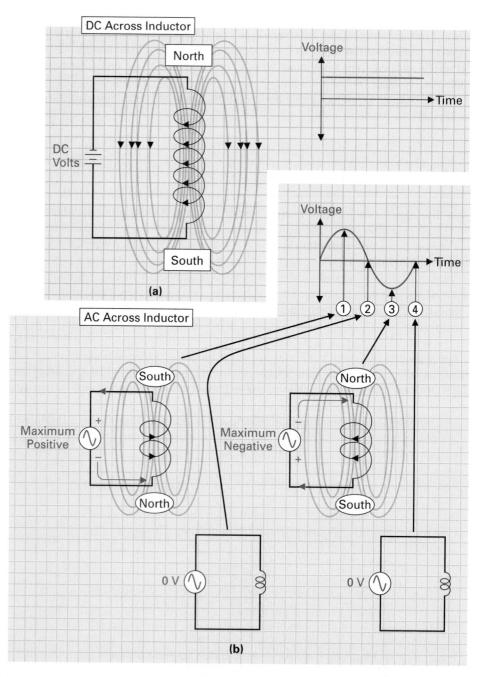

FIGURE 13-2   Magnetic Field Produced by (a) DC across an Inductor and (b) AC across an Inductor.

Use the following questions to test your understanding of Section 13-1.

1. Define mutual inductance and explain how it differs from self-inductance.

2. True or false: Mutual induction is possible only with direct current flow.

13-2   BASIC TRANSFORMER

Figure 13-3(a) illustrates a basic transformer, which consists of two coils within close prox-imity to one another, to ensure that the second coil will be cut by the magnetic flux lines pro-duced by the first coil, and thereby ensure mutual inductance. The ac voltage source is elec-trically connected (through wires) to the primary coil or winding, and the load (R_L) is electrically connected to the secondary coil or winding.

In Figure 13-3(b), the ac voltage source has produced current flow in the primary cir-cuit, as illustrated. This current flow produces a north pole at the top of the primary winding and, as the ac voltage input swings more negative, the current increase causes the magnetic field being developed by the primary winding to increase. This expanding magnetic field cuts the coils of the secondary winding and induces a voltage, and a subsequent current flows in the secondary circuit, which travels up through the load resistor. The ac voltage fol-lows a sinusoidal pattern and moves from a maximum negative to zero and then begins to build up toward a maximum positive.

In Figure 13-3(c), the current flow in the primary circuit is in the opposite direction due to the ac voltage increase in the positive direction. As voltage increases, current in-

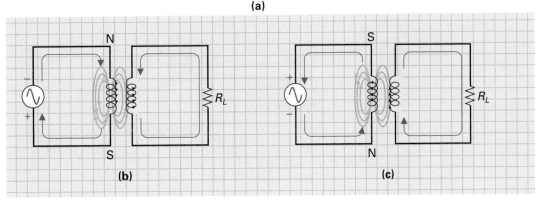

FIGURE 13-3   Transformer Action.

creases, and the magnetic field expands and cuts the secondary winding, inducing a voltage and causing current to flow in the reverse direction down through the load resistor.

You may have noticed a few interesting points about the basic transformer just discussed.

1. As primary current increases, secondary current increases, and as primary current decreases, secondary current also decreases. It can therefore be said that the frequency of the alternating current in the secondary is the same as the frequency of the alternating current in the primary.

2. Although the two coils are electrically isolated from one another, energy can be transferred from primary to secondary, because the primary converts electrical energy into magnetic energy, and the secondary converts magnetic energy back into electrical energy.

SELF-TEST EVALUATION POINT FOR SECTION 13-2

Use the following questions to test your understanding of Section 13-2.

1. True or false: A transformer achieves electrical isolation.
2. True or false: The transformer converts electrical energy to magnetic, and then from magnetic back to electrical.
3. What are the names given to the windings of a transformer?

13-3   TRANSFORMER LOADING

Let's now carry our discussion of the basic transformer a little further and see what occurs when the transformer is not connected to a load, as shown in Figure 13-4. Primary circuit current is determined by $I = V/Z$, where Z is the impedance of the primary coil (both its inductive reactance and resistance) and V is the applied voltage. Since no current can flow in the secondary, because an open in the circuit exists, the primary acts as a simple inductor, and the primary current is small due to the inductance of the primary winding. This small primary current lags the applied voltage due to the counter emf by approximately 90° because the coil is mainly inductive and has very little resistance.

When a load is connected across the secondary, as shown in Figure 13-5, a change in conditions occurs and the transformer acts differently. The important point that will be ob-

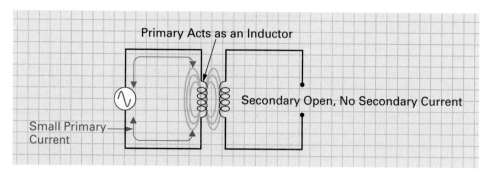

FIGURE 13-4   Unloaded Transformer.

FIGURE 13-5   Loaded Transformer.

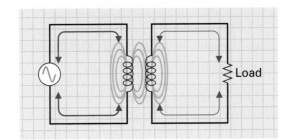

served is that as we go from a no-load to a load condition the primary current will increase due to mutual inductance. Let's follow the steps one by one.

1. The ac applied voltage sets up an alternating magnetic field in the primary winding.

2. The continually changing flux of this primary field induces and produces a counter emf into the primary to oppose the applied voltage.

3. The primary's magnetic field also induces a voltage in the secondary winding, which causes current to flow in the secondary circuit through the load.

4. The current in the secondary winding produces another magnetic field that is opposite to the field being produced by the primary.

5. This secondary magnetic field feeds back to the primary and induces a voltage that tends to cancel or weaken the counter emf that was set up in the primary by the primary current.

6. The primary's counter emf is therefore reduced, so primary current can now increase.

7. This increase in primary current is caused by the secondary's magnetic field; consequently, the greater the secondary current, the stronger the secondary magnetic field, which causes a reduction in the primary's counter emf, and therefore a primary current increase.

In summary, an increase in secondary current ($I_s\uparrow$) causes an increase in primary current ($I_p\uparrow$), and this effect in which the primary induces a voltage in the secondary (V_s) and the secondary induces a voltage into the primary (V_p) is known as *mutual inductance.*

SELF-TEST EVALUATION POINT FOR SECTION 13-3

Use the following questions to test your understanding of Section 13-3.

1. True or false: An increase in secondary current causes an increase in primary current.

2. True or false: The greater the secondary current, the greater the primary's counter emf.

13-4   TRANSFORMER RATIOS AND APPLICATIONS

Basically, transformers are used for one of three applications:

1. To step up (increase) or step down (decrease) voltage

2. To step up (increase) or step down (decrease) current

3. To match impedances

In all three cases, any of the applications can be achieved by changing the ratio of the number of turns in the primary winding compared to the number of turns in the secondary winding. This ratio is appropriately called the turns ratio.

13-4-1 *Turns Ratio*

The **turns ratio** is the ratio between the number of turns in the secondary winding (N_s) and the number of turns in the primary winding (N_p).

$$\text{turns ratio} = \frac{N_s}{N_p}$$

Let us use a few examples to see how the turns ratio can be calculated.

☐ EXAMPLE:

If the primary has 200 turns and the secondary has 600, what is the turns ratio (Figure 13-6)?

FIGURE 13-6 Step-Up Transformer Example.

■ *Solution:*

$$\text{turns ratio} = \frac{N_s}{N_p}$$
$$= \frac{600}{200}$$
$$= \frac{3(\text{secondary})}{1(\text{primary})}$$
$$= 3$$

This simply means that there are three windings in the secondary to every one winding in the primary. Moving from a small number (1) to a larger number (3) means that we *stepped up* in value. Stepping up always results in a turns ratio figure greater than 1, in this case, 3.

☐ EXAMPLE:

If the primary has 120 turns and the secondary has 30 turns, what is the turns ratio (Figure 13-7)?

FIGURE 13-7 Step-Down Transformer Example.

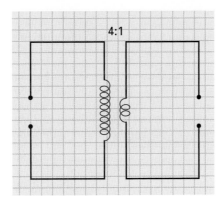

4:1

■ *Solution:*

$$\text{turns ratio} = \frac{N_s}{N_p}$$

$$= \frac{30}{120}$$

$$= \frac{1\,(\text{secondary})}{4\,(\text{primary})}$$

$$= 0.25$$

Said simply, there are four primary windings to every one secondary winding. Moving from a larger number (4) to a smaller number (1) means that we *stepped down* in value. Stepping down always results in a turns ratio figure of less than 1, in this case 0.25.

13-4-2 *Voltage Ratio*

Transformers are used within the power supply unit of almost every piece of electronic equipment to step up or step down the 115 V ac from the outlet. Some electronic circuits require lower-power supply voltages, while other devices may require higher-power supply voltages. The transformer is used in both instances to convert the 115 V ac to the required value of voltage.

Step Up

If the secondary voltage (V_s) is greater than the primary voltage (V_p), the transformer is called a **step-up transformer** ($V_s > V_p$), as shown in Figure 13-8. The voltage is stepped up or increased in much the same way as a generator voltage can be increased by increasing the number of turns.

If the ac primary voltage is 100 V and the turns ratio is a 1:5 step up, the secondary voltage will be five times that of the primary voltage, or 500 V, because the magnetic flux established by the primary cuts more turns in the secondary and therefore induces a larger voltage.

In this example, you can see that the ratio of the secondary voltage to the primary voltage is equal to the turns ratio; in other words,

Step-up Transformer
Transformer in which the ac voltage induced in the secondary is greater (due to more secondary windings) than the ac voltage applied to the primary.

$$\frac{V_s}{V_p} = \frac{N_s}{N_p}$$

or

$$\frac{500}{100} = \frac{500}{100}$$

FIGURE 13-8 **Step-Up Transformer.**

To calculate V_s, therefore, we can rearrange the formula and arrive at

$$V_s = \frac{N_s}{N_p} \times V_p$$

In our example, this is

$$V_s = \frac{500}{100} \times 100 \text{ V}$$
$$= 500 \text{ V}$$

Step Down

If the secondary voltage (V_s) is smaller than the primary voltage (V_p), the transformer is called a **step-down transformer** ($V_s < V_p$), as shown in Figure 13-9. The secondary voltage will be equal to

$$V_s = \frac{N_s}{N_p} \times V_p$$
$$= \frac{10}{100} \times 1000 \text{ V} = 100 \text{ V}$$

FIGURE 13-9 **Step-Down Transformer.**

CHAPTER 13 / TRANSFORMERS

Calculate the secondary voltage (V_s) if a $1:6$ step-up transformer has 24 V ac applied to the primary.

■ *Solution:*

$$V_s = \frac{N_s}{N_p} \times V_p$$

$$= \frac{6}{1} \times 24 \text{ V} = 144 \text{ V}$$

The coupling coefficient (k) in this formula is always assumed to be 1, which for most iron-core transformers is almost always the case. This means that all the primary magnetic flux is linking the secondary, and the secondary voltage is dependent on the number of secondary turns that are being cut by the primary magnetic flux.

The transformer can be used to transform the primary ac voltage into any other voltage, either up or down, merely by changing the transformer's turns ratio.

13-4-3   *Power and Current Ratio*

The power in the secondary of the transformer is equal to the power in the primary ($P_p = P_s$). Power, as we know, is equal to $P = V \times I$, and if voltage is stepped up or down, the current automatically is stepped down or up, respectively, in the opposite direction to voltage to maintain the power constant.

For example, if the secondary voltage is stepped up ($V_s\uparrow$), the secondary current is stepped down ($I_s\downarrow$), so the output power is the same as the input power.

$$P_s = V_s \uparrow \times I_s \downarrow$$

This is an equal but opposite change. Therefore, $P_s = P_p$; and you cannot get more power out than you put in. The current ratio is therefore inversely proportional to the voltage ratio:

$$\frac{V_s}{V_p} = \frac{I_p}{I_s}$$

If the secondary voltage is stepped up, the secondary current goes down:

$$\frac{V_s\uparrow}{V_p} = \frac{I_p}{I_s\downarrow}$$

If the secondary voltage is stepped down, the secondary current goes up:

$$\frac{V_s\downarrow}{V_p} = \frac{I_p}{I_s\uparrow}$$

If the current ratio is inversely proportional to the voltage ratio, it is also inversely proportional to the turns ratio:

$$\frac{I_p}{I_s} = \frac{V_s}{V_p} = \frac{N_s}{N_p}$$

By rearranging the current and turns ratio, we can arrive at a formula for secondary current, which is

$$I_s = \frac{N_p}{N_s} \times I_p$$

EXAMPLE:

The step-up transformer in Figure 13-10 has a turns ratio of 1 to 5. Calculate:

a. Secondary voltage (V_s)
b. Secondary current (I_s)
c. Primary power (P_p)
d. Secondary power (P_s)

FIGURE 13-10 Step-Up Transformer Example.

Solution:

The secondary has five times as many windings as the primary, and, consequently, the voltage will be stepped up by 5 between primary and secondary. If the secondary voltage is going to be five times that of the primary, the secondary current is going to decrease to one-fifth of the primary current.

a. $V_s = \dfrac{N_s}{N_p} \times V_p$

$\quad = \dfrac{5}{1} \times 100\ \text{V}$

$\quad = 500\text{V}$

b. $I_s = \dfrac{N_p}{N_s} \times I_p$

$\quad = \dfrac{1}{5} \times 10\ \text{A}$

$\quad = 2\ \text{A}$

c. $P_p = V_p \times I_p = 100\ \text{V} \times 10\ \text{A} = 1000\ \text{VA}$

d. $P_s = V_s \times I_s = 500\ \text{V} \times 2\ \text{A} = 1000\ \text{VA}$

Therefore, $P_p = P_s$.

13-4-4 *Impedance Ratio*

The **maximum power transfer theorem,** which has been discussed previously and is summarized in Figure 13-11, states that maximum power is transferred from source (ac generator) to load (equipment) when the impedance of the load is equal to the internal impedance of the source. If these impedances are different, a large amount of power could be wasted.

In most cases it is required to transfer maximum power from a source that has an internal impedance (Z_s) that is not equal to the load impedance (Z_L). In this situation, a transformer can be inserted between the source and the load to make the load impedance appear to equal the source's internal impedance.

For example, let's imagine that your car stereo system (source) has an internal impedance of 100 Ω and is driving a speaker (load) of 4 Ω impedance, as seen in Figure 13-12.

By choosing the correct turns ratio, the 4 Ω speaker can be made to appear as a 100 Ω load impedance, which will match the 100 Ω internal source impedance of the stereo system, resulting in maximum power transfer.

The turns ratio can be calculated by using the formula

$$\text{turns ratio} = \sqrt{\frac{Z_L}{Z_s}}$$

Z_L = Load impedance in ohms
Z_s = Source impedance in ohms

In our example, this will calculate out to be

$$\begin{aligned}
\text{turns ratio} &= \sqrt{\frac{Z_L}{Z_s}} \\
&= \sqrt{\frac{4}{100}} = \frac{\sqrt{4}}{\sqrt{100}} \\
&= \frac{2}{10} = \frac{1}{5} \\
&= 0.2
\end{aligned}$$

If the turns ratio is less than 1, a step-down transformer is required. A turns ratio of 0.2 means a step-down transformer is needed with a turns ratio of 5:1 ($\frac{1}{5} = 0.2$).

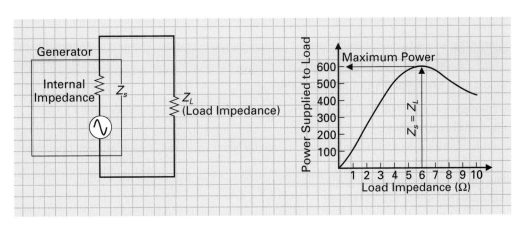

FIGURE 13-11   Maximum Power Transfer Theorem.

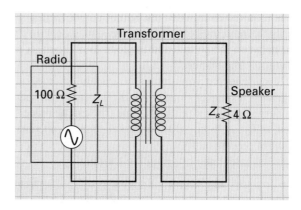

FIGURE 13-12   Impedance Matching.

EXAMPLE:

Calculate the turns ratio needed to match the 22.2 Ω output impedance of an amplifier to two 16 Ω speakers connected in parallel.

Solution:

The total load impedance of two 16 Ω speakers in parallel will be

$$Z_L = \frac{\text{product}}{\text{sum}} = \frac{16 \times 16}{16 + 16} = 8\ \Omega$$

The turns ratio will be

$$\text{turns ratio} = \sqrt{\frac{Z_L}{Z_s}}$$
$$= \sqrt{\frac{8\ \Omega}{22.2\ \Omega}}$$
$$= \sqrt{0.36} = 0.6$$

Therefore, a step-down transformer is needed with a turns ratio of $1.67:1$.

SELF-TEST EVALUATION POINT FOR SECTION 13-4

Use the following questions to test your understanding of Section 13-4.

1. What is the turns ratio of a 402 turn primary and 1608 turn secondary, and is this transformer step up or step down?
2. State the formula for calculating the secondary voltage (V_s).
3. True or false: A transformer can, by adjusting the turns ratio, be made to step up both current and voltage between primary and secondary.
4. State the formula for calculating the secondary current (I_s).
5. What turns ratio is needed to match a 25 Ω source to a 75 Ω load?
6. Calculate V_s if $N_s = 200$, $N_p = 112$, and $V_p = 115$ V.

13-5 WINDINGS AND PHASE

The way in which the primary and secondary coils are wound around the core determines the polarity of the voltage induced into the secondary relative to the polarity of the primary. In Figure 13-13, you can see that if the primary and secondary windings are both wound in a clockwise direction around the core, the voltage induced in the primary will be in phase with the voltage induced in the secondary. Both the input and output will also be in phase with one another if the primary and secondary are both wound in a counterclockwise direction.

In Figure 13-14, you will notice that in this case we have wound the primary and secondary in opposite directions, the primary being wound in a clockwise direction and the secondary in a counterclockwise direction. In this situation, the output ac sine-wave voltage is 180° out of phase with respect to the input ac voltage.

In a schematic diagram, there has to be a way of indicating to the technician that the secondary voltage will be in phase or 180° out of phase with the input. The **dot convention** is a standard used with transformers and is illustrated in Figure 13-15(a) and (b).

A positive on the primary dot causes a positive on the secondary dot, and similarly, since ac is applied, a negative on the primary dot will cause a negative on the secondary dot. In Figure 13-15(a), you can see that when the top of the primary swings positive or negative, the top of the secondary will also follow suit and swing positive or negative, respectively, and you will now be able to determine that the transformer secondary voltage is in phase with the primary voltage.

Dot Convention

A standard used with transformer symbols to indicate whether the secondary voltage will be in phase or out of phase with the primary voltage.

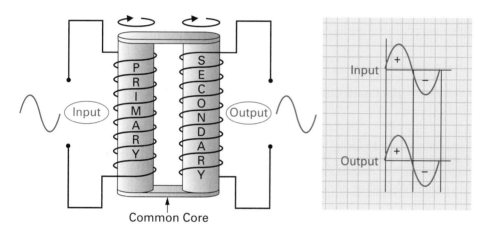

FIGURE 13-13   Primary and Secondary in Phase.

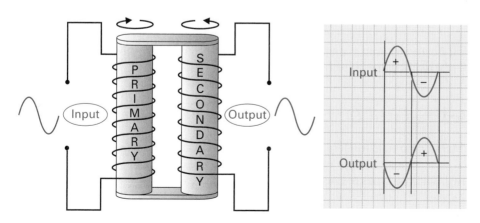

FIGURE 13-14   Primary and Secondary Out of Phase.

In Figure 13-15(b), however, the dots are on top and bottom, which means that as the top of the primary winding swings positive, the bottom of the secondary will go positive, which means the top of the secondary will actually go negative, resulting in the secondary ac voltage being out of phase with the primary ac voltage.

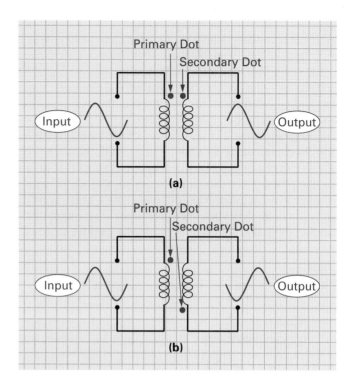

FIGURE 13-15   Dot Convention. (a) In-Phase Dots. (b) Out-of-Phase Dots.

SELF-TEST EVALUATION POINT FOR SECTION 13-5

Use the following questions to test your understanding of Section 13-5.

1. How does the winding of the primary and secondary affect the phase of the output with respect to the input?

2. Describe the transformer dot convention used on schematics.

13-6   TRANSFORMER TYPES

We will treat transformers the same as every other electronic component and classify them as either fixed or variable turns ratio.

13-6-1   *Fixed Turns Ratio Transformers*

Fixed turns ratio transformers have a turns ratio that cannot be varied. They are generally wound on a common core to ensure a high k and can be classified by the type of core material used. The two types of fixed transformers are:

1. Air core [Figure 13-16(a)]
2. Iron or ferrite core [Figure 13-16(b)]

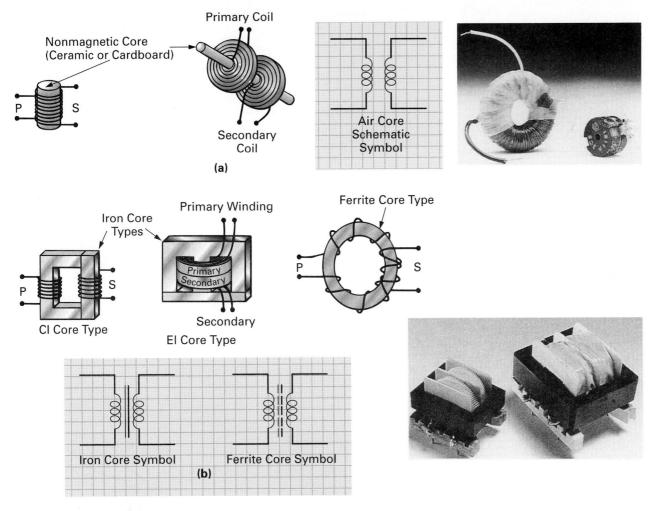

FIGURE 13-16 **Fixed Turns Ratio Transformers. (a) Air-Core Transformers. (Courtesy of Sencore.) (b) Iron- or Ferrite-Core Transformers. (Courtesy of Coilcraft.)**

The air-core transformers typically have a nonmagnetic core, such as ceramic or a cardboard hollow shell, and are used in high-frequency applications. The electronic circuit symbol just shows the primary and secondary coils. The more common iron- or ferrite-core transformers concentrate the magnetic lines of force, resulting in improved transformer performance; they are symbolized by two lines running between the primary and secondary. The iron-core transformer's lines are solid, while the ferrite-core transformer's lines are dashed.

Figure 13-17 illustrates the physical appearance of a few types of fixed transformers.

13-6-2 *Variable Turns Ratio Transformers*

Variable turns ratio transformers have a turns ratio that can be varied. They can be classified as follows:

1. Center-tapped secondary
2. Multiple-tapped secondary
3. Multiple windings
4. Single winding

Figure 13-18 illustrates the first two of these, the center-tapped and the multiple-tapped secondary types. Transformers can often have tapped secondaries. A tapped winding will have a lead connected to one of the loops other than the two end connections.

FIGURE 13-18   Secondary Tapped Transformers. (a) Center Tapped. (Courtesy of Sencore.) (b) Multiple Tapped.

FIGURE 13-19 **Center-Tapped Secondary Transformer.**

Center-Tapped Secondary

If the tapped lead is in the exact center of the secondary, the transformer is said to have a center-tapped secondary, as shown in Figure 13-19. With the **center-tapped transformer,** the two secondary voltages are each half of the total secondary voltage. If we assume a 1:1 turns ratio (primary turns = secondary turns) and a 20 V ac primary voltage and therefore secondary voltage, each of the output voltages between either end of the secondary and the center tap will be 10 V waveforms, as shown in Figure 13-19. The two secondary outputs will be 180° out of phase with one another from the center tap. When the top of the secondary swings positive (+), the bottom of the secondary will be negative, and vice versa.

Figure 13-20 illustrates a typical power company utility pole, which as you can see makes use of a center-tapped secondary transformer. These transformers are designed to step down the high 4.8 kV from the power line to 120 V/240 V for commercial and residential customers. Within the United States, the actual secondary voltage may be anywhere between 225 and 245 V, depending on the demand. The demand is actually higher during the day and in the winter months (source voltage is pulled down by smaller customer load resistance). The multiple taps on the primary are used to change the turns ratio and compensate for power line voltage differences.

The center tap of the secondary winding is connected to a copper earth grounding rod. This center tap wire is called the **neutral wire,** and within the building it is color coded with a white insulation. Residential or commercial appliances or loads that are designed to operate at 240 V are connected between the two black wires, while loads that are designed to operate at 120 V are connected between a black wire and the white wire (neutral).

Multiple-Tapped Secondary

If several leads are attached to the secondary, the transformer is said to have a multiple-tapped secondary. This multitapped transformer will tap or pick off different values of ac voltage, as seen in Figure 13-21.

Multiple Windings

The simple transformer has only one primary and one secondary. In some applications, transformers can have multiple primaries and even multiple secondaries.

Multiple primaries. The major application of this arrangement is to switch between two primary voltages and obtain the same secondary voltage. In Figure 13-22(a), the two primaries are connected in parallel, so the primary only has an equivalent 100 turns;

Center-Tapped Transformer

A transformer that has a connection at the electrical center of the secondary winding.

Neutral Wire

The conductor of a polyphase circuit or of a single-phase three-wire circuit that is intended to have a ground potential. The potential differences between the neutral and each of the other conductors are approximately equal in magnitude and are also equally spaced in phase.

FIGURE 13-20   Center-Tapped Utility Pole Transformer.

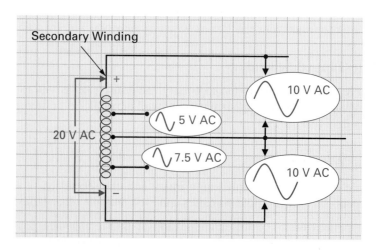

FIGURE 13-21   Multiple-Tapped Secondary Transformer.

and since the secondary has 10 turns, this 10:1 step-down ratio will result in a secondary voltage of

$$V_s = \frac{N_s}{N_p} \times V_p$$

$$= \frac{10}{100} \times 120 \text{ V}$$

$$= 12 \text{ V}$$

If the two primaries are connected in series, as shown in Figure 13-22(b), the now 200 turn primary will result in a secondary voltage of

$$V_s = \frac{N_s}{N_p} \times V_p$$

$$= \frac{10}{120} \times 240 \text{ V}$$

$$= 12 \text{ V}$$

Some portable electrical or electronic equipment, such as radios or shavers, have a switch that allows you to switch between 120 V ac (U.S. wall outlet voltage) and 240 V ac

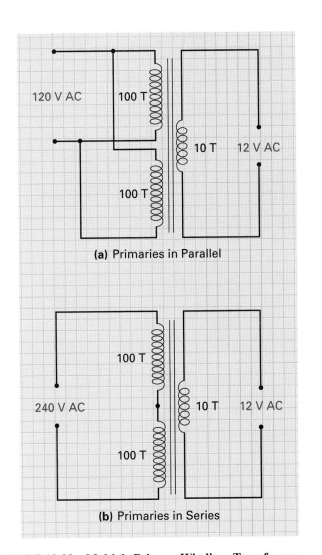

(a) Primaries in Parallel

(b) Primaries in Series

(c)

FIGURE 13-22   Multiple Primary Windings Transformer.

(European wall outlet voltage), as shown in Figure 13-22(c). By activating the switch, you can always obtain the correct voltage to operate the equipment, whether the wall socket is supplying 240 or 120 V.

Multiple secondaries. In some applications, more than one secondary is wound onto a common primary and core. The advantage of this arrangement can be seen in Figure 13-23, where many larger and smaller voltages can be acquired from one primary voltage.

Single Winding (Autotransformer)

The single winding transformer or autotransformer is used in the automobile ignition system (switched dc pulse system) to raise the low 12 V battery voltage up to about 20 kV for the spark plugs, as shown in Figure 13-24(a). Autotransformers are constructed by winding one continuous coil onto a core, which acts as both the primary and secondary. This yields three advantages for the single winding transformer:

1. They are smaller.
2. They are cheaper.
3. They are lighter than the normal separated primary/secondary transformer types.

Its disadvantage, however, is that no electrical isolation exists between primary and secondary. It is normally used as a step-up transformer for the automobile, or in televisions to obtain 20 kV. This transformer type can also be arranged to step down voltage, as shown in Figure 13-24(b).

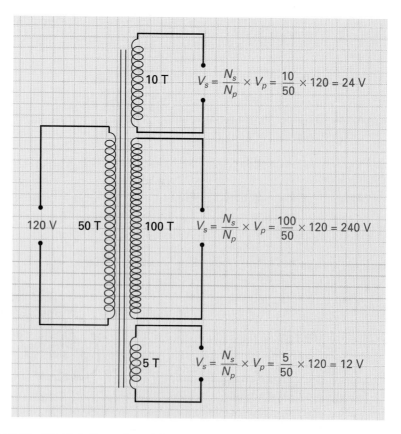

FIGURE 13-23 Multiple Secondary Windings Transformer.

FIGURE 13-24 Single-Winding Transformer. (a) Step-Up. (b) Step-Down.

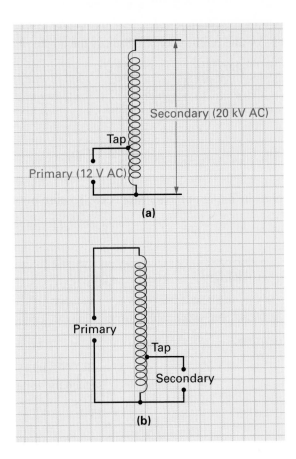

Secondary (20 kV AC)

Tap

Primary (12 V AC)

(a)

Primary

Tap

Secondary

(b)

SELF-TEST EVALUATION POINT FOR SECTION 13-6

Use the following questions to test your understanding of Section 13-6.

1. List the two types of fixed transformers.
2. List the four basic classifications for variable transformers.
3. True or false: The two outputs of a center-tapped secondary transformer will always be 90° out of phase with one another.
4. In what application would a multiple primary transformer be used?
5. In what application would a single-winding transformer be used?
6. What are the advantages of a single-winding transformer?

13-7 TRANSFORMER RATINGS

A typical transformer rating could read 1 kVA, 500/100, 60 Hz. The 500 normally specifies the maximum primary voltage, the 100 normally specifies the maximum secondary voltage, and the 1 kVA is the apparent power rating. In this example, the maximum load current will equal

$$I_s = \frac{\text{apparent power } (P_A)}{\text{secondary voltage } (V_s)} \qquad \left(\begin{array}{c} P = V \times I; \text{ therefore,} \\ I = \dfrac{P}{V} \end{array} \right)$$

$$= \frac{1\text{ kVA}}{100\text{ V}} = 10\text{ A}$$

With this secondary voltage at 100 V and a maximum current of 10 A, the smallest load resistor that can be connected across the output of the secondary is

$$R_L = \frac{V_s}{I_s}$$

$$= \frac{100 \text{ V}}{10 \text{ A}}$$

$$= 10 \ \Omega$$

Exceeding the rating of the transformer will cause overheating and even burning out of the windings.

EXAMPLE:

Calculate the smallest value of load resistance that can be connected across a 3 kVA, 600/200, 60 Hz step-down transformer.

Solution:

$$I = \frac{\text{apparent power } (P_A)}{\text{secondary voltage } (V_s)}$$

$$\frac{3 \text{ kVA}}{200 \text{ V}} = 15 \text{ A}$$

$$R_L = \frac{V_s}{I_s} = \frac{200 \text{ V}}{15 \text{ A}} = 13.3 \ \Omega$$

SELF-TEST EVALUATION POINT FOR SECTION 13-7

Use the following questions to test your understanding of Section 13-7.

1. What do each of the values mean when a transformer is rated as a 10 kVA, 200/100, 60 Hz?

2. If a 100 Ω resistor is connected across the secondary of a transformer that is to supply 1 kV and is rated at a maximum current of 8 A, will the transformer overheat and possibly burn out?

13-8 TESTING TRANSFORMERS

Like inductors, transformers, which are basically two or more inductors, can develop one of three problems:

1. An open winding
2. A complete short in a winding
3. A short in a section of a winding

13-8-1 *Open Primary or Secondary Winding*

An open in the primary winding will prevent any primary current, and therefore there will be no induced voltage in the secondary and therefore no voltage will be present across the load.

An open secondary winding will prevent the flow of secondary current, and once again, no voltage will be present across the load.

An open in the primary or secondary winding is easily detected by disconnecting the transformer from the circuit and testing the resistance of the windings with an ohmmeter. Like an inductor, a transformer winding should have a low resistance, in the tens to hundreds range. An open will easily be recognized because of its infinite (maximum) resistance.

13-8-2 *Complete Short or Section Short in Primary or Secondary Winding*

A partial or complete short in the primary winding of a transformer will result in an excessive source current that will probably blow the circuit's fuse or trip the circuit's breaker. A partial or complete short in the secondary winding will cause an excessive secondary current, which in turn will result in an excessive primary current.

In both instances, a short in the primary or secondary will generally burn out the primary winding unless the fuse or circuit breaker opens the excessive current path. An ohmmeter can be used to test for partial or complete shorts in transformer windings; however, all coils have a naturally low resistance which can be mistaken for a short. An inductive or reactive analyzer can accurately test transformer windings, checking the inductance value of each coil.

It is virtually impossible to repair an open, partially shorted, or completely shorted transformer winding, and therefore defective transformers are always replaced with a transformer that has an identical rating.

SELF-TEST EVALUATION POINT FOR SECTION 13-8

Use the following questions to test your understanding of Section 13-8.

1. Which test instrument could you use to test for a suspected open primary winding?
2. If a secondary winding had a resistance of 50 Ω, would this value indicate a problem?

REVIEW QUESTIONS

Multiple-Choice Questions

1. Transformer action is based on:
   a. Self-inductance
   b. Air between the coils
   c. Mutual capacitance
   d. Mutual inductance

2. An increase in transformer secondary current will cause a(an) _____ in primary current.
   a. Decrease   b. Increase

3. If 50% of the magnetic lines of force produced by the primary were to cut the secondary coil, the coefficient of coupling would be:
   a. 75   b. 50   c. 0.5   d. 0.005

4. A step-up transformer will always have a turns ratio _____, while a step-down transformer has a turns ratio _____.
   a. $< 1, > 1$ b. $> 1, > 1$ c. $> 1, < 1$ d. $< 1, < 1$

5. With an 80-V ac secondary voltage center-tapped transformer, what would be the voltage at each output, and what would be the phase relationship between the two secondary voltages?
   a. 20 V, in phase with one another
   b. 30 V, 180° out of phase
   c. 40 V, in phase
   d. 40 V, 180° out of phase

6. One application of the autotransformer would be:
   a. To obtain the final anode high-voltage supply for the cathode-ray tube in a television
   b. To obtain two outputs, 180° out of phase with one another
   c. To tap several different voltages from the secondary
   d. To obtain the same secondary voltage for different voltages

7. Alternating current can be used only with transformers because:
   a. It produces an alternating magnetic field.
   b. It produces a fixed magnetic field.
   c. Its magnetic field is greater than that of dc.
   d. Its rms is 0.707 of the peak.

8. The dot convention is used to indicate

   a. Primary/secondary phase
   b. Secondary power output
   c. Secondary load impedance
   d. Primary/secondary turns ratio

9. Assuming 100% efficiency, the output power, P_s, is always equal to:

   a. P_p **d.** Both (a) and (c)
   b. $V_s \times I_s$ **e.** Both (a) and (b)
   c. $0.5 \times P_p$

10. If the primary winding of a transformer were open, the result would be:

   a. No flux linkage between primary and secondary
   b. No primary or secondary current
   c. No induced voltage in the secondary
   d. All of the above

Practice Problems

11. Calculate the turns ratio of the following transformers and state whether they are step up or step down:

   a. P = 12T, S = 24T   **c.** P = 24T, S = 5T
   b. P = 3T, S = 250T   **d.** P = 240T, S = 120T

12. Calculate the secondary ac voltage for all the examples in Question 11 if the primary voltage equals 100 V.

13. Calculate the secondary ac current for all the examples in Question 11 if the primary current equals 100 mA.

14. What turns ratio would be needed to match a source impedance of 24 Ω to a load impedance of 8 Ω?

15. What turns ratio would be needed to step:

   a. 120 V to 240 V   **c.** 30 V to 14 V
   b. 240 V to 720 V   **d.** 24 V to 6 V

16. For a 24 V, 12-turn primary, and 16-, 2-, 1-, and 4-turn multiple secondary transformer, calculate each of the secondary voltages.

17. If a 2:1 step-down transformer has a primary input voltage of 120 V, 60 Hz, and a 2 kVA rating, calculate the maximum secondary current and smallest load resistor that can be connected across the output.

18. Indicate the polarity of the secondary voltages in Figure 13-25(a) and (b).

19. Referring to Figure 13-26(a) and (b), sketch the outputs, showing polarity and amplitude with respect to the inputs.

20. If a transformer is rated at 500 VA, 60 Hz, and the primary voltage is 240 V ac, and the secondary voltage is 600 V ac, calculate:

   a. Maximum load current   **b.** Smallest value of R_L

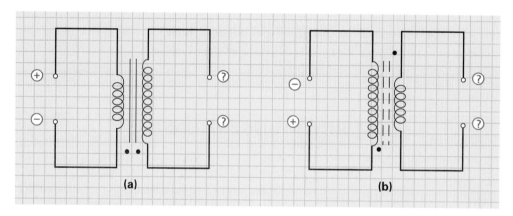

(a)   **(b)**

FIGURE 13-25   **Dot Convention Examples.**

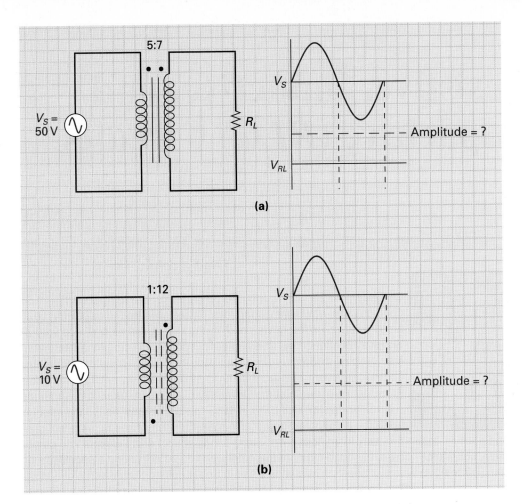

FIGURE 13-26 **Input/Output Polarity Examples.**

Resistive, Inductive, and Capacitive (*RLC*) Circuits

14

The Fairchildren

On December 23, 1947, John Bardeen, Walter Brattain, and William Shockley first demonstrated how a semiconductor device, named the *transistor,* could be made to amplify. However, the device had mysterious problems and was very unpredictable. Shockley continued his investigations, and in 1951 he presented the world with the first reliable junction transistor. In 1956, the three shared the Nobel Prize in physics for their discovery, and much later, in 1972, Bardeen would win a rare second Nobel Prize for his research at the University of Illinois in the field of superconductivity.

Shockley left Bell Labs in 1955 to start his own semiconductor company near his home in Palo Alto and began recruiting personnel. He was, however, very selective, only hiring those who were bright, young, and talented. The company was a success, although many of the employees could not tolerate Shockley's eccentricities, such as posting everyone's salary and requiring that the employees rate one another.

William Shockley, John Bardeen, and Walter Brattain (left to right)

Two years later, eight of Shockley's most talented defected. The "traitorous eight," as Shockley called them, started their own company only a dozen blocks away, named Fairchild Semiconductor.

More than 50 companies would be founded by former Fairchild employees. One of the largest was started by Robert Noyce and two other colleagues from the group of eight Shockley defectors; they named their company Intel, which was short for "intelligence."

Introduction

In this chapter we will combine resistors (*R*), inductors (*L*), and capacitors (*C*) into series and parallel ac circuits. Resistors, as we have discovered, operate and react to voltage and current in a very straightforward way, in that the voltage across a resistor is in phase with the resistor current.

Inductors and capacitors operate in essentially the same way, in that they both store energy and then return it back to the circuit. However, they have completely opposite reactions to voltage and current. To help you remember the phase relationships between voltage and current for capacitors and inductors, you may wish to use the following memory phrase:

ELI the ICE man

This phrase states that voltage (symbolized E) leads current (I) in an inductive (L) circuit (abbreviated by the word "ELI"), while current (I) leads voltage (E) in a capacitive (C) circuit (abbreviated by the word "ICE").

In this chapter we will study the relationships between voltage, current, impedance, and power in both series and parallel *RLC* circuits. We will also examine the important *RLC* circuit characteristic called resonance, and see how *RLC* circuits can be made to operate as filters. In the final section we will discuss how complex numbers can be used to analyze series and parallel ac circuits containing resistors, inductors, and capacitors.

14-1   SERIES *RLC* CIRCUIT

Figure 14-1 begins our analysis of series *RLC* circuits by illustrating the current and voltage relationships. The circuit current is always the same throughout a series circuit and can therefore be used as a reference. Studying the waveforms and vector diagrams shown alongside the components, you can see that the voltage across a resistor is always in phase with the current, while the voltage across the inductor leads the current by 90° and the voltage across the capacitor lags the current by 90°.

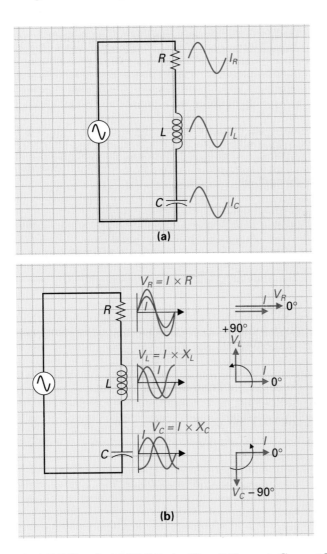

FIGURE 14-1   Series *RLC* Circuit. (a) *RLC* Series Circuit Current: Current Flow Is Always the Same in All Parts of a Series Circuit. (b) *RLC* Series Circuit Voltages: I Is in Phase with V_R, I Lags V_L by 90°, and I Leads V_C by 90°.

Now let's analyze the impedance, current, voltage, and power distribution of this circuit in a little more detail.

14-1-1   *Impedance*

Impedance is the total opposition to current flow and is a combination of both reactance (X_L, X_C) and resistance (R).

Capacitive reactance can be calculated by using the formula

$$X_C = \frac{1}{2\pi fC}$$

In the example,

$$X_C = \frac{1}{2\pi \times 60 \times 10 \text{ } \mu\text{F}} = 265.3 \text{ } \Omega$$

Inductive reactance is calculated by using the formula

$$X_L = 2\pi fL$$

In the example,

$$X_L = 2\pi \times 60 \times 20 \text{ mH} = 7.5 \text{ } \Omega$$

The difference between X_L and X_C is equal to 257.8, and since X_C is greater than X_L, the circuit is said to act capacitively. Reactance, however, is not in phase with resistance, and impedance is the vector sum of the reactive (X) and resistive (R) vectors. The formula, based on the Pythagorean theorem, is

$$Z = \sqrt{R^2 + X^2}$$

In this example, therefore, the circuit impedance will be equal to

$$Z = \sqrt{R^2 + X^2} = \sqrt{33^2 + 257.8^2} = 260 \text{ } \Omega$$

Since reactance (X) is equal to the difference between (symbolized $\sim$) X_L and X_C ($X_L \sim X_C$), the impedance formula can be modified slightly to incorporate the calculation to determine the difference between X_L and X_C.

$$Z = \sqrt{R^2 + (X_L \sim X_C)^2}$$

Using our example with this new formula, we arrive at the same value of impedance, and since the difference between X_L and X_C resulted in a capacitive vector, the circuit is said to act capacitively.

$$\begin{aligned} Z &= \sqrt{R^2 + (X_L \sim X_C)^2} \\ &= \sqrt{33^2 + (7.5 \sim 265.3)^2} \\ &= \sqrt{33^2 + 257.8^2} \\ &= 260 \text{ } \Omega \end{aligned}$$

If, on the other hand, the component values were such that the difference was inductive, then the circuit would be said to act inductively.

CHAPTER 14 / RESISTIVE, INDUCTIVE, AND CAPACITIVE (*RLC*) CIRCUITS

14-1-2   *Current*

The current in a series circuit is the same at all points throughout the circuit, and therefore

$$I = I_R = I_L = I_C$$

Once the total impedance of the circuit is known, Ohm's law can be applied to calculate the circuit current:

$$I = \frac{V_S}{Z}$$

In the example, circuit current is equal to

$$
\begin{aligned}
I &= \frac{V_S}{Z} \\
&= \frac{115\ \text{V}}{260\ \Omega} \\
&= 0.44\ \text{A} \quad \text{or} \quad 440\ \text{mA}
\end{aligned}
$$

14-1-3   *Voltage*

Now that you know the value of current flowing in the series circuit, you can calculate the voltage drops across each component, as shown in Figure 14-2.

$$V_R = I \times R$$

$$V_L = I \times X_L$$

$$V_C = I \times X_C$$

Since none of these voltages are in phase with one another, they must be added vectorially to obtain the applied voltage.

$$V_S = \sqrt{V_R^2 + (V_L \sim V_C)^2}$$

FIGURE 14-2   **Series Voltage Drops.**

In the example circuit, the applied voltage is, as we already know, 115 V.

$$V_S = \sqrt{15^2 + (3 \sim 117)^2}$$
$$= \sqrt{225 + 12{,}996}$$
$$= \sqrt{13{,}221}$$
$$= 115 \text{ V}$$

14-1-4   *Phase Angle*

As can be seen in Figure 14-1(b), there is a phase difference between the source voltage (V_S) and the circuit current (I). This phase difference can be calculated with either of the following formulas.

$$\theta = \arctan \frac{V_L \sim V_C}{V_R}$$

$$\theta = \arctan \frac{X_L \sim X_C}{R}$$

In the example circuit, θ is

$$\theta = \arctan \frac{3.3 \text{ V} \sim 116.732 \text{ V}}{14.52 \text{ V}}$$

$$= \arctan 7.812$$

$$= 82.7°$$

Since the example circuit is capacitive (ICE), the phase angle will be $-82.7°$ since V_S lags I in a circuit that acts capacitively.

14-1-5   *Power*

The true power or resistive power (P_R) dissipated by a circuit can be calculated using the formula

$$P_R = I^2 \times R$$

which in our example will be

$$P_R = 0.44^2 \times 33 \ \Omega = 6.4 \text{ W}$$

The apparent power (P_A) consumed by the circuit is calculated by

$$P_A = V_S \times I$$

which in our example will be

$$P_A = 115 \text{ V} \times 0.44 = 50.6 \text{ volt-amperes (VA)}$$

The true or actual power dissipated by the resistor is, as expected, smaller than the apparent power that appears to be being used.

The power factor can be calculated, as usual, by

$$PF = \cos \theta = \frac{R}{Z} = \frac{P_R}{P_A}$$

PF of 0 = reactive circuit

PF of 1 = resistive circuit

In the example circuit, PF = 0.126, indicating that the circuit is mainly reactive.

EXAMPLE:

For a series circuit where $R = 10\,\Omega$, $L = 5$ mH, $C = 0.05\,\mu F$, and $V_S = 100$ V/2 kHz, calculate:

a. X_C f. Apparent power

b. X_L g. True power

c. Z h. Power factor

d. I i. Phase angle

e. V_R, V_C, and V_L

Solution:

a. $X_C = \dfrac{1}{2\pi f C} = 1.6\ \text{k}\Omega$

b. $X_L = 2\pi f L = 62.8\ \Omega$

c. $Z = \sqrt{R^2 + (X_L \sim X_C)^2}$

$\quad = \sqrt{(10\ \Omega)^2 + (1.6\ \text{k}\Omega \sim 62.8\ \Omega)^2}$

$\quad = \sqrt{10\ \Omega^2 + 1.54\ \text{k}\Omega^2}$

$\quad = 1.54\ \text{k}\Omega$ (capacitive circuit due to high X_C)

d. $I = \dfrac{V_S}{Z} = \dfrac{100\ \text{V}}{1.54\ \text{k}\Omega} = 64.9\ \text{mA}$

e. $V_R = I \times R = 64.9\ \text{mA} \times 10\ \Omega = 0.65\ \text{V}$

$\quad V_C = I \times X_C = 64.9\ \text{mA} \times 1.6\ \text{k}\Omega = 103.9\ \text{V}$

$\quad V_L = I \times X_L = 64.9\ \text{mA} \times 62.8\ \Omega = 4.1\ \text{V}$

f. Apparent power $= V_S \times I = 100\ \text{V} \times 64.9\ \text{mA} = 6.49\ \text{VA}$

g. True power $= I^2 \times R = 64.9^2 \times 10\ \Omega = 42.17\ \text{mW}$

h. $\text{PF} = \dfrac{R}{Z} = \dfrac{10\ \Omega}{1.5\ \text{k}\Omega} = 0.006$ (reactive circuit)

i. $\theta = \arctan \dfrac{V_L \sim V_C}{V_R}$

$\quad = \arctan \dfrac{4.1\ \text{V} \sim 103.9\ \text{V}}{0.65\ \text{V}}$

$\quad = \arctan 153.54$

$\quad = 89.63°$

Capacitive circuit (ICE); therefore, V_S lags I by $-89.63°$.

SELF-TEST EVALUATION POINT FOR SECTION 14-1

Use the following questions to test your understanding of Section 14-1.

1. List in order the procedure that should be followed to fully analyze a series *RLC* circuit.
2. State the formulas for calculating the following in relation to a series *RLC* circuit.

    a. Impedance   **c.** Apparent power   **e.** True power   **g.** V_L **i.** Phase angle (θ)

    b. Current   **d.** V_S **f.** V_R **h.** V_C **j.** Power factor (PF)

14-2　PARALLEL *RLC* CIRCUIT

Now that the characteristics of a series circuit are understood, let us connect a resistor, inductor, and capacitor in parallel with one another. Figure 14-3(a) and (b) show the current and voltage relationships of a parallel *RLC* circuit.

14-2-1　*Voltage*

As can be seen in Figure 14-3(a), the voltage across any parallel circuit will be equal and in phase. Therefore,

$$V_R = V_L = V_C = V_S$$

14-2-2　*Current*

With current, we must first calculate the individual branch currents (I_R, I_L, and I_C) and then calculate the total circuit current (I_T). An example circuit is illustrated in Figure 14-4, and the branch currents can be calculated by using the formulas

$$I_R = \frac{V}{R}$$

$$I_L = \frac{V}{X_L}$$

$$I_C = \frac{V}{X_C}$$

The 180° phase difference between I_C and I_L results in a cancellation.

The total current (I_T) can be calculated by using the Pythagorean theorem on the right triangle.

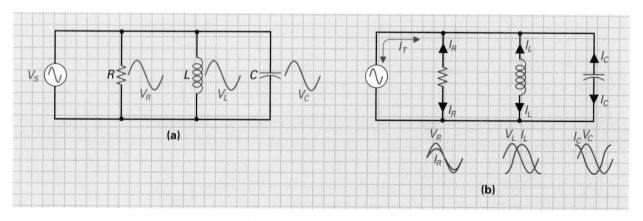

(a)

(b)

FIGURE 14-3　Parallel *RLC* Circuit. (a) *RLC* Parallel Circuit Voltage: Voltages across Each Component Are All Equal and in Phase with One Another in a Parallel Circuit. (b) *RLC* Parallel Circuit Currents: I_R Is in Phase with V_R, I_L Lags V_L by 90°, and I_C Leads V_C by 90°.

$$I_R = \frac{V}{R} = \frac{115 \text{ V}}{33 \text{ }\Omega} = 3.5 \text{ A}$$

$$I_L = \frac{V}{X_L} = \frac{115 \text{ V}}{7.5 \text{ }\Omega} = 15.3 \text{ A}$$

$$X_L = 2\pi fL$$
$$= 2\pi \times 60 \times 20 \text{ mH} = 7.5 \text{ }\Omega$$

$$I_C = \frac{V}{X_C} = \frac{115 \text{ V}}{265.3 \text{ }\Omega} = 0.43 \text{ A}$$

$$X_C = \frac{1}{2\pi fL}$$
$$= \frac{1}{2\pi \times 60 \times 10 \text{ }\mu\text{F}} = 265.3 \text{ }\Omega$$

FIGURE 14-4 Example *RLC* Parallel Circuit.

$$I_T = \sqrt{I_R^{\,2} + I_X^{\,2}}$$

$$(I_X = I_L \sim I_C)$$

$$= \sqrt{3.5^2 + 14.9^2}$$
$$= 15.3 \text{ A}$$

14-2-3 *Phase Angle*

There is a phase difference between the source voltage (V_S) and the circuit current (I_T). This phase difference can be calculated using the formula

$$\theta = \arctan \frac{I_L \sim I_C}{I_R}$$

$$= \arctan \frac{15.3 \text{ A} - 0.43 \text{ A}}{3.5 \text{ A}}$$

$$= \arctan 4.25$$
$$= 76.7°$$

Since this is an inductive circuit (ELI), the total current (I_T) will lag the source voltage (V_S) by $-76.7°$.

14-2-4 *Impedance*

With the total current (I_T) known, the impedance of all three components in parallel can be calculated by the formula

$$Z = \frac{V}{I_T}$$

$$= \frac{115 \text{ V}}{15.3 \text{ A}}$$
$$= 7.5 \text{ }\Omega$$

14-2-5 Power

The true power dissipated can be calculated using

$$P_R = I_R^2 \times R$$

$$= 3.5^2 \times 33\ \Omega$$
$$= 404.3\ \text{W}$$

The apparent power consumed by the circuit is calculated by

$$P_A = V_S \times I_T$$

$$= 115\ \text{V} \times 15.3\ \text{A}$$
$$= 1759.5\ \text{volt-amperes (VA)}$$

Finally, the power factor can be calculated, as usual, with

$$\text{PF} = \cos\theta = \frac{P_R}{P_A}$$

In the example circuit, PF $= 0.23$.

SELF-TEST EVALUATION POINT FOR SECTION 14-2

Use the following questions to test your understanding of Section 14-2.

1. State the formulas for calculating the following in relation to a parallel *RLC* circuit:
   a. I_R
   b. I_T
   c. I_C
   d. I_L
   e. θ
   f. Z

2. State the formulas for:
   a. P_R
   b. P_X
   c. P_A
   d. PF

14-3 RESONANCE

Resonance

Circuit condition that occurs when the inductive reactance (X_L) is equal to the capacitive reactance (X_C).

Resonance is a circuit condition that occurs when the inductive reactance (X_L) and the capacitive reactance (X_C) have been balanced. Figure 14-5 illustrates a parallel- and a series-connected *LC* circuit. If a dc voltage is applied to the input of either circuit, the capacitor will act as an open (X_C = infinite Ω) and the inductor will act as a short ($X_L = 0\ \Omega$).

If a low-frequency ac is now applied to the input, X_C will decrease from maximum, and X_L will increase from zero. As the ac frequency is increased further, the capacitive reactance will continue to fall ($X_C\downarrow \propto 1/f\uparrow$) and the inductive reactance to rise ($X_L\uparrow \propto f\uparrow$), as shown in Figure 14-6.

As the input ac frequency is increased further, a point will be reached where X_L will equal X_C, and this condition is known as *resonance*. The frequency at which $X_L = X_C$ in either a parallel or a series *LC* circuit is known as the *resonant frequency* (f_0) and can be cal-

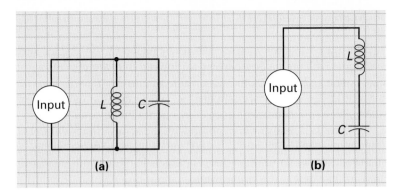

FIGURE 14-5   **Resonance. (a) Parallel *LC* Circuit. (b) Series *LC* Circuit.**

culated by the following formula, which has been derived from the capacitive and inductive reactance formulas:

$$f_0 = \frac{1}{2\pi\sqrt{LC}}$$

where f_0 = resonant frequency, in hertz (Hz)
L = inductance, in henrys (H)
C = capacitance, in farads (F)

☐   **EXAMPLE:**

Calculate the resonant frequency (f_0) of a series *LC* circuit if L = 750 mH and C = 47 μF.

■   *Solution:*

$$f_0 = \frac{1}{2\pi\sqrt{L \times C}}$$

$$= \frac{1}{2\pi\sqrt{(750 \times 10^{-3}) \times (47 \times 10^{-6})}}$$

$$= 26.8 \text{ Hz}$$

FIGURE 14-6   **Frequency versus Reactance.**

14-3-1   *Series Resonance*

Figure 14-7(a) illustrates a series *RLC* circuit at resonance ($X_L = X_C$), or **series resonant circuit.** The ac input voltage causes current to flow around the circuit, and since all the components are connected in series, the same value of current (I_S) will flow through all the components. Since R, X_L, and X_C are all equal to 100 Ω and the current flow is the same throughout, the voltage dropped across each component will be equal.

The voltage across the resistor is in phase with the series circuit current (I_S); however, since the voltage across the inductor (V_L) is 180° out of phase with the voltage across the capacitor (V_C), and both are equal to one another, V_L cancels V_C, when both are measured in series.

Three unusual characteristics occur when a circuit is at resonance, which do not occur at any other frequency.

(1) The first is that if V_L and V_C cancel, the voltage across L and C will measure 0 V on a voltmeter. Since there is effectively no voltage being dropped across these two components, all the voltage must be across the resistor ($V_R = 12$ V). This is true; however, since the same current flows throughout the series circuit, a voltmeter will measure 12 V across C, 12 V across L, and 12 V across R, as shown in Figure 14-7(b). It now appears that the voltage drops around the series circuit (36 V) do not equal the voltage applied (12 V). This is not true, as V_L and V_C cancel, because they are out of phase with one another, so Kirchhoff's voltage law is still valid.

FIGURE 14-7   Series Resonant Circuit.

CHAPTER 14 / RESISTIVE, INDUCTIVE, AND CAPACITIVE (*RLC*) CIRCUITS

(2) The second unusual characteristic of resonance is that because the total opposition or impedance (Z) is equal to

$$Z = \sqrt{R^2 + (X_L \sim X_C)^2}$$

and the difference between X_L and X_C is 0 ($Z = \sqrt{R^2 + 0}$), the impedance of a series circuit at resonance is equal to the resistance value R ($Z = \sqrt{R^2} = R$). As a result, the applied ac voltage of 12 V is forcing current to flow through this series *RLC* circuit. Since current is equal to $I_S = V/Z$ and $Z = R$, the circuit current at resonance is dependent only on the value of resistance. The capacitor and inductor are invisible and are seen by the source as simply a piece of conducting wire with no resistance, as illustrated in Figure 14-7(c). Since only resistance exists in the circuit, current (I_S) and voltage (V_S) are in phase with one another, and as expected for a purely resistive circuit, the power factor will be equal to 1.

(3) To emphasize the third strange characteristic of series resonance, we will take another example, shown in Figure 14-8. In this example, R is made smaller (10 Ω) than X_L and X_C (100 Ω each).The circuit current in this example is equal to $I = V/R = 12\,V/10\,Ω = 1.2\,A$, as $Z = R$ at resonance. Since the same current flows throughout a series circuit, the voltage across each component can be calculated.

$$V_R = I \times R = 1.2\,A \times 10 = 12\,V$$
$$V_L = I \times X_L = 1.2\,A \times 100 = 120\,V$$
$$V_C = I \times X_C = 1.2\,A \times 100 = 120\,V$$

As V_L is 180° out of phase with V_C, the 120 V across the capacitor cancels with the 120 V across the inductor, resulting in 0 V across L and C combined, as shown in Figure 14-8(b).

FIGURE 14-8   **Circuit Effects at Resonance.**

Since L and C have the ability to store energy, the voltage across them individually will appear larger than the applied voltage.

If the resistance in the circuit is removed completely, as shown in Figure 14-8(c), the circuit current, which is determined by the resistance only, will increase to a maximum $(I\uparrow = V/R\downarrow)$ and, consequently, cause an infinitely high voltage across the inductor and capacitor $(V\uparrow = I\uparrow \times R)$. In reality, the ac source will have some value of internal resistance, and the inductor, which is a long length of thin wire $(R\uparrow)$, will have some value of resistance, as shown in Figure 14-8(d), which limits the series resonant circuit current.

In summary, we can say that in a series resonant circuit:

1. The inductor and capacitor electrically disappear due to their equal but opposite effect, resulting in a 0 V drop across the series combination, and the circuit seems purely resistive.
2. The current flow is large because the impedance of the circuit is low and equal to the series resistance (R), which has the source voltage developed across it.
3. The individual voltage drops across the inductor or capacitor can be larger than the source voltage if R is smaller than X_L and X_C.

Quality Factor

As discussed previously in the inductance chapter, the Q factor is a ratio of inductive reactance to resistance and is used to express how efficiently an inductor will store rather than dissipate energy. In a series resonant circuit, the Q factor indicates the quality of the series resonant circuit, or is the ratio of the reactance to the resistance.

$$Q = \frac{X_L}{R}$$

or, since $X_L = X_C$,

$$Q = \frac{X_C}{R}$$

Another way to calculate the Q of a series resonant circuit is by using the formula

$$Q = \frac{V_L}{V_R} = \frac{V_C}{V_R}$$ (at resonance only)

or, since $V_R = V_S$,

$$Q = \frac{V_L}{V_S} = \frac{V_C}{V_S}$$ (at resonance only)

If the Q and source voltage are known, the voltage across the inductor or capacitor can be found by transposition of the formula, as can be seen in the example in Figure 14-9.

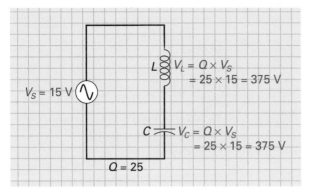

FIGURE 14-9   Quality Factor at Resonance.

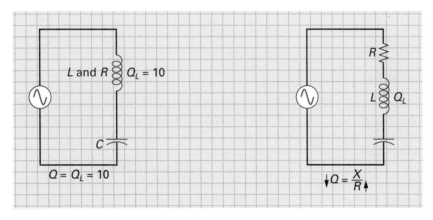

FIGURE 14-10   Resistance within Inductor.

The Q of a resonant circuit is almost entirely dependent on the inductor's coil resistance, because capacitors tend to have almost no resistance at all, only reactance, which makes them very efficient.

The inductor has a Q value of its own, and if only L and C are connected in series with one another, the Q of the series resonant circuit will be equal to the Q of the inductor, as shown in Figure 14-10. If the resistance is added in with L and C, the Q of the series resonant circuit will be less than that of the inductor's Q.

☐ **EXAMPLE:**

Calculate the resistance of the series resonant circuit illustrated in Figure 14-11.

FIGURE 14-11   Series Resonant Circuit Example.

$$Q = \frac{V_L}{V_S} = \frac{100 \text{ V}}{10 \text{ V}} = 10$$

Since $Q = X_L/R$ and $R = X_L/Q$, if the inductive reactance can be found, the R can be determined.

$$X_L = 2\pi \times f \times L$$
$$= 2\pi \times 100 \times 8 \text{ mH}$$
$$= 5 \ \Omega$$

R will therefore equal

$$R = \frac{X_L}{Q}$$
$$= \frac{5}{10}$$
$$= 0.5 \ \Omega$$

Bandwidth

Bandwidth
Width of the group or band of frequencies between the half-power points.

Frequency Response Curve
A graph indicating a circuit's response to different frequencies.

Cutoff Frequency
Frequency at which the gain of the circuit falls below 0.707 of the maximum current or half-power (−3 dB).

A series resonant circuit is selective in that frequencies at resonance or slightly above or below will cause a larger current than frequencies well above or below the circuit's resonant frequency. The group or band of frequencies that causes the larger current is called the circuit's **bandwidth.**

Figure 14-12 illustrates a series resonant circuit and its bandwidth. The × marks on the curve illustrate where different frequencies were applied to the circuit and the resulting value of current measured in the circuit. The resulting curve produced is called a **frequency response curve,** as it illustrates the circuit's response to different frequencies. At resonance, $X_L = X_C$ and the two cancel, which is why maximum current was present in the circuit (100 mA) when the resonant frequency (100 Hz) was applied.

The bandwidth includes the group or band of frequencies that cause 70.7% or more of the maximum current to flow within the series resonant circuit. In this example, frequencies from 90 to 110 Hz cause 70.7 mA or more, which is 70.7% of maximum (100 mA), to flow. The bandwidth in this example is equal to

$$\text{BW} = 110 - 90 = 20 \text{ Hz}$$

(110 Hz and 90 Hz are known as **cutoff frequencies**).

FIGURE 14-12   Series Resonant Circuit Bandwidth. (a) Circuit. (b) Frequency Response Curve.

Referring to the bandwidth curve in Figure 14-12(b), you may notice that 70.7% is also called the **half-power point,** although it does not exist halfway between 0 and maximum. This value of 70.7% is not the half-current point but the half-power point, as we can prove with a simple example.

Half-Power Point

A point at which power is 50%. This half-power point corresponds to 70.7% of the total current.

■ **EXAMPLE:**

$R = 2$ kΩ and $I = 100$ mA; therefore, power $= I^2 \times R = 100$ mA$^2 \times 2$ kΩ $= 20$ W. If the current is now reduced so that it is 70.7% of its original value, calculate the power dissipated.

■ *Solution:*

$$P = I^2 \times R = 70.7 \text{ mA}^2 \times 2 \text{ kΩ} = 10 \text{ W}$$

In summary, the 70.7% current points are equal to the 50% or half-power points. A circuit's bandwidth is the band of frequencies that exists between the 70.7% current points or half-power points.

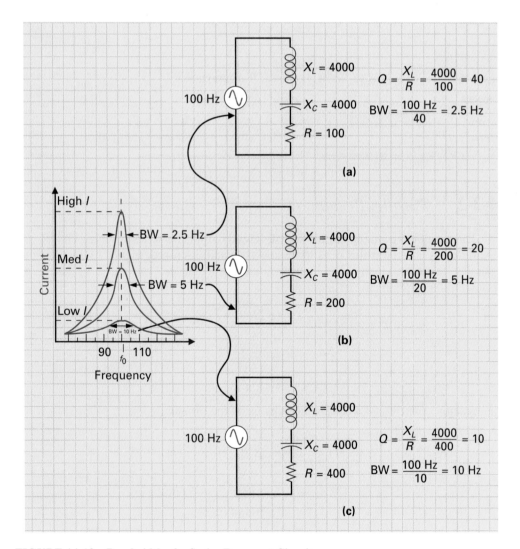

FIGURE 14-13   Bandwidth of a Series Resonant Circuit.

The bandwidth of a series resonant circuit can also be calculated by use of the formula

$$BW = \frac{f_0}{Q_{f_0}}$$

f_0 = Resonant frequency

Q_{f_0} = Quality factor at resonance

This formula states that the BW is proportional to the resonant frequency of the circuit and inversely proportional to the Q of the circuit.

Figure 14-13 illustrates three example response curves. In these three examples, the value of R is changed from 100 Ω to 200 Ω to 400 Ω. This does not vary the resonant frequency, but simply alters the Q and therefore the BW. The resistance value will determine the Q of the circuit, and since Q is inversely proportional to resistance, Q is proportional to current; consequently, a high value of Q will cause a high value of current.

In summary, the bandwidth of a series resonant circuit will increase as the Q of the circuit decreases (BW↑ = f_0/Q↓), and vice versa.

SELF-TEST EVALUATION POINT FOR SECTION 14-3-1

Use the following questions to test your understanding of Section 14-3-1.

1. Define *resonance.*

2. What is series resonance?

3. In a series resonant circuit, what are the three rather unusual circuit phenomena that take place?

4. How does Q relate to series resonance?

5. Define *bandwidth.*

6. Calculate BW if f_0 = 12 kHz and Q = 1000.

14-3-2 *Parallel Resonance*

Parallel Resonant Circuit

Circuit having an inductor and capacitor in parallel with one another, offering a high impedance at the frequency of resonance.

The **parallel resonant circuit** acts differently from the series resonant circuit, and these different characteristics need to be analyzed and discussed. Figure 14-14 illustrates a parallel resonant circuit. The inductive current could be calculated by using the formula

$$I_L = \frac{V_L}{X_L}$$
$$= \frac{10 \text{ V}}{1 \text{ k}\Omega}$$
$$= 10 \text{ mA}$$

The capacitive current could be calculated by using the formula

$$I_C = \frac{V_C}{X_C}$$
$$= \frac{10 \text{ V}}{1 \text{ k}\Omega}$$
$$= 10 \text{ mA}$$

Looking at the vector diagram in Figure 14-14(b), you can see that I_C leads the source voltage by 90° (ICE) and I_L lags the source voltage by 90° (ELI), creating a 180° phase differ-

FIGURE 14-14   Parallel Resonant Circuit. (a) Circuit. (b) Vector Diagram.

ence between I_C and I_L. This means that when 10 mA of current flows up through the inductor, 10 mA of current will flow in the opposite direction down through the capacitor, as shown in Figure 14-15(a). During the opposite alternation, 10 mA will flow down through the inductor and 10 mA will travel up through the capacitor, as shown in Figure 14-15(b).

If 10 mA arrives into point X and 10 mA of current leaves point X, no current can be flowing from the source (V_S) to the parallel LC circuit; the current is simply swinging or oscillating back and forth between the capacitor and inductor.

The source voltage (V_S) is needed initially to supply power to the LC circuit and start the oscillations; but once the oscillating process is in progress (assuming the ideal case), current is only flowing back and forth between inductor and capacitor, and no current is flowing from the source. So the LC circuit appears as an infinite impedance and the source can be disconnected, as shown in Figure 14-15(c).

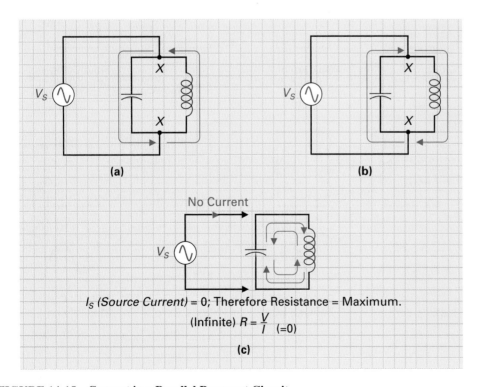

FIGURE 14-15   Current in a Parallel Resonant Circuit.

Flywheel Action

Flywheel Action

Sustaining effect of oscillation in an LC circuit due to the charging and discharging of the capacitor and the expansion and contraction of the magnetic field around the inductor.

Let's discuss this oscillating effect, called **flywheel action,** in a little more detail. The name is derived from the fact that it resembles a mechanical flywheel, which, once started, will keep going continually until friction reduces the magnitude of the rotations to zero.

The electronic equivalent of the mechanical flywheel is a resonant parallel-connected *LC* circuit. Figure 14-16(a) through (h) illustrate the continual energy transfer between capacitor and inductor, and vice versa. The direction of the circulating current reverses each half-cycle at the frequency of resonance. Energy is stored in the capacitor in the form of an electric field between the plates on one half-cycle, and then the capacitor discharges, supplying current to build up a magnetic field on the other half-cycle. The inductor stores its energy in the form of a magnetic field, which will collapse, supplying a current to charge the capacitor, which will then discharge, supplying a current back to the inductor, and so on. Due to the "storing action" of this circuit, it is sometimes related to the fluid analogy and referred to as a **tank circuit.**

Tank Circuit

Circuit made up of a coil and capacitor that is capable of storing electric energy.

The Reality of Tanks

Under ideal conditions a tank circuit should oscillate indefinitely if no losses occur within the circuit. In reality, the resistance of the coil reduces that 100% efficiency, as does friction with the mechanical flywheel. This coil resistance is illustrated in Figure

FIGURE 14-16   Energy and Current in an *LC* Parallel Circuit at Resonance.

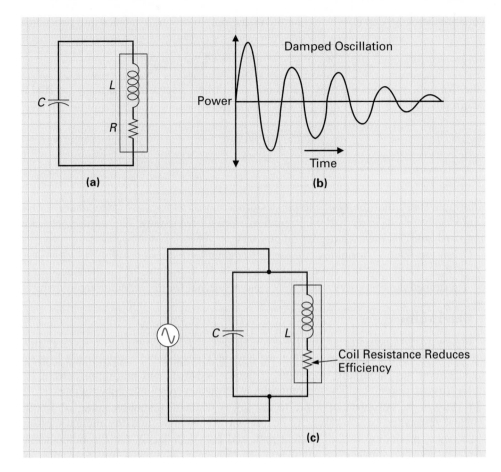

FIGURE 14-17   Losses in Tanks.

14-17(a), and, unlike reactance, resistance is the opposition to current flow, with the dissipation of energy in the form of heat. As a small part of the energy is dissipated with each cycle, the oscillations will be reduced in size and eventually fall to zero, as shown in Figure 14-17(b).

If the ac source is reconnected to the tank, as shown in Figure 14-17(c), a small amount of current will flow from the source to the tank to top up the tank or replace the dissipated power. The higher the coil resistance is, the higher the loss and the larger the current flow from source to tank to replace the loss.

Quality Factor

In the series resonant circuit, we were concerned with voltage drops since current remains the same throughout a series circuit, so

$$Q = \frac{V_C \text{ or } V_L}{V_S} \quad \text{(at resonance only)}$$

In a parallel resonant circuit, we are concerned with circuit currents rather than voltage, so

$$Q = \frac{I_{\text{tank}}}{I_S}$$

(at resonance only)

The quality factor, Q, can also be expressed as the ratio between reactance and resistance:

$$Q = \frac{X_L}{R} \quad \text{(at any frequency)}$$

Another formula, which is the most frequently used when discussing and using parallel resonant circuits, is

$$Q = \frac{Z_{\text{tank}}}{X_L}$$

(at resonance only)

This formula states that the Q of the tank is proportional to the tank impedance. A higher tank impedance results in a smaller current flow from source to tank. This assures less power is dissipated, and that means a higher-quality tank.

Of all the three Q formulas for parallel resonant circuits, $Q = I_{\text{tank}}/I_S$, $Q = X_L/R$, and $Q = Z_{\text{tank}}/X_L$, the latter is the easiest to use as both X_L and the tank impedance can easily be determined in most cases where C, L, and R internal for the inductor are known.

Bandwidth

Figure 14-18 illustrates a parallel resonant circuit and two typical response curves. These response curves summarize what we have described previously, in that a parallel resonant circuit has maximum impedance [Figure 14-18(b)] and minimum current [Figure 14-18(c)] at resonance. The current versus frequency response curve shown in Figure 14-18(c) is the complete opposite to the series resonant response curve. At frequencies below resonance (< 10 kHz), X_L is low and X_C is high, and the inductor offers a low reactance, producing a high current path and low impedance. On the other hand, at frequencies above resonance (> 10 kHz), the capacitor displays a low reactance, producing a high current path and low impedance. The parallel resonant circuit is like the series resonant circuit in that it responds to a band of frequencies close to its resonant frequency.

The bandwidth (BW) can be calculated by use of the formula

$$BW = \frac{f_0}{Q_{f_0}}$$

f_0 = Resonant frequency
Q_{f_0} = Quality factor at resonance

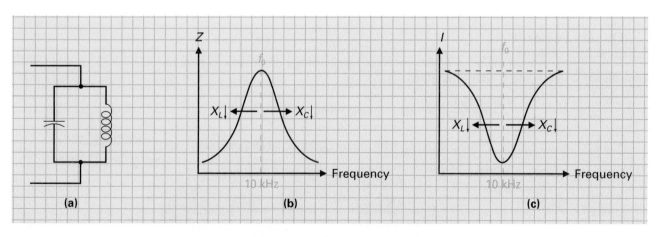

FIGURE 14-18   Parallel Resonant Circuit Bandwidth.

CHAPTER 14 / RESISTIVE, INDUCTIVE, AND CAPACITIVE (*RLC*) CIRCUITS

□ EXAMPLE:

Calculate the bandwidth of the circuit illustrated in Figure 14-19.

FIGURE 14-19 Bandwidth Example.

$f_0 = 125$ kHz

$Q_{f_0} = 75$

■ *Solution:*

$$\text{BW} = \frac{f_0}{Q_{f_0}}$$

$$= \frac{125 \text{ kHz}}{75}$$

$$= 1.7 \text{ kHz}$$

$$\frac{1.7 \text{ kHz}}{2} = 0.85 \text{ kHz}$$

therefore the bandwidth extends from

$$f_0 + 0.85 \text{ kHz} = 125.85 \text{ kHz}$$
$$f_0 - 0.85 \text{ kHz} = 124.15 \text{ kHz}$$
$$\text{BW} = 124.15 \text{ kHz to } 125.85 \text{ kHz}$$

Selectivity

Circuits containing inductance and capacitance are often referred to as **tuned circuits** since they can be adjusted to make the circuit responsive to a particular frequency (the resonant frequency). **Selectivity,** by definition, is the ability of a tuned circuit to respond to a desired frequency and ignore all others. Parallel resonant *LC* circuits are sometimes too selective, as the *Q* is too large, producing too narrow a bandwidth, as shown in Figure 14-20(a) $(\text{BW}\!\downarrow = f_0/Q\!\uparrow)$.

In this situation, because of the very narrow response curve, a high resistance value can be placed in parallel with the *LC* circuit to provide an alternative path for source current. This process is known as *loading* or *damping* the tank and will cause an increase in source current and decrease in *Q* $(Q\!\downarrow = I_{\text{tank}}/I_{\text{source}}\!\uparrow)$. The decrease in *Q* will cause a corresponding increase in BW $(\text{BW}\!\uparrow = f_0/Q\!\downarrow)$, as shown by the examples in Figure 14-20, which illustrates a 1000 Ω loading resistor [Figure 14-20(b)] and a 100 Ω loading resistor [Figure 14-20(c)].

In summary, a parallel resonant circuit can be made less selective with a broader bandwidth if a resistor is added in parallel, providing an increase in current and a decrease in impedance, which widens the bandwidth.

Tuned Circuit

Circuit that can have its components' values varied so that the circuit responds to one selected frequency yet heavily attenuates all other frequencies.

Selectivity

Characteristic of a circuit to discriminate between the wanted signal and the unwanted signal.

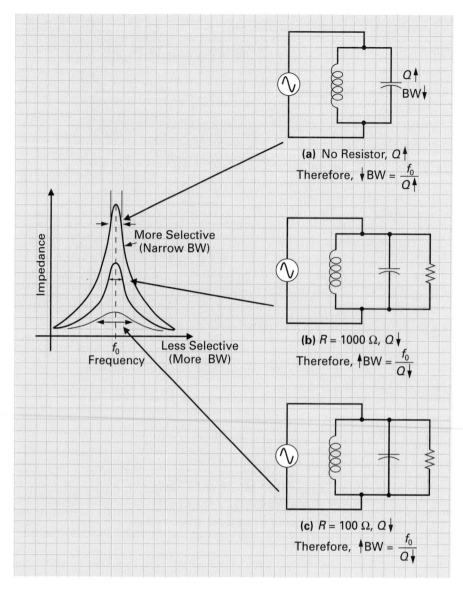

FIGURE 14-20 Varying Bandwidth by Loading the Tank Circuit.

Use the following questions to test your understanding of Section 14-3-2.

1. What are the differences between a series and a parallel resonant circuit?
2. Describe flywheel action.
3. Calculate the value of Q of a tank if $X_L = 50\ \Omega$ and $R = 25\ \Omega$.
4. When calculating bandwidth for a parallel resonant circuit, can the series resonant bandwidth formula be used?
5. What is selectivity?

14-4 APPLICATIONS OF *RLC* CIRCUITS

In the previous chapters, you saw how *RC* and *RL* filter circuits are used as low- or high-pass filters to pass some frequencies and block others. There are basically four types of filters:

1. Low-pass filter, which passes frequencies below a cutoff frequency
2. High-pass filter, which passes frequencies above a cutoff frequency
3. Bandpass filter, which passes a band of frequencies
4. Band-stop filter, which stops a band of frequencies

14-4-1 Low-Pass Filter

Figure 14-21(a) illustrates how an inductor and capacitor can be connected to act as a low-pass filter. At low frequencies, X_L has a small value compared to the load resistor (R_L), so nearly all the low-frequency input is developed and appears at the output across R_L. Since X_C is high at low frequencies, nearly all the current passes through R_L rather than C.

At high frequencies, X_L increases and drops more of the applied input across the inductor rather than the load. The capacitive reactance, X_C, aids this low-output-at-high-frequency effect by decreasing its reactance and providing an alternative path for current to flow.

Since the inductor basically blocks alternating current and the capacitor shunts alternating current, the net result is to prevent high-frequency signals from reaching the load. The way in which this low-pass filter responds to frequencies is graphically illustrated in Figure 14-21(b).

14-4-2 High-Pass Filter

Figure 14-22(a) illustrates how an inductor and capacitor can be connected to act as a high-pass filter. At high frequencies, the reactance of the capacitor (X_C) is low while the reactance of the inductor (X_L) is high, so all the high frequencies are easily passed by the capacitor and blocked by the inductor, so they all are routed through to the output and load.

At low frequencies, the reverse condition exists, resulting in a low X_L and a high X_C. The capacitor drops nearly all the input, and the inductor shunts the signal current away from the output load.

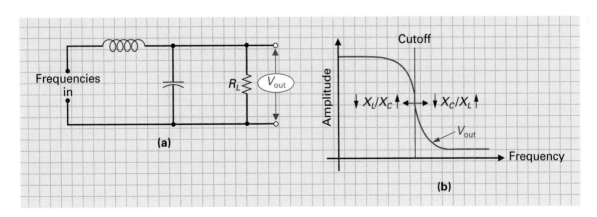

FIGURE 14-21 Low-Pass Filter. (a) Circuit. (b) Frequency Response.

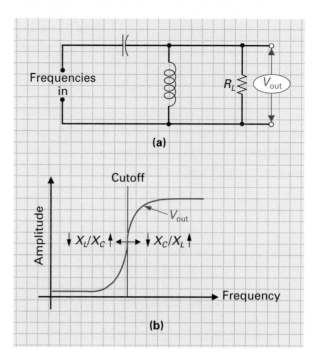

FIGURE 14-22 **High-Pass Filter. (a) Circuit. (b) Frequency Response.**

14-4-3 *Bandpass Filter*

Figure 14-23(a) illustrates a series resonant **bandpass filter,** and Figure 14-23(b) shows a parallel resonant bandpass filter. Figure 14-23(c) shows the frequency response curve produced by the bandpass filter. At resonance, the series resonant LC circuit has a very low impedance and will consequently pass the resonant frequency to the load with very little drop across the L and C components.

Below resonance, X_C is high, and the capacitor drops a large amount of the input signal; above resonance, X_L is high and the inductor drops most of the input frequency voltage. This circuit will therefore pass a band of frequencies centered around the resonant frequency of the series LC circuit and block all other frequencies above and below this resonant frequency.

Figure 14-23(b) illustrates how a parallel resonant LC circuit can be used to provide a bandpass response. The series resonant circuit was placed in series with the output, whereas the parallel resonant circuit will have to be placed in parallel with the output to provide the same results. At resonance, the parallel resonant circuit or tank has a high impedance, so very little current will be shunted away from the output; it will be passed on to the output, and almost all the input will appear at the output across the load.

Above resonance, X_C is small, so most of the input is shunted away from the output by the capacitor; below resonance, X_L is small, and the shunting action occurs again, but this time through the inductor.

Figure 14-24 illustrates how a transformer can be used to replace the inductor to produce a bandpass filter. At resonance, maximum flywheel current flows within the parallel circuit made up of the capacitor and the primary of the transformer (L), which is known as a *tuned transformer.* With maximum flywheel current, there will be a maximum magnetic field, which means that there will be maximum power transfer between primary and secondary. So nearly all the input will be coupled to the output (coupling coefficient $k = 1$) and appear across the load at and around a small band of frequencies centered on resonance.

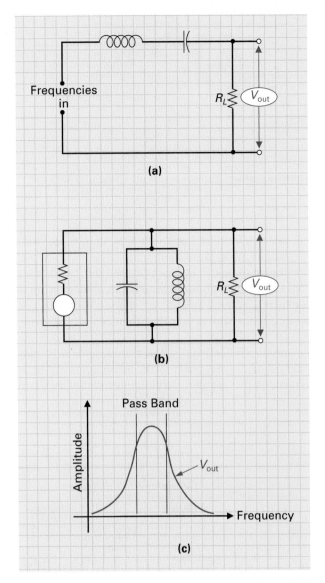

FIGURE 14-23 **Bandpass Filter. (a) Series Resonant Bandpass Filter. (b) Parallel Resonant Bandpass Filter. (c) Frequency Response.**

Above and below resonance, current within the parallel resonant circuit will be smaller. So the power transfer ability will be less, effectively keeping the frequencies outside the pass-band from appearing at the output.

14-4-4 *Band-Stop Filter*

Figure 14-25(a) illustrates a series resonant and Figure 14-25(b) a parallel resonant **band-stop filter.** Figure 14-25(c) shows the frequency response curve produced by a band-stop filter. The band-stop filter operates exactly the opposite to a bandpass filter in that it blocks or attenuates a band of frequencies centered on the resonant frequency of the *LC* circuit.

In the series resonant circuit in Figure 14-25(a), the *LC* impedance is very low at and around resonance, so these frequencies are rejected or shunted away from the output. Above

Band-stop Filter

A filter that attenuates alternating currents whose frequencies are between given upper and lower cutoff values while passing frequencies above and below this band.

FIGURE 14-24   Parallel Resonant Bandpass Circuit Using a Transformer.

and below resonance, the series circuit has a very high impedance, which results in almost no shunting of the signal away from the output.

In the parallel resonant circuit in Figure 14-25(b), the *LC* circuit is in series with the load and output. At resonance, the impedance of a parallel resonant circuit will be very high, and the band of frequencies centered around resonance will be blocked. Above and below resonance, the impedance of the tank is very low, so nearly all the input is developed across the output.

Filters are necessary in applications such as television or radio, where we need to tune in (select or pass) one frequency that contains the information we desire, yet block all the millions of other frequencies that are also carrying information, as shown in Figure 14-26.

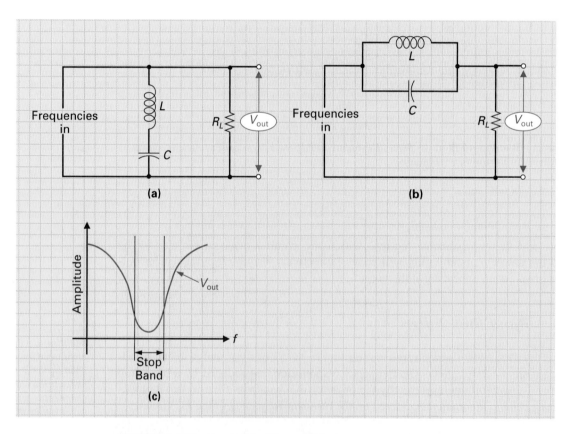

FIGURE 14-25   Band-Stop Filter. (a) Series Resonant Band-Stop Filter. (b) Parallel Resonant Band-Stop Filter. (c) Frequency Response.

FIGURE 14-26 Tuning in of Station by Use of a Bandpass Filter.

Use the following questions to test your understanding of Section 14-4.

1. Of the four types of filters, which:
   a. Would utilize the inductor as a shunt?
   b. Would utilize the capacitor as a shunt?
   c. Would use a series resonant circuit as a shunt?
   d. Would use a parallel resonant circuit as a shunt?
2. In what applications can filters be found?

REVIEW QUESTIONS

Multiple-Choice Questions

1. Capacitive reactance is _____ to frequency and capacitance, while inductive reactance is _____ to frequency and inductance.

   a. Proportional, inversely proportional
   b. Inversely proportional, proportional
   c. Proportional, proportional
   d. Inversely proportional, inversely proportional

2. Resonance is a circuit condition that occurs when:

   a. V_L equals V_C d. Both (a) and (c)
   b. X_L equals X_C e. Both (a) and (b)
   c. L equals C

3. As frequency is increased, X_L will _____, while X_C will _____.

   a. Decrease, increase
   b. Increase, decrease
   c. Remain the same, decrease
   d. Increase, remain the same

4. In an RLC series resonant circuit, with $R = 500\ \Omega$ and $X_L = 250\ \Omega$, what would be the value of X_C?

   a. $2\ \Omega$ c. $250\ \Omega$
   b. $125\ \Omega$ d. $500\ \Omega$

5. At resonance, the voltage drop across both a series-connected inductor and a capacitor will equal:

   a. 70.7 V   c. 10 V
   b. 50% of the source   d. Zero

6. In a series resonant circuit the current flow is _____, as the impedance is _____ and equal to _____.

   a. Large, small, R c. Large, small, X
   b. Small, large, X d. Small, large, R

7. A circuit's bandwidth includes a group or band of frequencies that cause _____ or more of the maximum current, or more than _____ of the maximum power to appear at the output.

   a. 110, 90   c. 70.7%, 50%
   b. 50%, 70.7%   d. Both (a) and (c)

8. The bandwidth of a circuit is proportional to the:

   a. Frequency of resonance   c. Tank current
   b. Q of the tank   d. Two of the above

9. Series or parallel resonant circuits can be used to create:

   a. Low-pass filters
   b. Low-pass and high-pass filters
   c. Bandpass and band-stop filters
   d. All of the above

10. Flywheel action occurs in:

    a. A tank circuit   **d.** Two of the above

    b. A parallel *LC* circuit   **e.** None of the above

    c. A series *LC* circuit

Practice Problems

11. Calculate the values of capacitive and inductive reactance for the following when connected across a 60 Hz source:

    a. 0.02 µF   **e.** 4 mH

    b. 18 µF   **f.** 8.18 H

    c. 360 pF   **g.** 150 mH

    d. 2700 nF   **h.** 2H

12. If a 1.2 kΩ resistor, a 4 mH inductor, and an 8 µF capacitor are connected in series across a 120 V/60 Hz source, calculate:

    a. X_C **g.** V_C

    b. X_L **h.** Apparent power

    c. Z **i.** True power

    d. I **j.** Resonant frequency

    e. V_R **k.** Circuit quality factor

    f. V_L **l.** Bandwidth

13. If a 270 Ω resistor, a 150 mH inductor, and a 20 µF capacitor are all connected in parallel with one another across a 120 V/60 Hz source, calculate:

    a. X_L **f.** I_T

    b. X_C **g.** Z

    c. I_R **h.** Resonant frequency

    d. I_L **i.** Q factor

    e. I_C **j.** Bandwidth

14. Calculate the impedance of a series circuit if $R = 750\ \Omega$, $X_L = 25\ \Omega$, and $X_C = 160\ \Omega$.

15. Calculate the impedance of a parallel circuit with the same values as those of Question 14 when a 1 V source voltage is applied.

Answers to Self-Test Evaluation Points

STEP 1-1
1. Elements are made up of similar atoms; compounds are made up of similar molecules.
2. Protons, neutrons, electrons
3. Copper
4. Like charges repel; unlike charges attract.

STEP 1-2
1. Amp
2. Current = Q/t (number of coulombs divided by time in seconds).
3. The direction of flow is different: negative to positive is termed electron current flow; positive to negative is known as conventional current flow.
4. The ammeter

STEP 1-3
1. Volts
2. 3000 kV
3. The voltmeter
4. They are directly proportional.

STEP 1-4
1. False
2. Copper
3. 28.6 millisiemens

STEP 1-5
1. True
2. Mica
3. The voltage needed to cause current to flow through a material
4. Small

STEP 1-6
1. No
2. Yes
3. A short circuit provides an unintentional path for current to flow. A closed circuit has a complete path for current.
4. An open switch opens a path in a closed circuit. A closed switch closes a path in a closed circuit.

STEP 2-1
1. It is the opposition to current flow.
2. The difference is in how much or how little current will flow.
3. They are inversely proportional.
4. Small

STEP 2-2
1. A circuit is said to have a resistance of 1 ohm when 1 volt produces a current of 1 ampere.
2. $I = V/R = 24\,\text{V}/6\,\Omega = 4\,\text{A}$.

3. A memory aid to help remember Ohm's law:

4. Current is proportional to voltage and inversely proportional to resistance.
5. $V = I \times R = 25\,\text{mA} \times 1\,\text{k}\Omega = (25 \times 10^{-3}) \times (1 \times 10^{3}) = 25\,\text{V}$.
6. $R = V/I = 12\,\text{V}/100\,\mu\text{A} = 12\,\text{V}/100 \times 10^{-6} = 120\,\text{k}\Omega$.

STEP 2-3
1. Type of conducting material used, cross-sectional area, length, temperature
2. The resistance of a conductor is proportional to its length and resistivity, and inversely proportional to its cross-sectional area.
3. True
4. True

STEP 2-4
1. Light, heat, magnetic, chemical, electrical, mechanical
2. When energy is transformed, work is done. Power is the rate at which work is done or energy is transformed.
3. $W = Q \times V, P = V \times I$
4. When 1 kW of power is used in 1 hour

STEP 3-1
1. Carbon composition, carbon film, metal film, wirewound, metal oxide, thick film
2. SIPs have one row of connecting pins, while DIPs have two rows of connecting pins.
3. Rheostat has two terminals; potentiometer has three terminals.
4. Linear means that the resistance changes in direct proportion to the amount of change of the input, while a tapered potentiometer varies nonuniformly.

STEP 3-2
1. Ohmmeter
2. See Figure 3-13.

STEP 3-3
1. General purpose are ±5% or greater; precision are ±2% or less.
2. $222 \times 0.1, \pm0.25\% = 22.2\,\Omega, \pm0.25\%$
3. Yellow, violet, red, silver
4. True (General purpose, ±20% tolerance)

STEP 3-4 and 3-5
1. $P = I^2 \times R$
2. Small, large

STEP 3-6
1. Opens
2. Ohmmeter

STEP 4-1
1. Load resistance is the resistance of a device, circuit, or system. Load current is the amount of current drawn from the source by the device, circuit, or system. A dc voltage source is one that will supply a constant polarity output voltage, while a dc current is a unidirectional (one direction) current.
2. False
3. They are inversely proportional.
4. A load that will draw a large amount of current.

STEP 4-2
1. A device that will generate a voltage between the opposite faces of a crystal as a result of pressure being applied.
2. A device that will generate a voltage when heat is applied.
3. A photovoltaic cell.
4. A generator
5. Negative plate, positive plate, electrolyte.

STEP 4-3 and 4-4
1. Indicate the maximum current that can flow without blowing the fuse and the maximum voltage value that will not cause an arc.
2. Fast-blow fuses open almost instantly after an excessive current occurs, whereas a slow-blow fuse will have a certain time delay until it opens.
3. Thermal, magnetic, thermomagnetic
4. SPST, SPDT, DPST, DPDT, NOPB, NCPB, rotary, and DIP

STEP 4-5
1. It is a solderless breadboard used to accommodate experimental circuits.
2. Power.

STEP 5-1 and 5-2
1. A circuit in which current has only one path
2. 8 A

STEP 5-3
1. $R_T = R_1 + R_2 + R_3 + \cdots$
2. $R_T = R_1 + R_2 + R_3 = 2\text{ k}\Omega + 3\text{ k}\Omega + 4700 = 9.7\text{ k}\Omega$

STEP 5-4
1. True
2. True
3. $R_T = R_1 + R_2 = 6 + 12 = 18\ \Omega;\ I_T = V_S/R_T = 18/18 = 1\text{ A}$
   $V_{R_1} = 1\text{ A} \times 6\ \Omega = 6\text{ V}$
   $V_{R_2} = 1\text{ A} \times 12\ \Omega = 12\text{ V}$
4. $V_X = (R_X/R_T) \times V_S$
5. Potentiometer
6. No

STEP 5-5
1. $P = I \times V$ or $P = V^2/R$ or $P = I^2 \times R$
2. $P = V^2/R = 12^2/12 = 144/12 = 12\text{ W}$

3. Wirewound, at least 12 W, ideally a 15 W
4. $P_T = P_1 + P_2 = 25\text{ W} + 3800\text{ mW} = 28.8\text{ W}$

STEP 5-6
1. Component will open, component's value will change, and component will short.
2. No current will flow; source voltage dropped across it.
3. False
4. No voltage drop across component, yet current still flows in circuit; resistance equals zero for component.

STEP 6-1 and 6-2
1. When two or more components are connected across the same voltage source so that current can branch out over two or more paths
2. False
3. $V_{R_1} = V_S = 12\text{ V}$
4. No

STEP 6-3
1. The sum of all currents entering a junction is equal to sum of all currents leaving that junction.
2. $I_2 = I_T - I_1 = 4\text{ A} - 2.7\text{ A} = 1.3\text{ A}$
3. $I_X = (R_T/R_X) \times I_T$
4. $I_T = V_T/R_T = 12/1\text{ k}\Omega = 12\text{ mA};\ I_1 = 1\text{ k}\Omega/2\text{ k}\Omega \times 12\text{ mA}$
   $= 6\text{ mA}$

STEP 6-4
1. $R_T = \dfrac{R_1 \times R_2}{R_1 + R_2}$
2. $R_T = \dfrac{1}{(1/R_1) + (1/R_2) + (1/R_3)} + \cdots$
3. $R_T = \dfrac{\text{common value of resistors }(R)}{\text{number of parallel resistors }(n)}$
4. $R_T = \dfrac{1}{(1/2.7\text{ k}\Omega) + (1/24\text{ k}\Omega) + (1/1\text{ M}\Omega)} = 2.421\text{ k}\Omega$

STEP 6-5
1. True
2. $P_1 = I_1 \times V = 2\text{ mA} \times 24\text{ V} = 48\text{ mW}$
3. $P_T = P_1 + P_2 = 22\text{ mW} + 6400\ \mu\text{W} = 28.4\text{ mW}$
4. Yes

STEP 6-6
1. No current will flow in the open branch; total current will decrease.
2. Maximum current is through shorted branch; total current will increase.
3. Will cause a corresponding opposite change in branch current and total current
4. False

STEP 7-1 and 7-2
1. By tracing current to see if it has one path (series connection) or more than one path (parallel connection)
2. $R_{1,2} = R_1 + R_2 = 12\text{ k}\Omega + 12\text{ k}\Omega = 24\text{ k}\Omega$
   $= R_{1,2,3} = \dfrac{R_{1,2} \times R_3}{R_{1,2} + R_3} = \dfrac{24\text{ k}\Omega \times 6\text{ k}\Omega}{24\text{ k}\Omega + 6\text{ k}\Omega}$
   $= \dfrac{144\text{ k}\Omega}{30\text{ k}\Omega} = 4.8\text{ k}\Omega$
3. STEP A: Find equivalent resistances of series-connected resistors. STEP B: Find equivalent resistances of parallel-connected combinations. STEP C: Find equivalent resistances of remaining series-connected resistances.

4. $R_{1,2} = R_1 + R_2 = 470 + 330 = 800\ \Omega$

$R_{1,2,3} = \dfrac{R_{1,2} \times R_3}{R_{1,2} + R_3} = \dfrac{800 \times 270}{800 + 270} = \dfrac{216\ \text{k}\Omega}{1.07\ \text{k}\Omega} = 201.9\ \Omega$

FIGURE 7-1, 4

STEP 7-3

1. Step 1: Find total resistance (Steps A, B and C). Step 2: Find total current. Step 3: Find voltage drop with $I_T \times R_X$.
2. The voltage drops previously calculated would not change since the ratio of the series resistor to the series equivalent resistors remains the same, and therefore the voltage division will remain the same.

STEP 7-4, 7-5, and 7-6

1. Find total resistance; find total current; find the voltage across each series and parallel combination resistors; find the current through each branch of parallel resistors; find the total and individual power dissipated.
2. This is a do-it-yourself question; each answer will vary.

STEP 7-7

1. When a load resistance changes the circuit and lowers output voltage

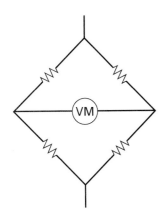

FIGURE 7-7, 2

2. Used to check for an unknown resistor's resistance

STEP 7-8

1. If component is in series with circuit, no current will flow and source voltage will be across bad component. If component is parallel, no current will flow in that branch.
2. Total resistance will decrease and bad component will have 0 V dropped across it.
3. Will cause the circuit's behavior to vary

STEP 8-1

1. **a.** Alternating current
   b. Direct current
2. AC

3. DC
4. DC flows in one direction, whereas ac first flows in one direction and then in the opposite direction.

STEP 8-2-1

1. AC generators can be larger, less complex, and cheaper to run; transformers can be used with ac to step up/down, so low-current power lines can be used; easy to change ac to dc, but hard the other way around.
2. False
3. $P = I^2 \times R$
4. A device that can step up or down ac voltages
5. 120 V ac
6. AC, dc

STEP 8-2-2 and 8-2-3

1. Sound, electromagnetic, electrical
2. 1133 feet per second; 186,000 miles per second
3. Sound, electromagnetic, electrical
4. Electronic
5. Electrical

STEP 8-3

1.

2. (Sound wave); $\lambda(\text{mm}) = \dfrac{344.4\ \text{m/s}}{f(\text{Hz})}$;

   (Electromagnetic wave): $\lambda(\text{m}) = \dfrac{3 \times 10^8\ \text{m/s}}{f(\text{Hz})}$;

   different because sound waves travel at a different speed than do electromagnetic waves.

STEP 8-4-1

1. Yes
2. A device or circuit that converts ac to pulsating dc.

STEP 8-4-2

1. Cathode ray tube
2. Period = 80 μs
   Frequency = 12.5 kHz
3. Peak = 4 V
   Peak-to-Peak = 8 V
4. It allows us to make comparisons between the phase, amplitude, shape, and timing of two signals.

STEP 8-4-3

1. Sine, square, triangular, sawtooth
2. Analyze the frequency of any periodic wave.

STEP 9-1

1. Two plates, dielectric
2. Dielectric

STEP 9-2

1. True
2. None

STEP 9-3

1. The farad
2. Capacitance, C (farads) = charge, Q (coulombs)/voltage, V (volts)
3. $30,000 \times 10^{-6} = 0.03 \times 10^0 = 0.03$ F
4. $C = Q/V = 17.5/9 = 1.94$ F

STEP 9-4
1. Plate area, distance between plates, type of dielectric used
2. $C = \dfrac{(8.85 \times 10^{-12}) \times K \times A}{d}$
3. Double
4. Double

STEP 9-5
1. Glass
2. False

STEP 9-6
1. $C_T = \dfrac{1}{(1/C_1) + (1/C_2) + (1/C_3)}$

   $ = \dfrac{1}{(1/2 \ \mu F) + (1/3 \ \mu F) + (1/5 \ \mu F)}$

   $ = 0.968 \ \mu F$ or $1 \ \mu F$
2. $C_T = C_1 + C_2 + C_3 = 7 \ pF + 2 \ pF + 14 \ pF = 23 \ pF$
3. $V_{CX} = (C_T/C_X) \times V_T$
4. True

STEP 9-7
1. Mica, ceramic, paper, plastic, electrolytic
2. **a.** Electrolytic; **b.** ceramic
3. **a.** Air; **b.** mica, ceramic, plastic

STEP 9-8
1. 470 pF, 2% tolerance
2. 0.47 μF or 470 nF, 5% tolerance

STEP 9-9
1. 63.2%
2. 36.8%

STEP 9-10
1. Capacitive opposition is low for ac (short) and high for dc (open).
2. True

STEP 9-11
1. False
2. True

STEP 10-1
1. Opposition to current flow without the dissipation of energy
2. $X_C = 1/2\pi f C$
3. When frequency or capacitance goes up, there is more charge and discharge current, so X_C is lower.
4. $X_C = 1/2\pi f C = 1/2\pi \times 4 \ kHz \times 4 \ \mu F = 9.95 \ \Omega$

STEP 10-2
1. Current leads voltage by some phase angle less than 90%.
2. Z = total opposition to current flow; $Z = \sqrt{R^2 + X_C^2}$

STEP 10-3
1. Resistor current is in phase with voltage; capacitor current is 90° out of phase (leading) with voltage.
2. No

STEP 10-4
1. 0.5 μF
2. Capacitance meter or analyzer

STEP 10-5
1. **a.** Filter
   b. Voltage divider
   c. Differentiator
2. Capacitor, resistor
3. Long
4. Differentiator

STEP 11-1-1
1. False
2. Electron

STEP 11-1-2
1. Coil
2. True
3. DC
4. False
5. Used to determine the magnetic polarity; placing your left-hand fingers in the direction of the current, your thumb points in the direction of the north magnetic pole.
6. AC

STEP 11-1-3
1. Number of lines of force (or maxwells) in webers
2. Number of magnetic lines of flux per square meter
3. True
4. The state beyond which an electromagnet is incapable of further magnetic strength

STEP 11-1-4
1. Magnetic-type circuit breaker, relays
2. An NO relay is open when off; NC is closed when off.
3. A reed relay has an electromagnet around it, while a reed switch needs an external magnet to operate.
4. Home security circuits

STEP 11-2
1. The voltage induced or produced in a coil as the magnetic lines of force link with the turns of a coil
2. Faraday's: When the magnetic flux linking a coil is changing, an emf is induced.
   Lenz's: The current induced in a coil due to the change in the magnetic flux is such as to oppose the cause producing it.
3. Sine wave

STEP 12-1
1. When the current-carrying coil of a conductor induces a voltage within itself
2. The induced voltage, which opposes the applied emf;
   $V_{ind} = L \times (\Delta i/\Delta t)$
3. $V_{ind} = L \times (\Delta i/\Delta t) = 2 \ mH \times 4kA/s = 8 \ V$

STEP 12-2
1. Different applications: An electromagnet is used to generate a magnetic field; an inductor is used to oppose any changes of circuit current.
2. True

STEP 12-3
1. Number of turns, area of coil, length of coil, core material used
2. $L = \dfrac{N^2 \times A \times \mu}{l}$

STEP 12-4
1. False
2. **a.** $L_T = L_1 + L_2 + L_3 + \cdots$
   b. $L_T = 1/(1/L_1) + (1/L_2) + (1/L_3) + \cdots$

3. a. $L_T = L_1 + L_2 = 4 \text{ mH} + 2 \text{ mH} = 6 \text{ mH}$

b. $L_T = \dfrac{L_1 \times L_2}{L_1 + L_2}$ (using product over sum)

$= \dfrac{4 \text{ mH} \times 2 \text{ mH}}{4 \text{ mH} + 2 \text{ mH}} = \dfrac{8 \text{ mH}}{6 \text{ mH}} = 1.33 \text{ mH}$

STEP 12-5

1. Air core, iron core, ferrite core
2. Air core
3. Chemical compound, basically powdered iron oxide and ceramic
4. Toroidal-type inductors have a greater inductance.
5. Ferrite-core variable inductor
6. Core permeability

STEP 12-6

1. Current in an inductive circuit builds up in the same way that voltage does in a capacitive circuit, but the capacitive time constant is proportional to resistance, where the inductive time constant is inversely proportional to resistance.
2. True
3. False
4. It will continuously oppose the alternating current.

STEP 12-7

1. The opposition to current flow offered by an inductor without the dissipation of energy: $X_L = 2 \times \pi \times f \times L$
2. The higher the frequency (and therefore the faster the change in current) or the larger the inductance of the inductor, the larger the magnetic field created, the larger the counter emf will be to oppose applied emf; so the inductive reactance (opposition) will be large also.
3. Inductive reactance is an opposition, so it can be used in Ohm's law in place of resistance.
4. False

STEP 12-8

1. False
2. $V_S = \sqrt{V_R^2 + V_L^2} = \sqrt{4^2 + 2^2} = \sqrt{16 + 4} = \sqrt{20} = 4.47 \text{ V}$
3. The total opposition to current flow offered by a circuit with both resistance and reactance: $Z = \sqrt{R^2 + X_L^2}$
4. $+45°$
5. Power consumption above the zero line, caused by positive current and voltage or negative current and voltage
6. Quality factor of an inductor that is the ratio of the energy stored in the coil by its inductance to the energy dissipated in the coil by the resistance: $Q = X_L/R$
7. True power is energy dissipated and lost by resistance; reactive power is energy consumed and then returned by a reactive device.
8. $\text{PF} = $ true power (P_R)/apparent power (P_A) or $\text{PF} = R/Z$ or $\text{PF} = \cos\theta$

STEP 12-9

1. False
2. $I_T = \sqrt{I_R^2 + I_L^2}$

STEP 12-10

1. **a.** With an ohmmeter; resistance will be infinite instead of low.
   b. With an inductor analyzer to check its inductance value
2. An open

STEP 12-11

1. *RL* integrator, *RL* differentiator, *RL* filter

---

2. Can be made to function as either a high- or a low-pass filter
3. The integrator will approximate a dc level; the differentiator will output spikes of current.

STEP 13-1

1. Mutual inductance is the process by which an inductor induces a voltage in another inductor, whereas self-inductance is the process by which a coil induces a voltage within itself.
2. False

STEP 13-2

1. True
2. True
3. Primary and secondary

STEP 13-3

1. True
2. False

STEP 13-4

1. Turns ratio $= N_s/N_p = 1608/402 = 4$; step up
2. $V_s = N_s/N_p \times V_p$
3. False
4. $I_s = N_p/N_s \times I_p$
5. Turns ratio $= \sqrt{Z_L/Z_S} = \sqrt{75\ \Omega/25\ \Omega} = \sqrt{3} = 1.732$
6. $V_s = (N_s/N_p) \times V_p = (200/112) \times 115 = 205.4 \text{ V}$

STEP 13-5

1. If both primary and secondary are wound in the same direction, output is in phase with input; but if primary and secondary are wound in different directions, the output will be 180° out of phase with the input.
2. When there is a positive on the primary dot, there will be a positive on the secondary dot. If there is a negative on the primary dot, there will be a negative on the secondary dot.

STEP 13-6

1. Air core, iron or ferrite core
2. Center-tapped secondary, multiple-tapped secondary, multiple winding, single winding
3. False
4. To be able to switch between two primary voltages and obtain the same secondary voltage
5. Automobile ignition system
6. Smaller, cheaper, lighter than normal separated primary/secondary transformer types

STEP 13-7

1. 10 kVA is the apparent power rating, 200 V the maximum primary voltage, 100 V the maximum secondary voltage, at 60 cycles per second (Hz)
2. $R_L = V_s/I_s = 1 \text{ kV}/8 \text{ A} = 125\ \Omega$; $125 > 100$, so the transformer will overheat and possibly burn out.

STEP 13-8

1. An ohmmeter
2. Probably not since 50 Ω is a normal coil resistance

STEP 14-1

1. Calculate the inductive and capacitive reactance (X_L and X_C), the circuit impedance (Z), the circuit current (I), the component voltage drops (V_R, V_L, and V_C), and the power distribution and power factor (PF).
2. **a.** $Z = \sqrt{R^2 + (X_L \sim X_C)^2}$
   b. $I = V_s/Z$

c. Apparent power = $V_s \times I$ (volt-amperes)

d. $V_S = \sqrt{V_R^2 + (V_L \sim V_C)^2}$

e. True power = $I^2 \times R$(watts)

f. $V_R = I \times R$

g. $V_L = I \times X_L$

h. $V_C = I \times X_C$

i. $\theta = \arctan \dfrac{V_L \sim V_C}{V_R}$

j. PF = $\cos \theta$

STEP 14-2

1. a. $I_R = V/R$ **c.** $I_C = V/X_C$

   b. $I_T = \sqrt{I_R^2 + I_X^2}$ **d.** $I_L = V/X_L$

2. a. $P_R = I^2 \times R$ **c.** $P_A = V_S \times I_T$

   b. $P_X = I^2 \times X_L$ **d.** PF = $\cos \theta$

STEP 14-3-1

1. A circuit condition that occurs when the inductive reactance (X_L) and the capacitive reactance (X_C) have been balanced

2. A series *RLC* circuit that at resonance X_L equals X_C, so V_L and V_C will cancel, and $Z = R$

3. Voltage across *L* and *C* will measure 0; impedance only equals *R*; voltage drops across inductor or capacitor can be higher than source voltage

4. *Q* factor indicates the quality of the series resonant circuit, or is the ratio of the reactance to the resistance

5. Group or band of frequencies that causes the larger current flow

6. BW = f_0/Q = 12 kHz/1000 = 12 Hz

STEP 14-3-2

1. In a series resonant *RLC* circuit, source current is maximum; in a parallel resonant circuit, source current is minimum at resonance.

2. Oscillating effect with continual energy transfer between capacitor and inductor

3. $Q = X_L/R = 50\ \Omega/25\ \Omega = 2$

4. Yes

5. Ability of a tuned circuit to respond to a desired frequency and ignore all others

STEP 14-4

1. a. High pass   **c.** Band stop

   b. Low pass   **d.** Band pass

2. Television, radio, and other communications equipment

Answers to Odd-Numbered Problems

Chapter 1
1. b
3. d
5. a
7. a
9. b
11. d
13. a
15. d
17. $n = Q \times (6.24 \times 10^{18}) = 6.5 \times (6.24 \times 10^{18})$
   $= 4.056 \times 10^{19}$ electrons
19. **a.** 14 mA **c.** 776 nA
    b. 1.374 kA **d.** 910 μA
21. 0.945/1 = 0.945
23. 10 V/750 kV = 0.0000133 cm or 0.000133 mm
25. 35 × (2000 kV/cm ÷ 10) = 7000 kV

Chapter 2
1. a
3. a
5. d
7. a
9. c
11. **a.** $I = V/R = 120$ V/6 Ω = 20 amps

FIGURE Ch. 2, 11 (b)

13. $R = V/I = 120$ V/10 A = 12 Ω
15. $P = I \times V, I = P/V$
    a. 300/120 = 2.5 A   **c.** 60/120 = 0.5 A
    b. 100/120 = 833 mA   **d.** 25/120 = 208.3 mA
17. **a.** Volts   **c.** Kilowatts
    b. Milliamps   **d.** Megaohm
19. 13
21. $I = V/R = 12$ V/12 mA = 1 kΩ
23. $R = V/I = 120$ V/500 mA = 240 Ω
    $P = I \times V = 500$ mA × 120 V = 60 W
25. $P = V \times I$
    a. 120 V × 20 mA = 2.4 W   **c.** 9 V × 100 μA = 900 μW
    b. 12 V × 2 A = 24 W   **d.** 1.5 V × 4 mA = 6 mW
27. **a.** 1 kW   **c.** 1.25 MW
    b. 345 mW   **d.** 1250 μW
29. **a.** 7500 = 7.5 kW
    Energy Consumed = 7.5 kW × 1 hr
    = 7.5 kWh
    b. 25 W = 0.025 kW
    Energy Consumed = 0.025 kW × 6 hrs
    = 0.15 kWh
    c. 127,000 W = 127 kW
    Energy Consumed = 127 kW × 0.5 hr
    = 63.5 kWh

Chapter 3
1. b
3. c
5. b; $P = I^2R$
   $= 20$ mA² × 2 kΩ
   $= 0.8$ W
7. a
9. b
11. 5.6 kΩ × 0.1 = 560 Ω,
    5.6 kΩ ± 560 kΩ = 5.04 kΩ to 6.16 kΩ
13. $P = V^2/R = 12$ V²/12Ω = 144/12 = 12 W
15.

$$I = V/R = 50 \text{ mA}$$

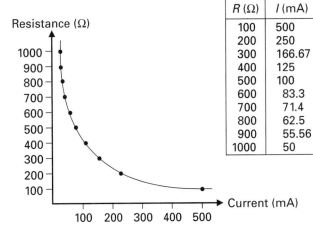

Resistance (Ω)

R (Ω)	I (mA)
100	500
200	250
300	166.67
400	125
500	100
600	83.3
700	71.4
800	62.5
900	55.56
1000	50

Current (mA)

17. See Section 3–6, #4   19. It will not burn up

Chapter 4
1. b   9. a
3. d   11. c
5. c   13. b
7. d   15. b
17. **a.** 150 Ah × 3600 = 54 × 10⁴ coulombs
    b. 150 Ah ÷ 5 A = 30 hours

Chapter 5
1. d   9. a
3. c   11. c
5. a   13. d
7. d   15. d
17. $I = \dfrac{V_S}{R_T}, R_T = R_1 + R_2 = 40 + 35 = 75$ Ω
    $I = 24/75 = 320$ mA; 150 Ω (double 75 Ω) needed to half current
19. 40 Ω, 20 Ω, 60 Ω. (Any values can be used, as long as the ratio remains the same.)

21. $I_{R1} = I_T = 6.5$ mA

23. $P_T = P_1 + P_2 + P_3 = 120 + 60 + 200 = 380$ W;
$I_T = P_T/V_S = 380/120 = 3.17$ A
$V_1 = P_1/I_T = 120$ W/3.17 A $= 38$ V;
$V_2 = P_2/I_T = 60$ W/3.17 A $= 18.9$ V
$V_3 = P_3/I_T = 200$ W/3.17 A $= 63.1$ V

25. a. $R_T = R_1 + R_2 + R_3$
$= 22$ kΩ + 3.7 kΩ + 18 kΩ = 43.7 kΩ
$I = V/R = 12$ V/43.7 kΩ = 274.6 μA

b. $R_T = V/I = 12$ V/10 mA $= 1.2$ kΩ
$P_T = V \times I = 12$ V $\times$ 10 mA $= 120$ mW

c. $R_T = R_1 + R_2 + R_3 + R_4 = 5 + 10 + 6 + 4 = 25$ Ω;
$V_S = I \times R_T = 100$ mA $\times$ 25 Ω = 2.5 V
$V_{R_1} = I \times R_1 = 100$ mA $\times$ 5 = 500 mV,
$V_{R_2} = I \times R_2 = 100$ mA $\times$ 10 = 1 V,
$V_{R_3} = I \times R_3 = 100$ mA $\times$ 6 = 600 mV
$V_{R_4} = I \times R_4 = 100$ mA $\times$ 4 = 400 mV,
$P_1 = I \times V_1 = 100$ mA $\times$ 500 mV = 50 mW
$P_2 = I \times V_2 = 100$ mA $\times$ 1 V = 100 mW,
$P_3 = I \times V_3 = 100$ mA $\times$ 600 mV = 60 mW
$P_4 = I \times V_4 = 100$ mA $\times$ 400 mV = 40 mW

d. $P_T = P_1 + P_2 + P_3 + P_4 = 12$ mW + 7 mW
$+ 16$ mW + 3 mW = 38 mW
$I = P_T/V_S = 38$ mW/12.5 V = 3.04 mA
$R_1 = P_1/I^2 = 12$ mW/(3.04 mA)^2 = 1.3 kΩ
$R_2 = P_2/I^2 = 7$ mW/(3.04 mA)^2 = 757.4 Ω
$R_3 = P_3/I^2 = 16$ mW/(3.04 mA)^2 = 1.73 kΩ
$R_4 = P_4/I^2 = 3$ mW/(3.04 mA)^2 = 324.6 Ω

27. Zero voltage drop across the shorted component, while there is also an increase in voltage across the others.

29. a. No current at all (zero)
b. Go to infinity (∞)
c. Measure source voltage
d. No voltage across any other component

Chapter 6

1. b   **7.** b
3. d   **9.** a
5. c
11. $R_T = R/$no. of R's $= 30$ k$\Omega/4 = 7.5$ kΩ
13. $R_T = R/$no. of R's $= 25/3 = 8.33$ Ω,
$I_T = V_S R_T = 10/8.33 = 1.2$ A
$I_1 = I_2 = I_3 = R_T/R_x \times I_T = 8.33/25 \times 1.2$
$= I_T/$no. of R's $= 1.2$ A/3 = 400 mA
15. $I_T = V_S/R_t = 14/700 = 20$ mA;
$I_X = I_T/$no. of $R = 20$ mA/3 = 6.67 mA
17. a. $R_T = \dfrac{R_1 \times R_2}{R_1 + R_2} = \dfrac{33 \times 22}{33 + 22} = \dfrac{726 \text{ k}\Omega}{55 \text{ k}\Omega} = 13.2$ kΩ
b. $I_T = V_S/R_T = 20/13.2$ kΩ = 1.5 mA
c. $I_1 = R_T/R_1 \times I_T = (13.2$ k$\Omega/33$ k$\Omega) \times 1.5$ mA = 600 μA
$I_2 = R_T/R_2 \times I_T = 13.2$ k$\Omega/22$ k$\Omega \times 1.5$ mA = 900 μA
d. $P_T = I_T \times V_S = 1.5$ mA $\times$ 20 = 30 mW
e. $P_1 = I_1 \times V_1$, $(V_1 = V_S = 20$ V),
600 μA $\times$ 20 = 12 mW
$P_2 = I_2 \times V_2$, $(V_2 = V_S = 20$ V), 900 μA $\times$ 20 = 18 mW
19. a. $R_T = \dfrac{R_1 \times R_2}{R_1 + R_2} = \dfrac{22 \text{ k}\Omega \times 33 \text{ k}\Omega}{22 \text{ k}\Omega + 33 \text{ k}\Omega}$
$= \dfrac{726 \text{ k}\Omega}{55 \text{ k}\Omega} = 13.2$ kΩ,
$I_T = \dfrac{V_S}{R_T} = \dfrac{10}{13.2 \text{ k}\Omega} = 757.6$ μA
$I_1 = \dfrac{R_T}{R_1} \times I_T = \dfrac{13.2 \text{ k}\Omega}{22 \text{ k}\Omega} \times 757.6$ μA = 454.56 μA
$I_2 = \dfrac{R_T}{R_1} \times I_T = \dfrac{13.2 \text{ k}\Omega}{33 \text{ k}\Omega} \times 757.6$ μA = 303.04 μA

b. $R_T = \dfrac{1}{(1/R_1) + (1/R_2) + (1/R_3)} = $
$\dfrac{1}{(1/220 \ \Omega) + (1/330 \ \Omega) + (1/470 \ \Omega)} = 103$ Ω
$I_T = V_S/R_T = 10/103 = 97$ mA, $I_1 = R_T/R_1 \times I_T$
$= 103/220 \times 97$ mA = 45.4 mA
$I_2 = (R_T/R_2) \times I_T = (103/330) \times 97$ mA = 30.3 mA
$I_3 = (R_T/R_2) \times I_T = (103/470) \times 97$ mA = 21.3 mA

21. a. $G_T = \dfrac{1}{R_1} + \dfrac{1}{R_2} + \dfrac{1}{R_3} = \dfrac{1}{5} + \dfrac{1}{5} + \dfrac{1}{5} = 0.6$ S,
$R_T = \dfrac{1}{G} = \dfrac{1}{0.6} = 1.67$ Ω

b. $G_T = \dfrac{1}{R_1} + \dfrac{1}{R_2} = \dfrac{1}{200} + \dfrac{1}{200} = 10$ mS,
$R_T = \dfrac{1}{G} = \dfrac{1}{10 \text{ mS}} = 100$ Ω

c. $G_T = \dfrac{1}{R_1} + \dfrac{1}{R_2} + \dfrac{1}{R_3} = \dfrac{1}{1 \text{ M}\Omega} + \dfrac{1}{500 \text{ M}\Omega} + \dfrac{1}{3.3 \text{ M}\Omega}$
$= 1.305$ μS, $R_T = \dfrac{1}{G} = \dfrac{1}{1.305 \ \mu\text{S}} = 766.3$ kΩ

d. $G_T = \dfrac{1}{R_1} + \dfrac{1}{R_2} + \dfrac{1}{R_3} = \dfrac{1}{5} + \dfrac{1}{3} + \dfrac{1}{2} = 1.033$ S,
$R_T = \dfrac{1}{G} = \dfrac{1}{1.033} = 967.7$ mΩ

23. a. $R_T = \dfrac{R_1 \times R_2}{R_1 + R_2} = \dfrac{15 \times 7}{15 + 7} = \dfrac{105}{22} = 4.77$ Ω

b. $R_T = \dfrac{1}{(1/R_1) + (1/R_2) + (1/R_3)}$
$= \dfrac{1}{(1/26 \ \Omega) + (1/15 \ \Omega) + (1/30 \ \Omega)} = 7.22$ Ω

c. $R_T = \dfrac{R_1 \times R_2}{R_1 + R_2} = \dfrac{5.6 \text{ k}\Omega \times 2.2 \text{ k}\Omega}{5.6 \text{ k}\Omega + 2.2 \text{ k}\Omega}$
$= \dfrac{12.32 \text{ M}\Omega}{7.8 \text{ k}\Omega} = 1.58$ kΩ

d. $R_T = \dfrac{1}{(1/R_1) + (1/R_2) + (1/R_3) + (1/R_4) + (1/R_5)} = $
$\dfrac{1}{(1/1 \text{ M}\Omega) + (1/3 \text{ M}\Omega) + (1/4.7 \text{ M}\Omega) + (1/10 \text{ M}\Omega) + (1/33 \text{ M}\Omega)}$
$= 596.5$ kΩ

25. a. $I_2 = I_T - I_1 - I_3 = 6$ mA $- 2$ mA $- 3.7$ mA $= 300$ μA
b. $I_T = I_1 + I_2 + I_3 = 6$ A $+ 4$ A $+ 3$ A $= 13$ A
c. $R_T = \dfrac{R_1 \times R_2}{R_1 + R_2} = \dfrac{5.6 \text{ M} \times 3.3 \text{ M}}{5.6 \text{ M} + 3.3 \text{ M}}$
$= \dfrac{18.48 \text{ M}\Omega}{8.9 \text{ M}\Omega} = 2.08$ MΩ
$V_S = I_T \times R_T = 100$ mA $\times$ 2.08 MΩ = 208 kV,
$I_1 = \dfrac{R_T}{R_1} \times I_T = \dfrac{2.08 \text{ M}\Omega}{5.6 \text{ M}\Omega} \times 100$ mA = 37 mA,
$I_2 = \dfrac{R_T}{R_2} \times I_T = \dfrac{2.08 \text{ m}\Omega}{3.3 \text{ M}\Omega} \times 100$ mA = 63 mA.
d. $I_1 = V_{R_1}/R_1 = 2/200$ k$\Omega = 10\mu$A,
$I_2 = I_T - I_1 = 100$ mA $- 10$ μA = 99.99 mA
$R_2 = \dfrac{V_{R_2}}{I_2} = \dfrac{2}{99.99 \text{ mA}} = 20.002$ Ω,
$P_T = I_T \times V_S = 100$ mA $\times$ 2 V = 200 mW

27. a
29. Total current would increase, and the branch current with the shorted resistor would increase to 20 V/1 Ω = 20A.

Chapter 7

1. c   7. d
3. b   9. c
5. d

11. **a.** $R_{1.2} = R_1 + R_2 = 2.5\ k\Omega + 10\ k\Omega = 12.5\ k\Omega$;
$R_{3.4} = R_3 + R_4 = 7.5\ k + 2.5\ k = 10\ k\Omega$

$$R_{3.4.5} = \frac{R_{3.4} \times R_5}{R_{3.4} + R_5} = \frac{10\ k\Omega \times 2.5\ M\Omega}{10\ k\Omega + 2.5\ M\Omega}$$

$$= \frac{256\ \Omega}{2.51\ M\Omega} = 9.96\ k\Omega$$

$R_{1.2.3.4.5} = R_{1.2} + R_{3.4.5} = 12.5\ k\Omega + 9.96\ k\Omega = 22.46\ k\Omega$

b. $I_T = V_S/R_T = 100/22.46\ k\Omega = 4.45\ mA$

c. $V_{R_1} = I_T \times R_1 = 4.45\ mA \times 2.5\ k\Omega = 11.125\ V$
$V_{R_2} = I_T \times R_2 = 4.45\ mA \times 10\ k\Omega = 44.5\ V$
$V_{R_{3.4.5}} = I_T \times R_{3.4.5} = 4.45\ mA \times 9.96\ k\Omega = 44.3\ V$

d. $I_{R_1} = I_T = 4.45\ mA$, $I_{R_2} = I_T = 4.45\ mA$

$$I_{R_3} = I_{R_4} = V_{3.4.5}/R_{3.4} = \frac{44.3}{10\ k\Omega} = 4.43\ mA$$

$I_{R_5} = V_{3.4.5}/R_5 = 44.3/2.5\ M\Omega = 17.73\ \mu A$

e. $P_T = I_T \times V_S = 4.45\ mA \times 100\ V = 445\ mW$
$P_{R_1} = I_{R_1} \times V_{R_1} = 4.45\ mA \times 11.125\ V = 49.5\ mW$
$P_{R_2} = I_{R_2} \times V_{R_2} = 4.45\ mA \times 44.5\ V = 198.025\ mW$
$P_{R_3} = I_3^2 \times R_3 = (4.43\ mA)^2 \times 7.5\ k\Omega = 147.2\ mW$
$R_{R_4} = I_4^2 \times R_4 = (4.43\ mA)^2 \times 2.5\ k\Omega = 49.1\ mW$
$P_{R_5} = V_{R_{3.4.5}} \times I_{R_5} = 44.3\ V \times 17.73\ \mu A = 785\ \mu W$

FIGURE Ch. 7

13. $R_{3.4} = \dfrac{R_3 \times R_4}{R_3 + R_4} = \dfrac{200 \times 300}{200 + 300} = \dfrac{60\ k\Omega}{500} = 120\ \Omega$

$R_{2.3.4} = R_2 + R_{3.4} = 100 + 120 = 220\ \Omega$

$R_{1.2.3.4} = \dfrac{R_1 \times R_{2.3.4}}{R_1 + R_{2.3.4}} = \dfrac{100 \times 220}{100 + 220} = \dfrac{22\ k\Omega}{320} = 68.75 = R_T$

$I_T = V_S/R_T = 10/68.75 = 145.45\ mA$
$V_{R_2} = I_{R_2} \times R_2,$

$I_{R_2} = I_{R_{2.3.4}} = \dfrac{R_T}{R_{2.3.4}} \times I_T = \dfrac{68.76}{220} \times 145.45\ mA$

$= 45.45\ mA = I_2$

$V_{R_2} = 45.45\ mA \times 100 = 4.545\ V,$
$V_{R_{3.4}} = I_{R_{3.4}} \times R_{3.4} = 45.45\ mA \times 120 = 5.45\ V$
$V_{R_1} = V_S = 10\ V$, $I_{R_1} = V_S/R_1 = 10/100 = 100\ mA$
$I_{R_2} = 45.45\ mA,$
$I_{R_3} = (R_{R_{3.4}}/R_3) \times I_{R_{3.4}} = (120/200) \times 45.45\ mA$
$= 27.27\ mA$
$I_{R_4} = I_{R_{3.4}} - I_{R_3} = 45.45\ mA - 27.27\ mA = 18.18\ mA$
$P_T = V_S \times I_T = 10 \times 145.45\ mA = 1.4545\ W$
$P_{R_1} = I_{R_1} \times V_{R_1} = 100\ mA \times 10\ V = 1\ W$
$P_{R_2} = I_{R_2} \times V_{R_2} = 45.45\ mA \times 4.545\ V = 207\ mW$

FIGURE Ch. 7

$P_{R_3} = I_{R_3} \times V_{R_3} = 27.27\ mA \times 5.45 = 149\ mW$
$P_{R_4} = I_{R_4} \times V_{r_4} = 18.18\ mA \times 5.45 = 99\ mW$
$V_A = V_S = 10\ V$, $V_B = V_A - V_{R_2} = 10 - 4.545 = 5.455$
$V_C = V_D = 0\ V$

15. This answer will vary with each person

Chapter 8

1. a   9. b
3. b   11. c
5. a   13. b
7. d   15. d

17. Frequency = 1/time:
  a. $1/16\ ms = 62.5\ Hz$;   **d.** $1/0.05s = 20\ Hz$;
  b. $1/1\ s = 1\ Hz$;   **e.** $1/200\ \mu s = 5\ kHz$;
  c. $1/15\ \mu s = 66.67\ kHz$;   **f.** $1/350\ ms = 2.86\ Hz$

19. **a.** Peak = $1.414 \times rms = 1.414 \times 40\ mA = 56.56\ mA$
  b. Peak to peak = $2 \times peak = 2 \times 56.56\ mA = 113.12\ mA$
  c. Average = $0.637 \times peak = 0.637 \times 56.56\ mA = 36\ mA$

21. **a.** peak to peak = $5\ cm \times 0.5\ V/cm = 2.5\ V_{p-p}$;
    $t = 4\ cm \times 20\ \mu s/cm = 80\ \mu s,$
    frequency = $1/t = 1/80\ \mu s = 12.5\ kHz$;
    period = $t = 80\ \mu s$; peak = $p-p/2 =$
    $2.5/2 = 1.25\ V_{pk}$;
    rms = $0.707 \times peak = 0.707 \times 1.25\ V = 883.75\ mV$;
    average = $0.637 \times peak = 0.637 \times 1.25\ V$
    $= 796.25\ mV_{avg}$

  b. peak to peak = $5\ cm \times 10\ V/cm = 50\ V_{p-p}$;
    $t = 4\ cm \times 10\ ms/cm = 40\ ms$;
    frequency = $1/t = 1/40\ ms = 25\ Hz$;
    period = $t = 40\ ms$, peak = $pk-pk/2 =$
    $50/2 = 25\ V_{pk}$,
    rms = $0.707 \times peak = 0.707 \times 25 = 17.675\ V$;
    average = $0.637 \times peak = 0.637 \times 25 = 15.925\ V_{avg}$

  c. peak to peak = $5\ cm \times 50\ mV/cm = 250\ mV_{p-p}$,
    $t = 4\ cm \times 0.2\ \mu s/cm =$
    $800\ ns$; frequency = $1/t = 1/800\ ns = 1.25\ MHz$;
    period = $t = 800\ ns$; peak =
    $pk-pk/2 = 250\ mV_{pk}/2 = 125\ mV_{pk}$;
    rms = $0.707 \times peak = 0.707 \times 125\ mV_{pk}$
    $= 88.375\ mV$; average = $0.637 \times peak$
    $= 0.637 \times 125\ mV_{pk} = 79.625\ mV$

23. $3.5\ cm \times 10\ V/cm = 35\ V\ p-p$
25. There would be no need to select a range setting since an
    autoset scopemeter would automatically adjust settings for
    best possible viewing.

Chapter 9

1. d   9. d
3. d   11. b
5. a   13. c
7. d

15. $Q = C \times V = 10\ \mu F \times 10\ V = 100\ \mu C$

17. $V = Q/C = 125 \times 10^{-6}/0.006\ \mu F = 20.83\ kV$

19. a. $C_T = C_1 + C_2 + C_3 + C_4 = 1.7\ \mu F + 2.6\ \mu F + 0.03\ \mu F +$
$1200\ pF = 4.3312\ \mu F$

   b. $C_T = \dfrac{1}{(1/C_1) + (1/C_2) + (1/C_3)} =$

$$\dfrac{1}{(1/1.6\ \mu F) + (1/1.4\ \mu F) + (1/4\ \mu F)} = 0.629\ \mu F$$

21. a. $10 + 4$ zeros $= 100,000\ pF$ or $0.1\ \mu F$

   b. $12 + 5$ zeros $= 1,200,000\ pF$ or $1.2\ \mu F$

   c. $0.01\ \mu F$

   d. $220\ pF$

23. a. $\tau = R \times C = 6\ k\Omega \times 14\ \mu F = 84\ ms$

   b. $\tau = R \times C = 12\ M\Omega \times 1400\ pF = 16.8\ ms$

   c. $\tau = R \times C = 170\ \Omega \times 24\ \mu F = 4.08\ ms$

   d. $\tau = R \times C = 140\ k\Omega \times 0.007\ \mu F = 980\ \mu s$

25. $C = Q/V = \dfrac{25\ \mu C}{125\ V} = 0.2\ \mu F$

Chapter 10

1. b

3. b

5. d

7. b

9. d

11. a

15. b

17. $V_S = \sqrt{V_R^2 + V_C^2} = \sqrt{6^2 + 12^2} = \sqrt{36 + 144} = \sqrt{180}$
$= 13.4\ V$

19. a. $I^R = V/R = 12\ V/4\ M\Omega = 3\ \mu A$

   b. $I_C = V/X_C = 12\ V/1.3\ k\Omega = 9.23\ mA$

   c. $I_T = \sqrt{I_R^2 + I_C^2} = \sqrt{(3\ \mu A)^2 + (9.23\ mA)^2}$
$= 9.23\ mA$

   d. $Z = V/I_T = 12/9.23\ mA = 1.3\ k\Omega$

   e. $\theta = \arctan(R/X_C) = \arctan(4\ M\Omega/1.3\ k\Omega) = 89.98°$

21. $C = \dfrac{1}{2\pi f X_C} = \dfrac{1}{2\pi(20\ kHz)\ 10\ k\Omega} = 795.8\ pF$

23.

24 V, $R = 40\ \Omega$, $X_C = 33\ \Omega$

FIGURE Ch. 10, 23(a)

   b. $Z = \sqrt{R^2 + X_C^2} = \sqrt{40^2 + 33^2} = 51.9\ \Omega$

$I = V/Z = 24\ V/51.9\ \Omega = 462.4\ mA$

$V_R = I \times R = 462.4\ mA \times 40\ \Omega = 18.496\ V$

$V_C = I \times X_C = 462.4\ mA \times 33\ \Omega = 15.2592\ V$

$I_R = I_C = I_T = 462.4\ mA$

$\theta = \arctan(X_C/R) = \arctan(33\ \Omega/40\ \Omega)\ 39.5°$

25. a. $Z = \sqrt{R^2 + X_C^2} = \sqrt{1\ M^2 + 2.5\ M^2} = 2.69\ M\Omega$

$I = V/Z = 12/2.69\ M\Omega = 4.5\ \mu A$

$V_R = I \times R = 4.5\ \mu A \times 1\ M\Omega = 4.5\ V$

$V_C = I \times X_C = 4.5\ \mu A \times 2.5\ M\Omega = 11.25\ V$

   b. $Z = \sqrt{R^2 + X_C^2} = \sqrt{300^2 + 200^2} = 360.6\ \Omega$

$I = V/Z = 50\ V/360.6\ \Omega = 138.7\ mA$

$V_R = I \times R = 138.7\ mA \times 300\ \Omega = 41.61\ V$

$V_C = I \times R = 138.7\ mA \times 200\ \Omega = 27.74\ V$

27. Lag, 90

---

29. $X_C = \dfrac{1}{2\pi f c} = \dfrac{1}{2\pi(35\ kHz)\ 10\ \mu F} = 455\ m\Omega.$

$Z = \sqrt{R^2 + X_C^2} = \sqrt{100\ k\Omega^2 + 455\ m\Omega^2} = 100\ k\Omega.$

$I = V/Z = 24\ V/100\ k\Omega = 240\ \mu A$

True power $= I^2 \times R = 240\ \mu A^2 \times 100\ k\Omega = 5.76\ mW$

Reactive power $= I^2 \times X_C = 240\ \mu A^2 \times 455\ m\Omega$
$= 26\ nVAR$

Apparent power $= \sqrt{P_R^2 + P_X^2} = 5.76\ mVA$

Power factor $= R/Z = 100\ k\Omega/100\ k\Omega = 1$

Chapter 11

1. a   **11.** a

3. b   **13.** c

5. b   **15.** b

7. e   **17.** a

9. c   **19.** b

Chapter 12

1. c

3. d

5. d

7. a

9. e

11. c

13. b

15. b

17. a. $22\ M\Omega$

   b. $78.6\ k\Omega$

   c. $314.2\ k\Omega$

19. a. $L_T = \dfrac{L_1 \times L_2}{L_1 + L_2} = \dfrac{12\ mH \times 8\ mH}{12\ mH + 8\ mH}$
$= \dfrac{96\ \mu H}{20\ mH} = 4.8\ mH$

   b. $L_T = \dfrac{1}{(1/L_1) + (1/L_2) + (1/L_3)}$

$$= \dfrac{1}{(1/75\ \mu H) + (1/34\ \mu H) + (1/27\ \mu H)} = 12.53\ \mu H$$

21. a. $V_S = \sqrt{V_R^2 + V_L^2} = \sqrt{12^2 + 6^2} = \sqrt{144 + 36}$
$= \sqrt{180} = 13.4\ V$

   b. $I = V_S/Z = 13.4\ V/14\ k\Omega = 957.1\ \mu A$

   c. $\angle = \arctan V_L/V_R = \arctan 2 = 63.4°.$

   d. $Q = V_L/V_R = 12/6 = 2.$

   e. $PF = \cos\theta = 0.448.$

23. $\tau = \dfrac{L}{R} = \dfrac{400\ mH}{2\ k\Omega} = 200\ \mu s$

V_L will start at 12 V and then exponentially drop to 0 V

Time	Factor	V_S	V_L
0	1.0	12	12.
$1\ T_C$	0.365	12	4.416
$2\ T_C$	0.135	12	1.62
$3\ T_C$	0.05	12	0.6
$4\ T_C$	0.018	12	0.216
$5\ T_C$	0.007	12	0.084

25. a. $R_T = R_1 + R_2 = 250 + 700 = 950\ \Omega$

   b. $L_T = L_1 + L_2 = 800\ \mu H + 1200\ \mu H = 2\ mH$

   c. $X_L = 2\pi f L = 2\pi(350\ Hz)\ 2\ mH = 4.4\ \Omega$

   d. $Z = \dfrac{R \times X_L}{\sqrt{R^2 + X_L^2}} = \dfrac{950 \times 4.4}{\sqrt{950^2 + 4.42^2}} = 4.39$

   e. $V_{RT} = V_{LT} = V_S = 20\ V$

   f. $I_{RT} = \dfrac{V_S}{R_T} = \dfrac{20\ V}{950\ \Omega} = 21\ mA,$

$I_{LT} = \dfrac{V_{S1}}{X_{LT}} = \dfrac{20\ V}{4.4\ \Omega} = 4.5\ A$

g. $I_T = \sqrt{I_R^2 + I_L^2} = \sqrt{21 \text{ mA}^2 + 4.5 \text{ A}^2} = 4.5 \text{ A}$

h. $\theta = \arctan(R/X_L) = \arctan 950/4.4 = 89.7°$

i. $P_R = I^2 \times R = 21 \text{ mA}^2 \times 950 = 418.95 \text{ mW}$

$P_X = I^2 \times X_L = 4.5 \text{ A}^2 \times 4.4 = 89.1 \text{ VAR}$

$P_A = \sqrt{P_R^2 + P_X^2} = \sqrt{418.95 \text{ mW}^2 + 89.1 \text{ A}^2} = 89.1 \text{ VA}$

j. $\text{PF} = P_R/P_A = 418.95 \text{ mW}/89.1 \text{ W} = 0.0047$

Chapter 13

1. d   **7.** a

3. c   **9.** a

5. d

11. a. Turns ratio $= N_S/N_P = 24/12 = 2$, step up

b. Turns ratio $= N_S/N_P = 250/3 = 83.3$, step up

c. Turns ratio $= N_S/N_P = 5/24 = 0.2083$, step down

d. Turns ratio $= N_S/N_P = 120/240 = 0.5$, step down

13. a. $I_S = (N_P/N_S) \times I_P = (12/24) \times 100 \text{ mA} = 50 \text{ mA}$

b. $I_S = (N_P/N_S) \times I_P = 3/250) \times 100 \text{ mA} = 1.2 \text{ mA}$

c. $I_S = (N_P/N_S) \times I_P = (24/5) \times 100 \text{ mA} = 480 \text{ mA}$

d. $I_S = (N_P/N_S) \times I_P = (240/120) \times 100 \text{ mA} = 200 \text{ mA}$

15. a. Turns ratio $= N_S/N_P = V_S/V_P = 240 \text{ V}/120 \text{ V}$

$= 2:1$ step up

b. Turns ratio $= N_S/N_P = V_S/V_P = 720 \text{ V}/240 \text{ V}$

$= 3:1$ step up

c. Turns ratio $= N_S/N_P = V_S/V_P = 14 \text{ V}/30 \text{ V}$

$= 0.467 \ (2.14:1)$ step down

d. Turns ratio $= N_S/N_P = V_S/V_P = 6 \text{ V}/24 \text{ V}$

$= 0.25 \ (4:1)$ step down

17. $I_S =$ apparent power/V_S,

$V_S =$ ratio $\times V_P = 1/2 \times 120 = 60 \text{ V}$

$I_S = 2 \text{ kVA}/60 \text{ V} = 33.3 \text{ A}$

$R_L = V_S/I_S = 60 \text{ V}/33.3 \text{ A} = 1.8 \ \Omega$

19. Follow polarity dots.

Chapter 14

1. b

3. b

5. d

7. c

9. c

11. a. $X_C = \dfrac{1}{2\pi fC} = \dfrac{1}{2\pi(60)0.02\mu\text{F}} = 132.6 \text{ k}\Omega$

b. $X_C = \dfrac{1}{2\pi fC} = \dfrac{1}{2\pi(60)18\mu\text{F}} = 147.4 \ \Omega$

c. $X_C = \dfrac{1}{2\pi fC} = \dfrac{1}{2\pi(60)360 \text{ pF}} = 7.37 \text{ M}\Omega$

d. $X_C = \dfrac{1}{2\pi fC} = \dfrac{1}{2\pi(60)2700 \text{ nF}} = 982.4 \ \Omega$

e. $X_L = 2\pi fL = 2\pi(60)4 \text{ mH} = 1.5 \ \Omega$

f. $X_L = 2\pi fL = 2\pi(60)8.18 \text{ H} = 3.08 \text{ k}\Omega$

g. $X_L = 2\pi fL = 2\pi(60)150 \text{ mH} = 56.5 \ \Omega$

h. $X_L = 2\pi fL = 2\pi(60)2 \text{ H} = 753.98 \ \Omega$

13. a. $X_L = 2\pi fL = 2\pi(60 \text{ Hz})150 \text{ mH} = 56.5 \ \Omega$

b. $X_C = \dfrac{1}{2\pi fc} = \dfrac{1}{2\pi(60 \text{ Hz})20 \ \mu\text{F}} = 132.6 \ \Omega$

c. $I_R = V/R = 120 \text{ V}/270 \ \Omega = 444.4 \text{ mA}$

d. $I_L = V/X_L = 120 \text{ V}/56.5 \ \Omega = 2.12 \text{ A}$

e. $I_C = V/X_C = 120 \text{ V}/132.6 \ \Omega = 905 \text{ mA}$

f. $I_T = \sqrt{I_R^2 + I_X^2} = \sqrt{(444.4 \text{ mA})^2 + (1.215)^2} = 1.29 \text{ A}$

g. $Z = V/I_T = 120 \text{ V}/1.29 \text{ A} = 93.02 \ \Omega$

h. Resonant frequency

$= \dfrac{1}{2\pi\sqrt{LC}} = \dfrac{1}{2\pi\sqrt{150 \text{ mH} \times 20 \ \mu\text{F}}} = 91.89 \text{ Hz}$

i. $X_L = 2\pi fL$

$= 6.28 \times 91.89 \text{ Hz} \times 150 \text{ mH}$

$= 86.54 \ \Omega$

$Q = \dfrac{X_L}{R} = \dfrac{86.54 \ \Omega}{270 \ \Omega} = 0.5769$

j. $\text{BW} = \dfrac{f_0}{Q} = \dfrac{91.89 \text{ Hz}}{0.5769} = 159.28 \text{ Hz}$

15. Using a source voltage of 1 volt:

$Z = V/I_T, I_T = \sqrt{I_R^2 + I_X^2}, I_R = V/R = 1/750 = 1.33 \text{ mA}$

$I_L = V/X_L = 1/25 = 40 \text{ mA}$,

$I_C = V/X_C = 1/160 = 6.25 \text{ mA}$

$I_X = I_L - I_C = 40 \text{ mA} - 6.25 \text{ mA} = 33.75 \text{ mA}$

$I_T = \sqrt{(1.33 \text{ mA})^2 + (33.75 \text{ mA})^2} = 33.78 \text{ mA}$

$Z = 1/33.78 \text{ mA} = 29.6 \ \Omega$

Index

A

AC charge and discharge, 253
AC meter, 218
Air core, 317
Alternating current, 193
American wire gauge, 42
Ammeter, 15
Amp, 12
Ampere, 12
Ampere, Andre, 12
Ampere-hours, 86
Amplitude, 201
Atom, 3
Atomic number, 3
Atomic weight, 5
Attenuate, 277
Average value, 203

B

Babbage, Charles, 230
Ballast resistor, 71
Bandpass filter, 394
Band-stop filter, 395
Bands, 6
Bandwidth, 384
Battery, 18
B-H curve, 291
Bleeder current, 176
Boot, Henry, 209
Branch current, 137
Breadboard, 95
Breakdown voltage, 27, 237

C

Cable, 45
Calibration, 222
Capacitance meter, 275
Capacitive reactance, 259
Capacitive time constant, 248

Capacitor, 230
Capacitor coding, 247
Capacitor phase relationship, 255
Capacitors in parallel, 239
Capacitors in series, 240
Center tapped transformer, 361
Christie, S. H., 160
Circuit breaker, 90
Closed circuit, 28
Coercive force, 294
Coil, 289
Cold resistance, 71
Communication, 197
Compound, 9
Conductance, 25
Conductor, 24
Conventional flow, 14
Coulomb, 11
Coulomb, Charles, 10
Counter EMF, 311
Cray, Seymour, 31
Current, 9
Current divider, 139
Current in parallel, 137
Current in series circuit, 103
Current in series-parallel, l69
Current ratio, 353
Cutoff frequency, 384

D

Dielectric strength, 27
Differentiator, 282
DIP, 59
Direct current, 193
Dissipation, 55
Dot convention, 357
Dry cell, 84
Du Fay, Charles, 25
Dual in-line package, 59
Duty cycle, 212